D0046550

# Colliding
## Worlds

*137: Jung, Pauli, and the Pursuit of a Scientific Obsession*

*Empire of the Stars: Obsession, Friendship,
and Betrayal in the Quest for Black Holes*

*Einstein, Picasso: Space, Time, and the Beauty
That Causes Havoc*

*Insights of Genius: Imagery and Creativity
in Science and Art*

*Imagery in Scientific Thought:
Creating 20th-Century Physics*

HOW CUTTING-EDGE

SCIENCE IS

# Colliding
# Worlds

REDEFINING

CONTEMPORARY ART

**ARTHUR I. MILLER**

 W. W. Norton & Company   New York • London

For information about permission to reproduce selections from this book, write to
Permissions, W. W. Norton & Company, Inc., 500 Fifth Avenue, New York, NY 10110

For information about special discounts for bulk purchases, please contact
W. W. Norton Special Sales at specialsales@wwnorton.com or 800-233-4830

Manufacturing by Courier Westford
Book design by Chris Welch
Production manager: Anna Oler

Library of Congress Cataloging-in-Publication Data

Miller, Arthur I.
Colliding worlds : how cutting-edge science is redefining contemporary art /
Arthur I. Miller. — First Edition.
pages cm
Includes bibliographical references and index.
ISBN 978-0-393-08336-1 (hardcover)
1. Science and the arts—History—20th century. 2. Science and the arts—History—
21st century. 3. Arts—Experimental methods—History—20th century. 4. Arts—
Experimental methods—History—21st century. I. Title.
NX180.S3M555 2014
700.1'05—dc23
2014005430

W. W. Norton & Company, Inc.
500 Fifth Avenue, New York, N.Y. 10110
www.wwnorton.com

W. W. Norton & Company Ltd.
Castle House, 75/76 Wells Street, London W1T 3QT

1 2 3 4 5 6 7 8 9 0

TO LESLEY

# CONTENTS

# LIST OF ILLUSTRATIONS

## INSERT

# ACKNOWLEDGMENTS

This book has been a voyage of discovery for me. I was struck by how today, more than ever before, the world around us is being represented in a manner that combines art, science, and technology. I immersed myself in this new movement, which I call artsci, spending over a year conducting interviews in addition to visiting some of the essential centers devoted to it.

The many artists and scientists I feature in this book graciously took time out from their busy schedules for my interviews and replied to my queries with "faster than light" responses, as did directors and staff at key centers.

My deepest appreciation to all those artists who provided me with images of their work.

I thank my agent Peter Tallack of the Science Factory for his encouragement, enthusiasm, and sagacious advice.

Huge thanks to my editor at W. W. Norton, Matt Weiland, his assistant, Sam MacLaughlin, and their team, for their invaluable criticisms and assistance in the production process.

My appreciation to Allegra Huston for her splendid copyediting. Any errors that remain are my own.

I am indebted to my wife, Lesley, who has provided support and encouragement at every stage of this book. Her comments on the manuscript were invaluable, as always. This book is dedicated to her.

Arthur I. Miller
London 2013
www.arthurimiller.com

Art does not reproduce the visible; rather, it makes visible.

—*Paul Klee*

On October 13, 1966, the New York glitterati and all those who basked in their glow descended on the cavernous 69th Regiment Armory on Lexington Avenue to celebrate the opening night of *9 Evenings: Theater and Engineering*, the first ever large-scale collaboration between artists, engineers, and scientists. Ten artists and thirty engineers took part and the technology they used was spectacularly new at the time.

Naturally Andy Warhol, in sunglasses and leather jacket, was there, surrounded by his entourage. He was heard to declare, "It's just great." The Beat poet Allen Ginsberg was approached by a young woman who whispered in his ear, "You probably don't remember me, but I'm Susan Sontag." Marcel Duchamp, who had kick-started the entire modern movement in art, was there too, no doubt remembering the moment fifty years earlier at the 1913 Armory show when his *Nude Descending a Staircase* had scandalized New York. The up-and-coming artist Chuck Close sat next to Duchamp. The fashion designer Tiger Morse wore a bare midriff outfit of white vinyl with a portable lamp which bathed her in a violet glow.

Everyone involved agreed that Robert Rauschenberg was the inspiration but John Cage was undoubtedly the star. Cage, the composer famous for his *4'33"*—four minutes and thirty-three seconds of intense silence—produced a collage of sounds randomly collected at that moment by telephones around the city and the Armory. As the performance went on, one by one members of the audience stepped onstage to add to the cacophony, playing with juicers and mixers installed there.

The following night, Rauschenberg, the celebrated iconoclast and artist, showed a piece called *Open Score*, basically a game of tennis with the rackets wired to transmit sound. Every time a racket hit a ball, an amplified "boing" resounded around the building and one of the forty-eight lights went out, until the audience was in total darkness.

The performances were by turns magical and chaotic. It was an Andy Warhol moment, although very much inspired by Duchamp. The two served, so to speak, as bookends, Warhol as the logical conclusion of Duchamp—from the ready-made to the Campbell's soup can. Everyone was sure what they were seeing was a brand new art movement that was going to blow a hole right through the middle of the traditional art scene—and they wanted to make sure they were in at the beginning. The opening night was a sellout, with 1,727 tickets sold; 1,500 people had to be turned away. In all, 11,000 people attended, with sellouts on three of the nine evenings. Everybody who was anybody, or had dreams of being somebody, was there to bask in the glow of the already famous. A *New York Times* reporter wrote of a sold-out performance, "A bomb dropped here would turn off the whole New York art scene."

In the nearly half century that has passed since that first explosion of excitement, art, science, and technology have rubbed up against one another in myriad ways. The resulting artworks have been sometimes beautiful, sometimes disturbing, sometimes subversive, sometimes downright crazy, but always interesting, new, and pushing the boundaries.

*Colliding Worlds* begins by taking the story back to the early days of the twentieth century, when inventions such as x-rays and photography transformed the way we see the world. Artists such as Picasso and Kandinsky took on board the latest scientific developments, while scientists found themselves driven by questions like the relevance of aesthetics to science and what makes a scientific theory beautiful. But it was not until the second half of the last century that the new movement, which has come to define the twenty-first century, really flowered, and it is this flowering that forms the bulk of my story. Its creators are artists and scientists working together to create images and objects of stunning beauty, along the way redefining the very concept of "aesthetic"—of what we mean by "art" and, eventually, by "science."

I started to write about how art interacts with science and technology in the 1980s, when few people other than the artists and scientists themselves were taking note. Over the years I watched as more and more artists emerged, along with more and more art fairs and more and more conferences. I watched as the movement grew from something underground to something far more mainstream that impinges on our daily life, the realm of what we all take for granted.

Full of curiosity, I began to track down and talk to those involved. I learned who these artists are, why they decided to become artists, what it meant to collaborate with scientists, and what their notions of aesthetics and beauty were in this strange and constantly evolving terrain—the avant-garde of the twenty-first century—and began to put together these dispatches from the edge of art and science. I discovered that the artists I spoke to are all engaged in the same quest: to find a way to unite art, science, and technology.

I looked for leading artists working in all the different areas of the new movement. I've limited myself to artists whose works illuminate science and might even contribute to scientific advances. I am less interested in those who simply use science to illustrate their themes. Although the results can be dazzling, they don't reflect back onto science or technology. Some of the artists I spoke to collaborate with scientists, others have learned at least some relevant scientific concepts, while others are both artists and scientists—artists who are also researchers.

To my surprise, collaboration between artists and scientists turned out to be a minefield. Is it always the artist who benefits, and not the scientist? Does a scientist's everyday research benefit from such collaborations? These are topics that came up again and again in the course of my research.

Initially I sought out these new-wave artists via galleries and museums. But the avant-garde has never been welcome in the traditional art world. Instead, these artists have created support networks of their own. They meet at international biennales and regular gatherings devoted to celebrating and exhibiting the latest creations in science-influenced art. Foremost among these are Ars Electronica in Linz, Austria, Zentrum für Kunst und Medientechnologie (ZKM) in Karlsruhe and Documenta in Kassel, both in Germany, the Science Gallery in Dublin, Le Laboratoire in Paris, CERN in Geneva, and the Wellcome Collection and GV Art in London. The School of Visual Arts (SVA) in New York focuses on science-influenced art, as do the MIT and NYU media labs, and there are departments devoted to it at the Slade School of Fine Art and Central Saint Martins in London, among others.

I also scoured scholarly papers and newspaper and magazine articles and books such as Edward Shanken's *Art and Electronic Media*, Bruce Wands's *Art of the Digital Age*, and Stephen Wilson's *Art + Science Now: How Scientific Research and Technological Innovation Are Becoming Key to 21st-Century Aesthetics*. They all provided interesting overviews of the subject but made no attempt to convey the people behind the art: their creativity and what drives them, their dreams, their struggles, the drama of developing a new art movement and what it is up against. To look deeper into all these topics I've chosen to interview some of the artists, scientists, and engineers who are actually involved.

One last problem is what to call this art form that is influenced by science or technology. Terms such as "artsci," "sciart," and "art-sci" seem inadequate to convey its beauty and subtleties, though I've opted for the first. I have no doubt that in the future these works will become known simply as "art."

# Colliding Worlds

# 1

# In Search of the Invisible

From the beginning of time, people have assumed there was an invisible world of some sort—of the spirits, the mind, or the world as explored by science. Artists have explored it on canvas and in sculpture, musicians in music, writers in words, and scientists with mathematics. The play between the visible and the invisible has always been at the heart of Western art and scientific thought.

In the fourteenth century, while the painter Giotto was exploring the geometry of planes with horizontal and vertical axes, thus moving toward an understanding of perspective, Nicole Oresme in France was inventing the graph, providing a visual image with which to investigate the laws governing falling objects. Then came the Renaissance. To its great masters, Leonardo da Vinci and Albrecht Dürer, there was no distinction between art and science. They carried out scientific investigations and painted pictures in the same spirit of inquiry and with the same creative fire.

In the seventeenth century, Newton was driven not only by the urge to make scientific developments but by alchemy, mysticism, and religion. His *Philosophiae Naturalis Principia Mathematica* (*Mathematical Principles of Natural Philosophy*), published in 1687, was the culmination of this quest. It gave no hint of its mystical origins, and seemed to offer a way to understand phenomena on the earth and in the heavens using only pure, cold logic. Hereafter, science was to be considered the serious pursuit of truth, while art was seen as merely decorative.

In fact, the flow of ideas between art and science never totally ceased, but it was not until the 1830s that it was renewed with great vigor. The English painter John Constable deeply admired the scientist Michael Faraday and wanted to be considered a scientist as well as an artist. He drew and classified cloud formations and was particularly inspired by the way Faraday used visual images, such as the ghostly lines of electric and magnetic force, to explore electric and magnetic phenomena.

In the 1850s, the Scottish scientist James Clerk Maxwell was beginning to formulate the equations necessary to understand Faraday's images. To do this, he had to deal with notions in physics that went beyond concepts basic to Newton's physics, such as weights and velocities. Art, meanwhile, was becoming more abstract, as in J. M. W. Turner's depictions of London, which astonished his contemporaries.

Then, at the end of the nineteenth century, just as the physics of Newton and Maxwell seemed unassailable, there were three discoveries that threatened to topple this magnificent edifice: x-rays, radioactivity, and the electron. For the first time, scientists had to directly confront phenomena beyond anything they could perceive with their senses—the world of the invisible.

Among much else, Newton had proposed a theory of color showing that white light was made up of red, orange, yellow, green, blue, and violet. Ever after, any discussion of color theory included the question of how we perceive color and whether there could be other theories based on a different set of primary colors. This question interested Johann Wolfgang von Goethe, who considered his own theory of color the best work he had ever done. Georges Seurat, too, was up to speed on the most current work, including the psychological effects of different colors—warm, cool, or arousing. He grouped different colors as dots such that, when viewed from a distance, they coalesced into a scene which conveyed the emotions he wanted to stir. In fact, his experiments with colored dots were experiments in the scientific sense. One of the results was to inspire Gabriel Lippmann to invent color photography.

Around the same time, Paul Cézanne was beginning to play with space in a way that broke with one of the Renaissance's most firmly established conventions: perspective. He pushed scenes up against the picture plane, thereby transforming a single perspective point into several, focusing on conception—what he knew was there— rather than perception—what he could see with his eyes. His still lifes appear to have been planned mathematically.

All this set the stage for the revolutionary developments of the twentieth century, led by two unparalleled masters: Einstein and Picasso.

## Reimagining space and time (1900-18)

One August day in 1914, at the train station in Avignon, two men awkwardly hugged each other. One, in French army uniform of dark coat, red trousers, and a kepi with a rifle slung over his shoulder, was a bear of a man. The other was shorter, solidly built, and swarthy, "a good looking bootblack," as Gertrude Stein wrote. Georges Braque was going to fight the war to end all wars, while Pablo Picasso had used his Spanish citizenship to avoid it. The two had worked as one. Fired by Picasso's brilliant ideas, they had created the most influential art movement of the twentieth century: Cubism.

Back in 1905, Picasso was so strapped for funds that he had taken to stealing milk and bread from people's doorways. But despite his penury, the young Spaniard, just twenty-five years old at the time, was sure that his destiny was to become the toast of the Parisian art scene.

At the time, Émile Loubet, an esteemed orator and honest statesman, was the president of France. France's great neighbor across the Channel still ruled the waves. Queen Victoria had died in 1901 after a reign of sixty-three years. Her son Edward VII had taken the throne and was now ruling over a decade of decadence.

Freud had recently published *The Interpretation of Dreams*. The Wright brothers had made their first flight, and France was enjoying the belle époque.

But there were seeds of discontent. Kaiser Wilhelm II in Germany was convinced that France, England, and Russia were plotting to encircle him.

Change was in the air. Europe was swept by an intellectual tidal wave whose proponents dubbed themselves the avant-garde. Their principal concern was to overthrow classical, intuitive notions of space and time—a mission which Picasso took very seriously. The momentum of change affected every aspect of life. The look of the city itself was transformed with the invention of prestressed concrete, which gave architects the freedom to open structures and play with space. Meanwhile, composers such as Erik Satie, Claude Debussy, and Arnold Schoenberg went on sonic adventures, going beyond the rigid structures of Germanic music to develop a new time frame of atonality. In poetry, Guillaume Apollinaire lifted words out of lines and sketched images with them.

Paris was electric with ideas. "We had no other preoccupation but what we were doing and saw nobody but each other. Apollinaire, Max Jacob, Salmon. Think of it, what an aristocracy!" Picasso recalled of this belle époque and of his closest friends who formed the core of his inner circle, his think tank. Their circle included Satie and Alfred Jarry, a literary fantasist who wrote about the fourth dimension and time travel. Jarry's speciality was demolishing bourgeois literary and social conventions; he was an intellectual agent provocateur.

Another member *en marge* was Maurice Princet. An insurance actuary by day, by night he hung out with Picasso's inner circle—*la bande à Picasso*; he had a penchant for philosophy and advanced mathematics, especially higher-dimensional geometries. "M. Maurice Princet preoccupies himself especially with painters who disdain ancient perspectives," recalled André Salmon, a close friend of Picasso. Princet was introduced into Picasso's circle by his notoriously unfaithful mistress, Alice Géry, who at one time had been involved with Picasso, preferring hot-blooded Spanish artists over gray insurance actuaries. Members of *la bande à Picasso* spoke of him as having a professorial manner but added that he "conceived of

mathematics like an artist. [He] was *le mathématicien du Cubisme*."
It was Princet who introduced Picasso to the great French scientist
Henri Poincaré's writings on geometry. Poincaré would turn out
to be the common denominator between Picasso and Einstein; his
writings influenced them both.

What impressed Picasso most about science was x-rays, with
their mind-bending message that what you saw was not necessarily what you got. Picasso's love of movies and photography provided
the technical background that enabled him to play with images, and
with reality.

He was especially interested in mathematics, and in particular the geometry of four dimensions—the three dimensions of our
everyday world (length, depth, and width) with an extra spatial
dimension. If the artist could glimpse the four-dimensional world,
he would be able to see all perspectives of a scene at once: a god's-
eye view; blooming, buzzing confusion. But how to project these
perspectives onto the two-dimensional canvas?

Some artists even attributed mystical attributes to the fourth
dimension, seeing it as a place where one could communicate with
the dead or tap one's powers of intuition; it was a world of ideas generated purely by thought, where the irrational moment of creativity
occurred.

Princet introduced Picasso to concepts from four-dimensional
geometry. He showed him a book on the subject written by Esprit
Jouffret, a keen amateur mathematician who was also a friend of
Poincaré. The equations meant nothing to Picasso, but the illustrations did. They were projections from the fourth dimension onto
the two-dimensional page, showing complex polyhedrons as if the
viewer was walking around them, viewing one perspective at a time.
This was the way Poincaré had suggested for showing the fourth
dimension, as Picasso knew from Princet's lectures on Poincaré's
best-selling book *Science and Hypothesis*. Picasso, however, disagreed. "Why not project all perspectives at once?" we can imagine
him asking Princet in a smoky bistro. He unveiled his response to
the conundrum in 1907, in his revolutionary and groundbreaking

*Les Demoiselles d'Avignon*, using all he had learned of mathematics, science, and technology to depict the fourth dimension in the face of one of the women.

*Les Demoiselles d'Avignon* was met with nothing but an embarrassed silence. Most of Picasso's friends and colleagues thought he had gone mad, and to make things worse his mistress, Fernande Olivier, walked out on him. For several months Picasso was severely depressed. The painting was not seen again for three years, when a photograph of it appeared in an architectural journal under the title "Study by Picasso." Commenting on the "terrible picture [that looms] through the chaos [in Picasso's atelier]," the author of the article, Gelett Burgess, asked Picasso whether he had used models: "'Where would I get them?' grinned Picasso."

The painting was not publicly exhibited until 1916, when art critics trashed it as "a nightmare," "overwhelming," and "yet another big daub by Picasso." Nine years passed before Picasso finally sold it to a wealthy Parisian art collector who promised to have it hung in the Louvre. The collector never did. On his death in 1938, his family sold it to the Museum of Modern Art in New York for the even then paltry sum of $28,000. For many years, only Picasso realized its worth and importance.

As we now know, *Les Demoiselles d'Avignon* was the foundation of Cubism.

IN 1905, WHILE Picasso was seeking his style, in a city not far in distance but light-years away in culture, another young man, two years older, had already found his. Intense and handsome, with a passion for the violin, Albert Einstein had the "kind of beauty that caused havoc," as a female admirer recalled. There were many similarities between the two, on both the personal and creative fronts, in particular the fact that the most creative periods of their lives spanned the same years: 1902–09.

At the time, both were involved in tempestuous relationships: Picasso with his flamboyant and flirtatious mistress, Fernande Olivier, and Einstein with his sultry and moody wife, Mileva Marić. And both struggled to make ends meet.

Born in 1879, in 1905 Einstein was a complete unknown. At his high school in Munich, his Greek teacher had told him he would not amount to anything—and indeed he didn't, at least in Greek. But Einstein did not waste his time. By the age of twelve, he had taught himself advanced mathematics and physics.

He then left Germany for Switzerland, which had managed to remain a peaceful place. Luckily for him, the Swiss Polytechnic Institute in Zürich did not require a high school diploma for entry. Nevertheless, when Einstein graduated in 1900, the Institute refused to give him a letter of recommendation. By now he had renounced his German citizenship.

Finally, in 1902, with the help of a friend's father, he got a job as a patent clerk third class at the Swiss federal patent office in Bern, an intellectual backwater. "[At the patent office I] was set free to produce my best creative work," he recalled. For culture, Einstein had a study group which consisted of himself and two pals, Conrad Habicht and Maurice Solovine. They called themselves the Olympia Academy, and took all knowledge as their province. Among the books they studied was Poincaré's *Science and Hypothesis*, which left them "spellbound," as a former member recalled. This was Einstein's think tank. "This group had a considerable influence on my development," he later recalled.

Three years later, at white heat, with no forewarning, Einstein published four papers that would change the course of science as well as of nations.

Since he was fifteen, Einstein had been mulling over the state of physics and had grown increasingly unhappy with it. He found certain of its inconsistencies "unbearable." In that era, he wrote, the equations of physics were interpreted in a way that "led to asymmetries that do not appear to be inherent in the phenomena."

He concluded that physics was riddled with redundant explanations and extraneous concepts that had led to these asymmetries, and pared all these away using a minimalist aesthetic. This led him to the discovery of his relativity theory in 1905. It was a response to aesthetic discontents. Thus he introduced the notion of symmetry into twentieth-century physics.

Einstein's theory opened the way to revealing a universe that was astounding, contradicting all our intuitions as experienced in daily life. There were no true times, just as there were no true lengths, for there was no one true perspective of any physical phenomenon. It was exactly what Picasso and the Cubists were discovering at almost the same moment.

"A consequence of the work on electrodynamics has suddenly occurred to me," Einstein wrote. This was that mass and energy were equivalent, that every mass is equivalent to a huge amount of energy. Thus the equation $E = mc^2$ made its appearance in September 1905. Einstein even suggested a way to release this energy by using radioactive processes.

Most physicists were convinced that Einstein's relativity theory was at odds with experimental data. Like Picasso, Einstein alone realized the worth of his work and persevered in claiming that the data were wrong, as indeed they turned out to be. Just about every physicist failed to spot the significance of what he had accomplished, asserting that Einstein had merely substantiated an already existing theory of the electron. Later, he recalled of the leading physicists of that era that "they were theorizing out of their depth," a rather audacious statement for a patent clerk. Like Picasso, he was confident of his destiny.

By the spring of 1909, Einstein's career had become meteoric. Walther Nernst, the impresario of physics in Berlin, visited him in Zürich, where he was an associate professor at the university, to consult him on his research into the way in which heat courses through matter, based on the still controversial quantum theory discovered barely ten years earlier by Nernst's colleague Max Planck. Einstein's research moved Planck's work from theory into the laboratory. At the same time, Einstein was working on expanding his theory of relativity to a point where scientists could use it to explore why the universe is as it is. The new so-called general theory of relativity began as one man's view of the cosmos. Ten years later, in 1919, its verification would result in his deification as genius incarnate.

The world meanwhile had suffered the cataclysm of war. Einstein used his immense prestige to advertise his long-held pacifist views, which had led him to renounce his German citizenship when he was living in Switzerland. He reacquired it in 1919 as a show of support for the liberal government of the Weimar Republic. He would renounce it again in 1933.

Picasso, too, was busy in his new atelier in Montmartre, creating paintings, photographs, and sculptures, and carrying out further researches into Cubism—reducing natural forms to geometry, the aesthetic he had discovered in 1907 with *Les Demoiselles d'Avignon*. He also photographed some of his Cubist paintings and laid the negatives one on top of the other to produce prints that were Cubism upon Cubism, Cubism to the nth degree, which he thought of as a higher order of Cubism. He made a multifaceted sculpture, *Head of Fernande*, and his portraits of her also had a true three-dimensionality. He was "as overworked" as God, he recalled.

New trends in both art and science were emerging, set in motion by the breakthroughs made by Picasso and Einstein. Ironically, both men would find it impossible to follow these through to their logical conclusions. At bottom Einstein was a classical physicist who clung to a view of the universe that was essentially an updated version of Newton's science of motion, and Picasso's bent was always toward an abstract figuratism. The two were determined to maintain a visual imagery in their work that connected with what we see in our daily lives.

## Beyond reality (1918–30)

Whereas Einstein and Picasso were reluctant to follow to the extreme the avenues that their magnificent breakthroughs had opened, other artists were not. Wassily Kandinsky, for one, had no problem moving totally away from reality as we see it with our eyes and stepping out into the void. "The work of art is born of the artist in a mysterious and secret way," he wrote in 1912.

Born in Moscow in 1866, Kandinsky was inspired to become an

artist when he went to an exhibition in Moscow in 1895 and saw how the French Impressionists played with color and light. He was particularly struck by Monet's *Haystacks*. "That it was a haystack the catalogue informed me. I could not recognize it. . . . I wondered if it would not be possible to go further in this direction," he wrote. He was twenty-nine at the time. Until then, he had had a successful law practice and even as a young painter he still looked like a lawyer—neatly dressed, hair combed, with wire-rimmed glasses, eager to provide a logical analysis of any situation at hand. This businesslike façade, however, concealed an intense belief in Russian mysticism which inspired his lifelong fascination with color and music.

In 1906 in belle époque Paris, Kandinsky met up with a group of artists famous for their extravagant use of pure colors and exaggeration of forms. They were known as Fauves, or "wild men," and among them was Henri Matisse. Kandinsky was dazzled. To him the works of Picasso and Braque, conversely, with their emphasis on objects, materials, and geometric representation, smacked of science. In these, he wrote, "natural form is often forcibly subjected to geometrical construction, a process which tends to hamper the abstract." He wanted to find the "spiritual existence," which meant exploring the unseen "absolute forces" behind material objects. The artist should strive "toward the abstract, the non-material," for only in this way could he express his inner emotions.

"The harmony of the new art," he continued, "appeals less to the eye and more to the soul." He used bright colors to "symbolize feelings" and evoke the "inner glory" of nature—to probe consciousness itself.

Nevertheless, he was impressed by Einstein's relativity theory and the equivalence of mass and energy, $E = mc^2$, which asserts that everything we see and touch is simply energy. Kandinsky interpreted this as the mystical aspect of modern science—that everything is amorphous. His 1910 painting *Improvisation* was completely abstract: it did not refer to any naturally occurring object or even any geometrical form. Kandinsky had discarded form altogether.

Around the same time, a group of Parisian artists was also look-ing for ways to probe emotion and consciousness using line and color, but in addition they wanted to bring in a scientific element. In particular, they wanted to communicate the dissolving of sub-stance into energy, to suggest that all forms were products of the mind. Picasso's friend the poet Apollinaire dubbed this movement Orphism after Orpheus, who sought pure music. Just as Picasso had emphasized conception over perception, Apollinaire did too, describing Orphism as the art of painting "with elements borrowed not from visual reality, but entirely created by the artist."

The Orphists met in the Parisian suburb of Puteaux, at the house of the three Duchamp brothers: Raymond Duchamp-Villon, Jacques Villon, and Marcel Duchamp, and thus were known as the Puteaux Group. Apollinaire was the poet of the group, which included Jean Metzinger and Albert Gleizes, who painted in a sort of academic style of Cubism and participated in group showings in salons. To Picasso's annoyance, Gleizes and Metzinger proclaimed themselves the founders of Cubism.

Like Kandinsky, these artists were very interested in develop-ments in science and technology. Electromagnetic waves emanate forcefully from Robert Delaunay's *Eiffel Tower*, while in his *Nude Descending a Staircase*, Duchamp explores movement in space and time.

Meanwhile, in Italy, a group of artists had sprung up who threw off all ties with the past and celebrated a new world grounded in modern technology, with the emphasis on the violence and beauty of speed. They called themselves the Futurists. Their images were stimulated by technology and fired by extremist rhetoric. "We want no part [of the past]. We spit every day on the Altar of Art," wrote their chief propagandist, Filippo Marinetti, who called himself "the caffeine of Europe." The Futurist manifesto of 1909, which he had a hand in writing, minced no words: "We will glorify war—the world's only hygiene." Then came the carnage of the First World War, which put a stop to the Futurists' belligerent propaganda. Marinetti would go on to write speeches for Mussolini.

Two years later, in 1911, the Futurists arrived in Paris. With their dapper dress, Italian good looks, and panache, they took the Parisian art community by storm, especially the women. Picasso had been looking for an excuse to dump Fernande, and found it when she had an affair with one of them. The Futurists revered Picasso and Braque, but had nothing but disdain for Picasso's followers—Duchamp and his friends, the Puteaux Group.

In 1916, yet another new art movement emerged, the brainchild of the Romanian expatriate, poet, and performance artist Tristan Tzara, this time in Zürich. Zürich had become a focal point for artists, writers, and political figures, including Lenin, as well as Picasso's longtime German dealer and friend, Daniel-Henry Kahnweiler, all seeking refuge from the war. The credo of the new movement was that it was time to put an end to the hierarchy that condoned the war and all the artistic trends associated with it, starting with good taste and rational thinking and moving on to Futurism and even Cubism, whose productions the new artists referred to as "Cathedrals made of shit." These iconoclasts were intellectual shock troops. They called their movement Dada and themselves Dadaists.

Meanwhile, Kandinsky's countryman Kazimir Malevich was striving to depict the cosmos in other ways. An intense, broad-shouldered, powerful man, in his early works he had been influenced by Fauvist colors and Cubist geometry, but wanted to go beyond this into purely abstract painting, as Kandinsky had. While designing the sets for the Cubist–Futurist opera *Victory over the Sun*, he was struck by the way in which three-dimensional stage props became basic two-dimensional forms under the spotlights. Perhaps, he thought, both the world about us and the unseen world might be built out of supremely one-dimensional forms—straight lines.

The first painting Malevich produced in this new experimental style—Suprematism—was a black square on a white background, in which the square, with its four straight sides, represented stability.

Next he painted a circle, reasoning that if a circle was expanded to fill the universe, its perimeter would become essentially a straight line. Then he painted a black cross constructed of five squares. To prove that these geometrical forms encompassed the whole cosmos, Malevich turned to science. In a series of writings and artworks focusing on magnetism, he convinced himself that geometrical forms had played a key role in James Clerk Maxwell's model of how magnets and wires carrying an electric current affected the space around them, which scientists in Maxwell's day believed was filled with a ghostly substance called the ether, another extraneous notion leading to asymmetries that Einstein got rid of using his minimalist aesthetic.

Like Picasso with Jouffret's book, Malevich could not understand the equations in Maxwell's writings. It was the images that dazzled him, among them the patterns formed by iron filings on a piece of paper placed near a magnet. Maxwell and his predecessor, Michael Faraday, had abstracted from these patterns to depict the unseen space around the magnet—the nonobjective world, which to Malevich was the only possible source of art. Inspired by these, he made a drawing of space using only straight lines.

By 1918, Malevich had moved beyond electromagnetism to depict the world about him as ultimately formless and colorless energy. That year he produced his first white-on-white painting, a white cross on a white canvas. The square had shed its material being and merged with infinity in a glare of pure whiteness. Malevich had entered the fourth dimension, achieving a sort of cosmic consciousness, a nirvana.

## Into the atom

For many years, scientists refused to embrace abstraction. They tried to describe the physical world in images drawn from everyday phenomena and dealt with objects in the atomic realm in the same way as those in our daily world, treating electrons as if they were charged billiard balls and obeyed the same laws of motion.

Then in 1905, Einstein suggested that light was emitted in indivisible bursts, or packets. He called these light quanta. Dazzlingly, he pointed out for the first time that the "profound formal distinction" made by scientists between waves and particles was unwarranted. Why assume that light was both particle and wave when, in specific phenomena, only one of these two was needed? He then applied his minimalist aesthetic to choose the particle aspect of light in order to deduce the law of the photoelectric effect, later put to practical use in developing devices that automatically open doors.

Most scientists dismissed Einstein's light quanta as seriously weird, useful only for doing calculations that could not be done using a theory of light as waves. How could a theory of light particles explain how two sources of light could interfere with each other? How could particles interfere with one another?

In fact, the whole issue of how electrons behaved in atoms baffled scientists until Niels Bohr investigated it. Einstein wrote of him, "He is like a sensitive child and walks about this world in a kind of hypnosis." Born in Copenhagen in 1885 into a highly intellectual family, Bohr was heavy-set, physically strong, yet clumsy—not one to get into a fight with—and very smart. He mumbled, speaking very softly and without enunciating clearly. From an early age he was exposed to the arcane world of Danish philosophy. He became steeped in Søren Kierkegaard's views on why we exist at all and Paul Martin Møller's tale of the boy who thinks about what he's thinking about and so on ad infinitum. Bohr's father, an eminent professor of physiology at the University of Copenhagen, had a wide intellectual circle which often met at his home, so Niels heard these topics explored in depth by Denmark's leading thinkers.

Bohr completed his PhD in 1911. That same year, laboratory experiments revealed that the atom seemed to behave like a miniature solar system with a positively charged nucleus at the center, like a sun, around which negatively charged electrons circulated like planets. This image of a minuscule solar system was most exciting, for it suggested that the atomic world was a reflection of the cosmos itself. Two years later, Bohr supplied the mathematics to back it up. His theory was immensely successful in making sense of many properties of atoms, in particular, of the sort of light they emitted, which formed their "fingerprint."

EIGHT YEARS LATER, as the First World War came to a close, science and art began to brush up against each other more closely than ever before.

In 1919 the English scientist Arthur Stanley Eddington set out to verify Einstein's generalization of relativity by testing the theory's startling prediction that starlight would bend near massive objects, i.e., other stars. He sailed to the tiny Portuguese island of Principe, off the west coast of Africa, and there carried out a spectacular experiment. His aim was to measure the extent that light from stars was bent in the sun's vicinity. To do so he photographed stars close to the edge of the sun during a total eclipse and compared their positions with those from star catalogues, with the sun removed, and found that the same stars were in distinctly different positions. He was thus able to measure how much a star's light was deflected by the sun's huge gravitational field. He concluded that Einstein was correct, and as a result of this dramatic proof relativity became a household word. Thanks to the explosion of interest, the public learned that we actually live in a world of four dimensions, where the fourth dimension is time. That same year, 1919, El Lissitzky, Malevich's colleague at the Artistic Institute in Vitebsk (Belarus), set out to incorporate science into his art.

Born in 1890 in a small Jewish community near Smolensk, Lazar Markovich Lissitzky, better known as El Lissitzky, was swarthy and handsome, with close-cropped hair and an intense intellectual stare. In 1917 he began to focus on graphic art and architecture and to make a name for himself. Two years later, Marc Chagall persuaded him and Malevich to come to his institute in Belarus, which specialized in graphic art. It was there that Lissitzky decided to rewrite Malevich's style of art—Suprematism—by incorporating relativity into it. He was particularly influenced by the way in which in relativity theory the three dimensions of space were combined with time into four-dimensional space-time, as the Russian mathematician Hermann Minkowski had demonstrated in 1907.

Minkowski had a stern Prussian bearing with a manicured handlebar mustache, high-collared shirt, and pince-nez glasses, and had been one of Einstein's teachers in Zürich. To describe Einstein's work, he used poetical statements such as, "Henceforth space by itself, and time by itself, are doomed to fade away into mere shadows, and only a kind of union of the two will preserve an independent reality." In his scientific papers Minkowski included two-dimensional diagrams of space-time. As in the four-dimensional world of the artist, where the fourth dimension was a spatial one, if a scientist could visualize this four-dimensional space he would be able to see every possible measurement of a phenomenon.

From 1921 onward, themes related to Minkowski's diagrams began to feature in Lissitzky's art, in particular in his collage *Vladimir Tatlin, Working on the Monument*. His choice of Tatlin was both symbolic and dramatic. Tatlin was a central figure in postrevolutionary Russian art. His first constructions had been abstract, like Picasso's, but after the 1917 October Revolution he began to incorporate rivets, iron, glass, and airplane wings to express the dynamic unfolding of forces. He argued that Socialist art should not be abstract and rejected the mysticism and abstraction of Malevich's work. Tatlin's art became known as Constructivism.

His most famous work, which inspired Lissitzky's collage, was his *Monument to the Third International*, a Constructivist tower designed to be erected in St. Petersburg after the Bolshevik Revolution of 1917, as the headquarters of the Comintern, the Third International. It was an extraordinary construction of industrial materials—iron, glass, steel—and a symbol of modernity. In shape it was a bit like Brueghel's *Tower of Babel*: a twin helix of girders which would spiral up 1,300 feet, higher than the Eiffel Tower, containing four geometric glass structures which would rotate at different speeds. It was conceived as a visual representation of Leninist dialectical materialism: but it was never built and was probably unbuildable.

Lissitzky believed that art was not meant to be merely art but to enhance society as a whole and that, to accomplish this, artists should take their place alongside scientists and engineers. Tatlin was his "new man," the architect of a new reality founded on relativity theory made concrete.

While Lissitzky was working on his Tatlin collage, he visited Weimar, in Germany, where he met the Dutch artist Theo van Doesburg. Van Doesburg had established an abstract art movement along with Piet Mondrian, among others. They produced a professional journal which they entitled *De Stijl* (The Style).

Van Doesburg intended De Stijl to be more than merely another art movement: it was to lead the way toward a new humanity and a new society. He conceived this lofty goal in 1915 when he saw Mondrian's abstract paintings and immediately contacted him.

Mondrian took a deep interest in science. "There are 'made' laws, 'discovered' laws, but also laws—a truth for all time," he wrote. "These are more or less hidden in the reality which surrounds us and do not change. Not only science but art also, shows us that reality, at first incomprehensible, gradually reveals itself, by the mutual relations that are inherent in things." This was a call for the artist and the scientist to work together to seek the true, unchanging laws of nature that lie in a realm beyond what we can perceive with our senses.

Van Doesburg admired the pureness of Mondrian's abstraction. But to his eyes it lacked modern scientific punch or scientific content. Its fourth dimension was static because it was a purely spatial dimension. Van Doesburg wanted to develop Mondrian's work and incorporate it into his De Stijl movement.

Whereas Mondrian was slow-moving and casually dressed, van Doesburg was the opposite. In his new approach to art and creativity there was no place for the romantic, bohemian image of the lonely, impoverished, poorly dressed artist starving in his freezing atelier. The new art van Doesburg envisioned was one that would integrate new technology and science and was practiced by men in suits and ties working in spotless offices.

In 1920, van Doesburg was already incorporating science into his art, using poems he called "X-images," where "X" stood both for "x-ray" and for the unknown in an equation. Then Lissitzky introduced him to relativity, with its four-dimensional space-time. This gave him the material to push Mondrian's style of art into modern times. He even proposed a plan for a university to be built in Amsterdam, with designs styled after Minkowski's space-time diagrams on its ceilings and walls, similar to Lissitzky's work in the Tatlin collage. It did affect the final design for the University of Amsterdam, built in 1923.

By the 1930s a battle was going on between exponents of nonfigurative and figurative art, centering on the term "relative." Mondrian wrote in an essay: "It is necessary to point out that the definitions 'figurative' and 'nonfigurative' are only approximate and relative. For every form, even every line, represents a figure; no form is absolutely neutral. Clearly, everything must be relative, but, since we need words to make our concepts understandable, we must keep to these terms." The term "relative" had taken on a life of its own, apart from its rigorous use in relativity theory. Here Mondrian wondered whether there can be a truly abstract art.

## Strange worlds (1930–50)

With the First World War over, the Roaring Twenties were in full swing. But there was industrial decline in Britain, the German economy was suffering rampant inflation, and the great crash of 1929 was looming on the horizon.

By now artists were releasing their hold on reality and creating fully abstract works; but scientists still tried to maintain either some visual imagery that could be abstracted from phenomena we can actually witness, or nothing at all. They were torn between figuration and abstraction. To make matters worse, by 1924 it had become clear that Bohr's theory could not accommodate the wealth of data on the atom that had accrued since he proposed it in 1913. It was totally apparent by now that atoms did not actually behave as if they were minuscule solar systems. The Austrian physicist Wolfgang Pauli wrote succinctly, "We do not want to clap the atom into the chains of our bias but on the contrary, we must adjust our concepts to experience." Instead of clinging to an outmoded image of the atom, scientists would be better off trying to understand the situation as it really was—listening to nature with a sympathetic ear attuned to the music of the spheres.

The man who finally found a way out was Werner Heisenberg, the German physicist who would later work on the German atomic bomb project. Heisenberg had the look of "a simple farm boy with short, fair hair, clear blue eyes, and a charming expression." He first encountered atomic physics in 1919, at the age of eighteen, while lying on a rooftop at the University of Munich, reading Plato's *Timaeus*. He was doing guard duty for the Free Corps, a hard-line conservative group made up mainly of disgruntled war veterans. A Communist uprising was in progress and Austria and Germany were in political turmoil.

Heisenberg was entranced by Plato's description of atoms as geometrical solids. He knew this was a fantasy, but he admired the way in which ancient Greek scientists had been prepared to consider even the most unlikely speculations. Perhaps he could do the same. A year later he formally entered the university and promptly began to make his mark in physics.

He set out to reshape Bohr's theory of the atom but finally decided that it was a dead end. Then, while recovering from hay fever in May 1925 on Heligoland, a small pollen-free archipelago in the North Sea, everything suddenly snapped into place. "I had the feeling that, through the surface of atomic phenomena, I was looking at a strangely beautiful interior, and felt almost giddy," he wrote. There and then Heisenberg rewrote atomic physics. Physicists dubbed his new theory "quantum mechanics."

Quantum mechanics solved many problems that had eluded the Bohr theory, but nevertheless physicists were uncomfortable because it could not be visualized. The mathematics it employed was unfamiliar and it was difficult to use. It was like trying to visualize infinity. To Heisenberg, this was not a problem. It meant that his theory was unencumbered with outdated visual imagery, like that of the solar system atom, which had led to confusion.

In quantum mechanics, electrons were treated as invisible particles. The year before, however, in 1924, the French physicist Louis de Broglie put forward convincing arguments that electrons could be waves as well as particles, just as Einstein had shown that light could be a particle as well as a wave.

At the end of 1925, the rakish Austrian physicist Erwin Schrödinger was on an illicit skiing weekend with a paramour when he had an epiphany. He worked out an equation and made it the basis of a new atomic physics in which electrons were waves surrounding the nucleus. He then carried out some of the calculations which had previously been done using Heisenberg's quantum mechanics but in a much simpler and more direct way using his theory, which employed visual imagery from classical physics—waves—and whose mathematics was familiar to all physicists, namely differential equations (in this case, his own—Schrödinger's—equation).

Schrödinger wanted to find a version of atomic theory that was different from Heisenberg's quantum mechanics. "Inspired by [de Broglie's suggestion that electrons were waves, I] felt discouraged not to say repelled by [quantum mechanics], which appeared very difficult to me and by the lack of visualizability," he wrote. Schrödinger's was an aesthetic choice—waves over particles; Heisenberg's choice of particles over waves was equally aesthetic. Schrödinger's wave mechanics took the world of physics by storm and overnight atomic physics went figurative again.

Heisenberg went apoplectic. "The more I reflect on the physical portion of Schrödinger's theory the more disgusting I find it. . . . What Schrödinger writes on the visualizability of his theory I consider crap," he wrote to that most incisive and critical of physicists, Wolfgang Pauli. Thus began what Heisenberg dubbed the battle between the waves and the particles.

In 1927, Bohr took the first steps toward resolving this knotty issue. He drew from his wanderings through the labyrinth of Danish philosophy and his knowledge of Eastern thought, bearing in mind the counterintuitive consequences of the high speed of light which means that time is relative. For Bohr, the key lay in the incredible minuteness of the atomic world, which is measured using an unimaginably tiny scale based on Planck's constant, $6.6 \cdot 10^{-34}$. (By contrast, the scale of relativity, used for high speeds and the vast universe, is based on the enormous speed of light, $3.0 \cdot 10^{8}$.)

In their atomic theories, Bohr argued, Heisenberg (quantum mechanics) and Schrödinger (wave mechanics) had each considered only one-half of the real world. Both the wave and the particle modes of existence were necessary in order to fully understand the ambiguous world of the atom. Bohr argued that it was necessary to opt for full symmetry, to take the electron as both wave and particle. The wave and particle aspects of an electron could be complementary in the yin-yang sense but at the same time mutually exclusive, which meant that only one side could ever be "seen" in an experiment. In other words, however you looked at an atomic entity, that was precisely what it was—either wave or particle, but never both at the same time. Bohr dubbed his new view "complementarity."

Bohr was widely read and highly cultured, with interests that went far beyond science into art. He seems to have had a particular interest in Cubism, as evidenced by the fact that, when he was given carte blanche by the Carlsberg Foundation to furnish his study in any way he wished, he chose a painting by Jean Metzinger, a member of the Puteaux Group. Cubism was in fact one root of his powerful notion of complementarity.

Perhaps he had read Metzinger and Gleizes's manifesto *On Cubism*. A Cubist painting, they wrote, represented a scene as if the observer were "moving around an object in order to seize it from several successive appearances." It was this that most impressed Bohr about Metzinger's painting. Mogens Anderson, a Danish artist and friend of Bohr, recollected Bohr's pleasure in giving "form to thoughts to an audience at first unable to see anything in it—They came with a preconceived idea of what art should be." Bohr's idea of complementarity offered a motif for the world of the atom that had striking parallels to the multiple perspectives offered by Cubism, providing a way of glimpsing beyond and behind the world of perceptions.

Heisenberg went on to conduct research into nuclear physics in 1932, still pursuing the correct imagery for the atomic domain, but it remained elusive. He did, however, come up with an important suggestion: to let the mathematics of the quantum theory decide the proper imagery.

IN 1921, THE French writer André Breton read a translation of Einstein's popularization *Relativity: The Special and the General Theory*. Breton was twenty-five at the time and his imagination was fired by what he read. "'One event can be the cause of another only if they both can be brought within the same point of space,' Einstein tells us. . . . I love at a certain altitude; what would I do even higher?" he wrote, playing with relativity theory—in which clocks at higher altitudes run slower than clocks at lower altitudes—as a way of looking into emotions and states of mind.

Breton and his circle were eager to examine states of mind using Freudian psychoanalysis and automatism to tap into the unconscious. They began to experiment with automatic writing, supposedly a key to the unconscious, in which artists drew spontaneously, without censoring their thoughts. To name the emerging movement and to describe his lyrical approach to his own creations in art and literature, Breton adopted a word coined in 1917: "sur-realism." It was a magical word.

Unlike the chaos and iconoclasm of Dada, Breton considered Surrealism to be constructive. Picasso thought otherwise. "The Surrealists never understood what I intended when I invented the word," he wrote. (In fact, the word had been Apollinaire's invention.) "Something more real than reality, a resemblance deeper and more real than the real, that is what constitutes the sur-real."

Picasso considered automatic writing and automatic drawing to be trivia. Although from time to time he used surrealistic effects, he detested being classified as a Surrealist. He was, however, on friendly terms with Breton throughout the 1920s and

1930s, although he never took him seriously as the almighty leader of a movement, as Breton liked to think of himself. Picasso always said that he was strange, and some letters that he purchased, from Breton to an ex-mistress, bore this out. Showing them to a friend, Picasso pointed to stains on one of them and asked, "What do you think this is?" "Hydrochloric acid?" the friend suggested. "No, it's sperm," Picasso told him. "That's how Breton was. A weird type."

But Picasso did agree on one aspect of Surrealism. "The Surrealists were right," he told his dealer and friend Daniel-Henry Kahnweiler. "Reality is more than the thing itself. I always look for its super-reality." Indeed, Picasso had done just that in *Les Demoiselles d'Avignon*, completed a decade and a half before the inception of Surrealism.

It was relativity, not quantum theory, that dazzled the Surrealists because there were still no readable popularizations of quantum theory in the 1920s and 1930s. Eddington's books on relativity theory, on the other hand, held them spellbound, with insightful and poetic explanations of abstruse scientific ideas such as, "Human personalities are not measurable by symbols any more than you can abstract the square root of a sonnet." Even to the Surrealists, this was surreal.

Einstein visited Spain in 1923 and translations of his works were published there. José Ortega y Gasset wrote an introduction to the printed version of Einstein's lecture in Madrid, positioning relativity theory in the history of Western thought, and also translated Freud's works into Spanish. Both Einstein and Freud greatly influenced the young Spanish artist Salvador Dalí.

"The difference between a madman and me is that I'm not mad," Dalí famously said.

"The new geometry of poetic thought demands a physical revision and accommodation of the order of those to which Einsteinian physics subjects all measurements," he wrote. Dalí painted figures and forms so precisely that they become otherworldly, placing them in highly imaginative renderings of a world oozing with flaccidity, sex, melancholy, and morbidity. What interested him most were the psychological aspects of relativity that he could draw out through psychoanalysis.

He demonstrated this in his 1931 painting *The Persistence of Memory*, with its drooping clocks, each in a different place and displaying a different time. He described it as "this harrowing and colossal question of Einsteinian space-time [as] a soft, extravagant, and solitary paranoiac-critical Camembert of time and space"—a typical Dalí-esque amalgam of relativity and psychoanalysis (and food). What he meant by "paranoiac-critical" was the Surrealist goal of extending the physical world, looking at one thing and seeing another, which Picasso referred to as superreality and had been striving for himself when he painted *Les Demoiselles d'Avignon*.

## Symmetry in art, symmetry in physics

"I think it is probable that negative protons can exist, since as far as the theory is yet definite, there is a complete and perfect symmetry between positive and negative electric charge, and if this symmetry is really fundamental in nature, it must be possible to reverse the charge on any kind of particle," said the English physicist Paul Dirac in 1933, when he received the Nobel Prize.

Dirac's greatest achievement was his prediction that every particle has a partner that is its antiparticle, opposite in charge but with the same mass, such that if the two came into contact they would annihilate in a blaze of light. He derived this startling conclusion from an equation he discovered which bears his name. The year before he won the Nobel Prize at the age of thirty-one, the antiparticle of the electron had actually been discovered in the laboratory. As Dirac said, it simply had to be the case. His certainty echoed Einstein's on receiving the news that Eddington had verified his general theory of relativity on his spectacular trip to the island of Principe in 1919: "I knew that the theory is correct. [If the results had come out otherwise] then I would have been sorry for the dear Lord. The theory is correct."

Four hundred years earlier, the astronomer and mystic Johannes Kepler had written about the "harmonies of the spheres." And in 1904, Poincaré wrote of scientists' "quest for this especial beauty, the sense of the harmony of the cosmos . . . just as the artist chooses from among the features of his model those which perfect the picture and give it character and life." "The scientist does not study nature because it is useful; he studies it because he delights in it, and he delights in it because it is beautiful," he went on. In his 1905 paper on relativity, Einstein had called for scientists to seek theories that revealed symmetries of nature. It was this that led scientists in the twentieth century to take up this quest.

Artists seldom defined symmetry as beauty, but this was not the case for scientists. In science, an equation has symmetry and is considered beautiful when it remains unchanged in form even if certain of its components are altered. If they are, for example, flipped from left to right, it is said that the equation exhibits mirror symmetry. A major symmetry occurs when an equation is unchanged in form even when its space and time coordinates are altered, as prescribed by Einstein's relativity theory. Such an equation is valid throughout the universe; there is a democracy between every laboratory, be it on the earth or in a star system a trillion miles away.

In art, on the other hand, symmetry is a balance that is pleasing to the eye between different elements on a canvas or in a sculpture. Artists have long understood the use of asymmetry. Picasso's *Three Musicians*, for example, is eye-catching due to Picasso's violation of just about every rule of symmetry.

## The Surrealists discover quantum physics

In the 1930s, popularizations of quantum mechanics began to appear. Surrealist artists were particularly influenced by *The New Scientific Spirit* by the French philosopher Gaston Bachelard. With a flowing white beard, long uncombed hair, and a generally disheveled appearance, Bachelard looked every inch the sage. Originally educated in physics and chemistry, he decided that the philosophy of science was his true calling, to which he brought theories of literature and psychoanalysis.

In *The New Scientific Spirit*, Bachelard wrote that developments in quantum theory "confront us with suggestive dilemmas." This was territory for the psychoanalyst because "psychologically the modern physicist is aware that the rational habits acquired from immediate knowledge and practical activity are crippling impediments of mind that must be overcome in order to regain the unfettered movement of discovery." He used the term "psychoanalysis" to mean an examination of the imagination in which the limitations imposed on it by logical thought were discarded.

Bachelard's writings led Breton to conclude that "modern scientific and artistic thought present us with identical structures." Breton went further in claiming that, according to Surrealism, "today, reason goes so far as to propose the continuous assimilation of the irrational."

The artist's freedom to change an object was stunningly used by the American artist Man Ray. His photographs, with their undulating forms, dazzlingly display the Surrealist style of playing with images from the world of perceptions, such as in *Violin d'Ingres*, his dazzling photograph of his lover Kiki de Montparnasse's back transformed into a violin, or his series of photographs entitled *Mathematical Objects*, depicting models of surfaces generated by mathematical equations. Ray's work influenced, among others, Max Ernst and Henry Moore.

Alongside Bachelard's provocative essays on quantum theory, Louis de Broglie—the French physicist who had discovered that electrons, like light, possessed a schizoid personality of both wave and particle—began writing in a reflective manner. In contrast to Bohr and Heisenberg, de Broglie continued to seek visual images of the atomic world that were abstractions from phenomena we see every day. But Bohr, Heisenberg, and particularly Pauli, all forceful and formidable adversaries, offered such strong criticism that it silenced de Broglie, a mild-mannered man. He did not argue his views again until after Pauli's death in 1958. Nevertheless, the Surrealists found the wave-particle duality with its counterintuitive ambiguities irresistible and read de Broglie assiduously.

In 1935 Christian Zervos, an influential French art collector, writer, and founder of the magazine *Cahiers d'Art,* wrote that the time had come to absorb quantum physics into art. His suggestion was taken up by younger Surrealists such as the Chilean artist Roberto Matta and the Austrian Wolfgang Paalen, who, in the 1940s, was to break with Surrealism for the very reason that it was not up-to-date on quantum theory. These artists looked for new images generated by the new physics, based in part on Freudian psychoanalysis, yet not abstract as Kandinsky had been. In this way they hoped to capture the tensions in physics (wave versus particle) as well as in psychoanalysis (id/ego/superego).

Matta was a notoriously difficult person but, nevertheless, remained faithful to the cause of Surrealism. His strongly expressed political views gave him the cachet of having been on the hit lists of both Franco in Spain and Pinochet in Chile. The American Abstract Expressionist Robert Motherwell was greatly influenced by Matta's art. He recalled, "Matta wanted to show the Surrealists up as middle-aged gray-haired men who weren't zeroed into contemporary reality."

Matta produced images, lines, and curves similar to Man Ray's photographs of mathematical surfaces but with added color. Duchamp put it well: "Matta's first and most important contribution to Surrealist painting was the discovery of regions of space hitherto unexplored in the realm of art."

Paalen was more critical than Matta of Surrealism. "It seems to me that we have to reach a potential concept of reality, based as much on the new directives of physics as on those of art," he wrote. Surrealism was not up to the task because "it tries to poeticize science, which can only lead to mysticism." Unlike Matta, Paalen wanted more concrete images along the lines of what de Broglie had described.

Like Matta, Paalen was both artist and thinker. He decided to

take on the challenge of representing the wave-particle duality head-on. He wanted to be seen as an advocate of the need for
science to influence art and for artist and scientist to seek a new
worldview together, as Mondrian had done. His depictions of the
wave-particle duality were, however, sadly, less than striking—
essentially swirls and vortices.

Dalí caught the mood when he wrote:

> In the Surrealist period I wanted to create the iconography of
> the interior world—the world of the marvelous, of my father
> Freud. I succeeded in doing it.
>
> Today the exterior world—that of physics—has transcended
> the one of psychology. My father today is Dr. Heisenberg.
>
> It is with pi-mesons and the most gelatinous and indeter-
> minate neutrinos that I want to paint the beauty of the angels
> and of reality.

Dalí realized that quantum theory, with its notions of probability and axioms such as Heisenberg's uncertainty principle,
had shattered the classical cosmos with its insistence that all processes developed in a continuous manner and could be described
with unlimited accuracy. But it was not until after 1945, when the
atomic bombs were dropped on Japan, that Dalí put his thoughts
onto canvas. In *The Disintegration of the Persistence of Memory*, he
produced dramatic images of a fragmented universe.

From the start, Dalí was a renegade who considered Breton
a dogmatic fool. Dalí espoused fascism instead of communism,
chose figuration over abstraction, and spoke glowingly of the
atomic bombs that had been dropped on Japan and the bright
future promised by nuclear energy. He pushed Breton and his
strong left-wing beliefs to his limit. In 1952, Breton threw him out
of the Surrealist movement. But by that time Surrealism had run
its course and Breton also would soon throw in the towel.

## Visualizing the invisible

By 1932, physicists had realized that an as yet undiscovered particle with no electrical charge at all existed inside the nucleus. They called it the neutron.

In a flight of inspired fancy, Heisenberg imagined that a neutron might be made up of an electron and a proton. Inside the nucleus the neutron could split into its component proton and electron. The electron would then move to another proton in the nucleus and the two fuse to become a new neutron. In this "exchange" the "migrating" electron would "transmit" an attractive force, which would explain why the nucleus was stable.

It soon became clear that in fact neutrons could not be made up of an electron and proton, but nevertheless some sort of exchange process would have to be dreamed up. The image of particles engaging with one another by exchanging other particles was extremely helpful. This unexpected scenario was exactly what Heisenberg had suggested as a way of finding the proper imagery for the atomic domain: to let the equations of quantum theory generate it—in other words, as Plato had suggested two thousand years earlier, to let mathematics guide the scientist into realms that could not be expressed in images derived from our everyday world.

Finally, in 1949, the American physicist Richard Feynman worked out the proper imagery for the atomic world: Feynman diagrams. "A half-assedly thought out semi-vision thing," was how he described his discovery. In the diagrams, Feynman replaced Bohr's iconic solar system image of the atom with lines indicating how an atom moves through space and time and how particles interact with one another by exchanging other particles, just as Heisenberg had suggested. The diagrams provide a glimpse of how objects that can be both a wave and a particle are able to interact with one another, in a visual form that seems to echo Mondrian's call for the "destruction of a particular form [and its replacement with] mutual forms of free lines."

Art and physics had reached the same juncture at about the

same time: representing the world in the fullest sense of the term, encompassing both seen and unseen. Although Mondrian kept up with events in science, Feynman's interest in art was limited to sketching naked women in pole-dancing establishments, so we must attribute his great advance solely to the increasingly abstract nature of theories of elementary particles.

# 2

# Montmartre in New York

Rarely can the onset of an astounding intellectual endeavor actually be located. The first collaborations between artists and scientists, however, can. They began in the early 1960s in lower Manhattan, in the vicinity of Fourth Avenue and East Tenth Street, in the East Village, a dilapidated area full of run-down tenements. It was Picasso's Montmartre transported to New York. The art critic Harold Rosenberg described it as the cradle of the new American abstract art, which was, as he pointed out, "the first art to appear here without a foreign address." Its stars were the de Koonings (Willem and Elaine), Mark Rothko, Joan Mitchell, Philip Guston, Robert Motherwell, and Angelo Ippolito.

East Tenth Street quickly became the new bohemia, the locus of the avant-garde, with happenings, impromptu jazz sessions, poetry readings, performance art, and discussions on just about anything and everything, including emerging styles of art and philosophies. Newcomers with nothing to lose opened small art galleries. The Brata Gallery on East Tenth hosted Jack Kerouac's first poetry readings, while the Tanager Gallery was lucky enough to exhibit Robert Rauschenberg and Jasper Johns.

All this was done on the fly, almost haphazardly, attended by those who were "around," as the saying went in New Yorkese, or who were attracted from elsewhere in the city by the magnetism of this déclassé neighborhood which, in any other context, would have been a no-go area. As in Montmartre at the time of Picasso, you went slumming. But in this magic domain there was freedom of thought, of action, of living—ideal conditions for the birth of new art movements such as Abstract Expressionism and its response, Pop Art, championed by Rauschenberg and Johns.

The improbable catalyst for all this was an electrical engineer from the Bell Telephone Laboratories at Murray Hill, New Jersey, called Billy Klüver.

## The bohemian engineer: Billy Klüver

In early 1960, an unlikely-looking pair of men were to be seen driving a Chevrolet convertible around the New Jersey garbage dumps, scavenging. Both were in their mid-thirties and European. One was tall with a long face reminiscent of the existentialist Albert Camus, with a cigarette firmly fixed between his lips and a Swiss accent. The other was smaller and neater, with a distinct Swedish accent. They collected scraps of old metal, bicycle parts, baby carriages, and other garbage. Later the Chevrolet reappeared in Manhattan, where the two heaved their load of garbage over the fence (lower in those days) into the sculpture garden of the Museum of Modern Art.

The result of their efforts was an explosive work that literally shook the New York art scene. It was called *Homage to New York* and, with a nod to that great and ever-changing city, was designed to self-destruct, which indeed it did in spectacular fashion. As for the two scavengers, one was the anarchic Swiss artist Jean Tinguely. His co-conspirator was an electrical engineer taking time off from his day job at Bell Labs: Billy Klüver. So who was this bohemian engineer?

Charismatic, with a high, domed forehead and receding hairline, Klüver was born on November 23, 1927, in Monaco, as Johan Wilhelm Klüver. Of Viking stock, the child of Swedish and Norwegian parents, he grew up in Stockholm where, in 1951, he graduated from the Royal Institute of Technology. He had two passions: electrical engineering and Swedish avant-garde cinema. In fact, he was president of the Stockholm University Film Society. In his senior thesis he combined the two by convincing his thesis supervisor, the Nobel laureate physicist Hannes Alfvén, to let him make an animated film showing how electrons move through electric and magnetic fields, a phenomenon on which Alfvén was an expert.

Klüver's interest in film was fueled by two friends, Pontus Hultén and Claes Oldenburg, both of whom would go on to have distinguished careers. In 1953, Hultén joined the Moderna Museet, Stockholm's museum of modern art. Seven years later, he became its director and began transforming it into one of the world's leading museums. Oldenburg's sculptures are known worldwide.

Shortly after graduating, Klüver moved to Paris where he worked for Thomson-Houston, a subsidiary of General Electric, as a nuts-and-bolts engineer. His projects included improving the antenna on the Eiffel Tower and designing underwater cameras for Jacques Cousteau. But America beckoned. His dream was to get a job at Bell Labs.

When American Telephone and Telegraph (AT&T) created Bell Telephone Laboratories in 1925, they envisaged it as a mecca for invention and also discovery. Early on in its existence, in 1927, the physicists Clinton Davisson and Lester Germer were working on a project exploring the structure of crystals using electrons when they noticed a peculiarity in the data. They had actually substantiated the schizoid nature of the electron as both wave and particle, for which Davisson won the Nobel Prize. Then in 1947, John Bardeen, Walter Brattain, and William Shockley invented a powerful amplifying device using solid-state materials, thus avoiding cumbersome and fragile glass vacuum tubes. This was the transistor, for which all three won Nobel Prizes. In 1965, Arno Penzias and Robert Wilson were puzzled by a noise in an antenna they were debugging. The noise seemed to come from everywhere, and persisted after all defects had been eliminated, including pigeon droppings. What they had discovered, as astrophysicists at nearby Princeton University informed them, was the echo from the Big Bang, 13.7 billion years ago; they were listening to the creation of our universe. More Nobel Prizes came. Claude Shannon went on to discover information theory at Bell, where he did his best thinking while riding his monocycle down one of the lab's long corridors and juggling three balls at the same time. Ken Knowlton, who also worked at Bell Labs, was one of the pioneers of computer art. Except for the transistor and Shannon's work, these discoveries had nothing to do with telephones.

As a place for making discoveries and inventions, fostering creativity at its highest level, it was inherent in the Bell Labs culture that great advances necessarily involved massive failures. Klüver, later, liked to point out that at Bell Labs, any scientist who didn't have a 90 percent failure rate with his experiments was no good. This adventurous, go-for-broke, anything-goes outlook served him well.

In 1954 Klüver was eager to join Bell Labs which was then one of the most innovative, exciting places for a scientist to work, and was certain he could get a position there. But the McCarthy hearings were going full steam, and foreigners at research centers were under scrutiny as security risks. So Klüver decided to lie low, and instead enrolled in the PhD program for electrical engineering at the University of California, Berkeley. Extraordinarily, he completed his degree in just over two years on a research topic that included theory and experiment, although he always said he preferred the hands-on approach of engineering over the theoretical side.

Afterward, he spent a year teaching at Berkeley. By this time Joseph McCarthy had been thoroughly discredited and Klüver was able to obtain the position he coveted at Bell Labs, the proper place for someone of his restless mind-set. He started off by exploring sound amplification devices as well as the possible uses for lasers.

By now Klüver's old pal Pontus Hultén had become an eminent curator. With the help of his introductions, Klüver made inroads into the East Village art scene, with its almost famous artists and almost famous taverns and coffeehouses populated by theorists like Clement Greenberg and Harold Rosenberg, who announced the onset of new movements and decided who the famous artists were. There was always an accompanying bevy of beautiful women. But Klüver wanted to do more than just hang around. He was interested in working with artists and bringing technology and art together—blurring boundaries. His passion for this quest had been fired some years earlier by C. P. Snow's electrifying 1959 Rede Lecture, "Two Cultures."

Trained as a chemist, Snow was also a novelist, gaining notoriety for his whodunits in university settings, and had held several senior civil service positions. After the war, he set himself the task of assessing Britain's future. It lay, he concluded, with science and technology, which offered hope and progress, whereas he saw the humanities as mired in the tragic condition of humankind. But they seemed to be diametrically opposed: scientists and engineers were woefully misinformed about the arts, and humanists (which included artists) were even more poorly informed about science. This would not do, especially in a postwar world driven by science and technology.

Klüver's utopian vision was to remove the boundary between science and the arts. As an engineer at Bell Labs, and using his connections in the art scene, he believed he could do it. He was a very unusual sort of engineer.

As a preliminary, he set up an art and science club at Bell Labs, telling his colleagues that it would make them better engineers. Sadly it "never went anywhere," he recalled.

Klüver's big break in his quest came when he met Tinguely, a well-established resident of the New York art scene. Tinguely's specialty was mechanical contrivances, usually made up of parts he collected from junkyards, powered with engines. To him, paintings were mere petrified objects.

Tinguely was a neo-Dadaist to the core. In 1960 he was working on his latest project, *Homage to New York*, in which he wanted to express his disgust with consumerism, materialism, and what he saw as a world gone amok with possessions. It was to be a slap in the face of fastidious cuckoo-clock Swiss technology.

Klüver came along just when Tinguely had begun thinking about this work, which was to become his most famous act of destruction. *Homage to New York* was an extraordinary contraption, a weird assemblage of small machines that would self-destruct one by one at preset times, sparking and smoking to an accompanying musical sound track while it rolled around haphazardly until it completely blew up in, as Klüver put it, "one glorious act of mechanical suicide."

Tinguely's problem was how to include the timers that would initiate the contraption's destruction. Klüver had the know-how. He also had a car, which put him in a privileged position among Tinguely's entourage. He and Tinguely scrounged suburban garbage dumps: years later, Klüver recalled the "stench sticking to your clothes. I can still smell it."

Tinguely also introduced Klüver to his friend Robert Rauschenberg. Klüver couldn't believe his luck. Not only had the avant-garde art world of New York opened up to him, he was working with two internationally acclaimed artists. Rauschenberg, much of whose own work also emerged from parts found in the street and junkyards, decided to come in with Tinguely, and the three men worked together.

Everything was ready by March 17 when the contraption was to perform its act of self-destruction in the sculpture garden of the Museum of Modern Art. Rauschenberg's contribution, dwarfed by Tinguely's, was entitled *Money Thrower*. It was a box filled with gunpowder and holding a dozen silver dollars. When the box exploded, the dollars would catapult into the crowd.

Everything was set. The performance, scheduled to last twenty-seven minutes, was witnessed by a chic invitation-only audience. As a reporter described it:

> During its short life, this artful contraption, programmed for a symphony of suicide, sent out smoke flashes, rang bells, played a piano with mechanical arms made of bicycle parts, poured gasoline on itself, set itself on fire, crushed bottles filled with evil-smelling gas, turned on a radio, spilled cans of paint on rolling scrolls, threw out silver dollars, melted its supports, sagged and nearly collapsed. The machine was supposed to have crawled to the museum pool and thrown itself in, but it didn't. . . . It was a direct and delightful assault on the belief that all art must be "lasting."

In the end, nothing went according to plan. The various timers did not detonate the charges on schedule. Rauschenberg's device, set to go off at a certain point in the program, exploded at an entirely different time.

"Art Goes Boom," was the *New York Journal-American* headline. The fire department had to be called in to douse the flames. The crowd loved it.

Tinguely was unperturbed. Klüver was in his element. Making mistakes was the credo of Bell Labs. Mistakes—or, better, unpredictability—became a part of performances based on technology. Expect the unexpected, was the byword.

Much of *Homage* had been built with Bell Labs equipment and in Bell Labs time. The day after *Homage* went off, Klüver's supervisor, John R. Pierce, barged into his office. Klüver assumed he was about to be fired. Instead Pierce demanded, "Why wasn't I invited?"

When asked to describe his art, Rauschenberg liked to say that he worked "in the gap between art and life," using everyday objects. He called these works "combines." He was inspired by Marcel Duchamp's legendary *Fountain* of 1917, a nonfunctioning upside-down urinal signed "R. Mutt," the precursor and epitome of Dada, the nonrational, nonconformist art movement in which it was the concept or idea that mattered, not the object, and which had completely transformed art. For Rauschenberg, a typical day began with an early morning stroll looking for junk. Often his assistants would widen the search beyond his immediate neighborhood.

In this quasi-realistic manner he explored the cosmos. But leaving it completely, which was what he interpreted Abstract Expressionism as doing, was not to his taste. Rauschenberg was a proponent of Pop Art, which would displace Abstract Expressionism from the art hierarchy while retaining its abstract meaning, in that Pop Art works were interpretable, if at all, only through signs, the subject matter of semiotics. Rauschenberg's work with Klüver opened up the possibility of using technology to put his ideas in motion. He wanted more. "I met Billy Klüver, the Bell Labs physicist," he recalled, "both here and in Sweden. He gave me the suggestion that the possibilities in technology were endless. Of course he was right. It was a difficult transition to make because I normally work very much by hand. I rely on the immediate sight and actuality of a piece. Moving on to theory and its possibilities was like being handed a ghost bouquet of promises."

Just days after *Homage to New York*, the two ran into each other at an exhibition of the Spanish artist Antonio Tàpies. Tàpies's decontextualized style bore similarities to Rauschenberg's, so it was no surprise that Klüver had shown up as well. The two decided to collaborate.

Rauschenberg had in mind a complex interactive piece that would respond to the viewer's movement, the ambient temperature, and light. In Klüver's opinion, the state of technology was not up to this. They decided to break it up into components. Several times they put together complete technical systems, then junked them. After more than three years of work, the final product was ready. Rauschenberg called it *Oracle*.

It was made up of five sculptures constructed from material found in the street: a rough bathtub with shower, a staircase, a window frame, a car door, and a pipe, fused together with galvanized sheet metal and mounted on casters. Inside each was an AM radio controlled by a central unit in one of the sculptures which acted like a command center. Rauschenberg didn't want any wires between the sculptures. This was where electrical engineering came in, in the innovative use of transmitters, receivers, amplifiers, and speakers. Klüver designed the command center that controlled the electronics, using fully transistorized FM wireless microphones, state-of-the-art technology at the time. The radios continually scanned all five available stations, producing a chance mélange of sounds—unpredictability. What emerged, Klüver recalled, was a "collage of ever-changing bits of music, talk and noise—loud, soft, clear or full of static," as random as the sounds heard floating from apartment windows on a walk through the Lower East Side. It was a multidimensional, multidisciplinary conglomeration of installation sound art.

*Oracle* had its debut in the cutting-edge Leo Castelli Gallery in May 1965. Only Rauschenberg's name was on it, not Klüver's. Reviews were uniformly excellent. The critic Lucy Pippard noticed that one of the five sculptures was the "most crowded and most business element; it lacks the absurdity of the other four and is the least individually beguiling." This was the one Klüver designed.

Rauschenberg went on to win the Norman Wait Harris silver medal and prize in 1968. *Oracle* is now on permanent display at the Pompidou Center in Paris. Unpredictability even broke into the work's installation for the grand opening there in 1977. As Klüver described it:

> Rauschenberg arrived and we worked frenetically to get *Oracle* going. An hour before the official opening, we plugged in the pump motor for the shower and blew all the fuses in our section of the museum. The electricians had all gone home for the day, and no one knew where to find the fuse box. When minister of culture Michel Guy and museum director Pontus Hultén arrived with President Giscard d'Estaing and Princess Grace, we were standing in semidarkness. The sound from a few lonely radio stations could be heard from *Oracle*, and with the refrigeration coils inoperative, the congealed fat was slowly melting—and creating a bad smell—in Joseph Beuys's piece *Plastischer Fuss, Elastischer Fuss* nearby.

Klüver's message was: no technology, no work of art. To ram it home, he insisted that the "technical elements involved in [these] works are just as much a part of the work of art as the paint in the painting." He continued, "The artist could not complete his intentions without the help of an engineer. The artist incorporates the work of the engineer in the painting or the sculpture or the performance ... Technology is well aware of its own beauty and does not need the artist to elaborate on this. I will argue that the use of the engineer by the artist is not only unavoidable but necessary."

The year 1966 was a watershed in his life. His supervisor's enthusiastic response to *Homage* reinforced his determination to collaborate with artists. He "felt it was a positive use of time." Besides Tinguely and Rauschenberg, Klüver went on to work with Andy Warhol, on *Silver Clouds*, metalized plastic film filled with helium and oxygen that wafted about Leo Castelli's gallery that year; John Cage and Merce Cunningham, in their production of *Variation V*, in which the dancers' movements triggered sounds and lights; and Jasper Johns, for whom he inserted battery-powered letters as lights into a painting.

Klüver insisted that technology pervades our existence. A favor-

2.1: *Variations V* (1965). John Cage, David Tudor, Gordon Mumma (foreground); Carolyn Brown, Merce Cunningham, Barbara Dilley (background). Photographer: Hervé Gloaguen.

ite example of his was the space program in which, so he said, it was sometimes difficult to know where the machine ended and the human being began. What was needed was the elimination of human error, a "new and maybe inhuman objective." At the time, many found it astounding that technology in the form of electronics and computers was being applied everywhere and seemed to promise improvements in growing food and providing shelter. Klüver liked to quote Marshall McLuhan, the self-proclaimed philosopher of communication theory: "Technology is the extension of our nervous system."

Perhaps, Klüver believed, he could achieve the dream fired by his reading of C. P. Snow: to remove the boundary between engineering and the arts in a way that benefited both artist and engineer, both of whom were essential for a true art-technology project. Klüver went even further: "I'm not so much interested in helping artists as

I am in seeing what effect the artist could have on technology. In the future, I see the artist having more and more impact, as he learns more and more about technical processes." Artists should create using technology because it is an inherent part of our lives.

"The artist's work is like that of a scientist. It is an investigation which may or may not yield meaningful results. . . . What I am suggesting is that the use of the engineer by the artist will stimulate new ways of looking at technology and dealing with life in the future." Like engineering, art was research, and both art and engineering achieved their fullest promise when the two worked hand in hand. Picasso, who was deeply influenced by the mathematics, science, and technology of his day, wrote similarly that "my studio is a sort of laboratory."

Klüver only occasionally used the word "science" alongside "art." In the passage above, this was a slip. He was quite explicit about this. Referring to an organization he was to form later that year, Experiments in Art and Technology (E.A.T.), he wrote, "Many people wanted E.A.T. to be about art and science, but I insisted it be art and technology. Art and science have really nothing to do with each other. Science is science and art is art. Technology is the material and the physicality." Julie Martin, Klüver's wife, told me that he often said, "What do an artist and a scientist have to talk about? But it's different with an engineer." With an engineer there could be something concrete, something hands-on.

The time had come for a major collaboration between artists and engineers.

Most of Klüver's colleagues had had no contact at all with the art community—or art for that matter, as Leonard Robinson, a fellow engineer at Bell Labs, humorously noted: "When I heard last May that we were going to Rauschenberg's I thought it was a Jewish delicatessen. That's how much I knew about art." Starting in summer 1965, Klüver, buoyed by encouragement from Bell, enlisted thirty of his colleagues to work with ten artists.

While Klüver extolled the virtues of working with artists, saying that it would make one a better engineer, Herb Schneider, a colleague, suspected there was another reason: "I watched Billy having fun with the artists. Billy was the link between Bell Labs and the artists. They were disorganized, but having a ball." Indeed, it was a great way to get away from suburbia and into the throbbing, swinging Greenwich Village art scene. Some engineers were never the same again.

At first, the artists and engineers met at Bell Labs. Senior figures such as John Pierce were behind the project, but nevertheless they insisted that the meetings be held late in the evening, well outside company time.

Klüver was concerned that the first meeting might not go well. In fact, everyone found it stimulating, although the artists began asking for technical miracles well beyond what the engineers could offer: walls made of warm air, materials that would decay and disappear during a performance, televisions that could revert to slow motion and back, balloons coming out of a hat like thought balloons, floating forms following people around, laser beams that could write on surfaces, artists suspended in midair, and large mirrors with continually changing curvatures. The artists appreciated that these ideas might not be possible, but it was a way to get the ball rolling.

The engineers replied with explanations that were far too technical. The problem was, they lacked a common language.

Herb Schneider was at first skeptical of these collaborations and then became a zealot: "At first artists were in control—nothing worked. They threw ideas to engineers expecting them to fix everything, but engineers could make no sense of what they wanted." So Schneider stepped in as translator. John Pierce also took an active role in the early deliberations, as did Manfred Schroeder, director of Bell's acoustics, speech, and mechanical research lab. After much discussion, the team agreed on various projects. Several weeks of rehearsals followed in a school gymnasium in New Jersey. The event was to be called *9 Evenings: Theater and Engineering.*

First, they had to choose the venue. Rauschenberg wanted a huge space to house the expected large audience. Radio City Music Hall and Yankee Stadium were suggested. Then Seymour Schweber, head of Schweber Electronics, one of the sponsors, suggested the 69th Regiment Armory on Lexington Avenue. He recalled the "nostalgia that surround[ed] it because of the original Cubist show in 1913 when Marcel Duchamp's *Nude Descending a Staircase* was exhibited and knocked the critics off their rockers." He believed that *9 Evenings* would represent a similar breakthrough, "affording the audience a taste of future shock." That had to be it.

In the event the staging was amateurish; no one had experience with a space of that size. Armories are cavernous. The Swedish multi-media artist Öyvind Fahlström, one of those involved, described it as more like a turn-of-the-century railroad station. The 69th Regiment Armory is 162 feet high with no facilities for attaching lighting equipment. Due to its size and the height of the ceiling the acoustics were terrible, added to which radio reception was not good and many stations were not available. Then the lighting expert became ill and had to be replaced by an inexperienced substitute. No one had any idea how much light would be needed to fill the space. Scaffolding had to be specially built to support the weight of 5,000- and 10,000-watt lights because the balcony railing was too weak. To add to all this, the team had seriously underestimated the equipment rental budget. And there was no production manager to handle staff and dismiss incompetents.

The Bell engineers worked huge hours for no pay and, says Cecil Stoker, one of the engineers, "their unhappy wives frequently came down to the Armory, children in tow, and made scenes." Klüver soldiered on.

But collaboration often broke down. While the artists told the engineers what they wanted and the engineers did their best to provide it, they did not easily understand each other. Finally Schneider realized that the common language of the engineers and artists was visual imagery, maps of their ideas: pictures, not words. He created wiring diagrams that both could understand and which were then turned into plugboards for electronic circuits that controlled lights and other equipment.

Least collaborative of all the artists was the composer John Cage, who "just said what he wanted," as Klüver's wife recalls, although he, too, pitched in at the end. Classically trained, for over twenty years he had experimented with how to mix Dada and Buddhism with music.

One of the most influential composers of the twentieth century, he set out to fracture the smooth flow of music into staccato bursts, often using new versions of instruments, such as a piano which he rejigged into a percussion instrument by inserting objects between the strings. He also pioneered the use of variable-speed turntables, in today's DJ style, as well as audio-frequency oscillators. Cage concluded that any sound—including silence—could be music.

Before the performance Klüver wrote a "for your eyes only" memo to the engineers. Here are his notes on Cage and Rauschenberg:

> <u>John Cage</u>: . . . Cage makes definite decisions about what he wants and what he does not want. The appearance before the concert is usually a big mess of equipment and wires. If something does not work it is okay. . . .
>
> <u>Robert Rauschenberg</u> is easy to work with. He will tell you exactly what he wants. His piece will probably not be ready until the first performance. If you have suggestions for him he will tell you if he likes them or not. It is great to watch him make choices. He will need a lot of help/the trouble is he looks like he does not need help, but that is not true. Due to his fame his piece will have great publicity value and every effort should be made to make it a success. He appreciates very much what you do for him.

At the end of his memo, Klüver urged engineers to reassure the artists that everything would work. They should make it clear that "We handle the technical end and they are not to worry about it. . . . If there are any problems, tell me about them rather than letting them interfere with the show." "Good luck," he concluded, as if laying the plans for a military campaign.

The engineers did their best to reassure the artists, though they must have had a tough time. As one engineer recalled, "Fuses were blowing like mad, weird flashes of sound and light would burst into the gym, occasionally the acrid smell and smoke of a burned out resistor would fill the air. But never did the engineers express doubt in front of the artists."

The advertising agency for the show was Ruder and Finn, Inc., aided by the artists and their friends, who spread the word that something fantastic and momentous was about to occur. Ruder and Finn divided the events into theater and concert pieces, promoting the show as avant-garde theater, a happening in which the audience would be welcome to participate. *9 Evenings* opened with a bang on October 13, 1966. *Newsweek* wrote on October 31:

> Half a millennium ago Leonardo da Vinci agonized over his attempt to integrate the logic of science and the inspiration of art. Things are better today—only recently a physicist spoke on the reversibility of time before a scientific meeting while playing the bongo drums. Or are they worse? In any case, today is today, and engineer Billy Klüver recalled the bongology incident last week as he introduced the largest and most organized attempt yet by artists to arm themselves with the ravishingly efficient implements of science.

The "bongology incident" was a lecture by the physicist Richard Feynman, which he enlivened by playing the bongo drums. The Armory event was a reversal of Feynman's anarchic act: artists armed with "implements of science."

For his piece, Cage and his crew moved around long tables jammed with electronics, playing with a collage of sounds collected by a bank of telephones from remote sources such as the ASPCA, restaurant kitchens, and police and marine radio channels, filling the Armory with random noises. Cage had wanted sounds from outer space as well. He might have had them, thanks to Penzias and Wilson's picking up an echo from the Big Bang just the previous year. During the opening night performance, a stagehand happened to see some telephones off the hook and hung them up. Silence ensued until the connections were reestablished. This was one silence Cage didn't want.

In his *Open Score* the following night, Rauschenberg used state-of-the-art electronics to subvert a game of tennis. Radios inside the racket handles transmitted sounds which were amplified by speak-

ers suspended by wires secured to the ceiling. The two players were
the artist Frank Stella and the tennis pro Mimi Kanarek. Every time
they hit the ball a light went out until there was total darkness. Individual members of the audience on the Armory floor were illuminated by infrared-sensitive television cameras and the images were
projected onto three large screens. The audience was aware of the
size of the crowd only through these projections.

The enthusiastic advertising campaign had led the audience to
expect polished performances with no glitches. But this could never
have been, owing to the vastness of the Armory and the lack of a production manager. It was the newness of it all. *9 Evenings: Theater
and Engineering* was experimental. It was a collaboration between
two groups of people who at first sight were poles apart, trying to
accommodate each other, to think along common lines, and to do
something that had never been attempted before.

The critic Brian O'Doherty rightly argued that technological malfunctions did not equal artistic failure:

> [*9 Evenings*] received, on the whole, an appalling press—based
> mainly on the justifiable irritation of interminable delays, techni-
> cal failures of the most basic sort, and long, dead spaces between,
> and sometimes in the middle of, pieces. Yet, as such irritations
> faded away, one is left with startlingly persistent residual images,
> and strong hints of an alternative theater that has been lagging in
> its post-Happenings penumbra between art and theater.

"Happenings" were impromptu performances in which the audience participated. They were unpredictable, unique, sometimes wild,
and occasionally dangerous. The best moments usually arose when
something "went wrong." The first took place in the 1950s and they
were a key part of the countercultural 1960s' scene. One of the most
notable was a performance by Yoko Ono in Tokyo in 1964, when she
appeared onstage swathed in fabric, presented the audience with a
pair of scissors and invited them to cut the fabric away. As O'Doherty
suggests, by 1966 happenings had probably passed their peak. *9 Evenings* had all their spontaneity and inspired anarchy.

With time, *9 Evenings: Theater and Engineering* has come to represent an inspired moment in the new combination of art and technology. The mechanical age was over. The electronic age had begun.

## A call to arms

It was in the air in the 1960s: artists seeking to bridge the gap between art and technology. Klüver spoke of "a new mode of interaction between science and technology on the one hand and art and life on the other. To use a scientific jargon that is currently in, I will try to define a new interface between these two areas." There had been sightings years before, ever since the coming of movies—an art form that emerged directly out of artists' interest in technology. Already people were pointing out that in order for a collaboration to be meaningful, the artist would have to be technologically literate, perhaps even a trained engineer, which was a bit of a stretch at that stage. Bell Labs opened their computer facilities to musicians who wanted to learn programming, a fairly basic tool; the revered avant-garde composers Edgard Varèse and Milton Babbitt were among the first to use them.

Klüver was eager to create an organization that would catch the swell of the tidal wave of collaboration between artists and engineers. Shortly before the first performance of *9 Evenings: Theater and Engineering,* he had met with Rauschenberg, Robert Whitman (an artist who had been involved in some of the earliest happenings), and Fred Waldhauer (an engineer at Bell known for his work on hearing aids). His aim was to ensure that whatever momentum was generated by *9 Evenings* would not be lost. The four founded Experiments in Art and Technology (E.A.T.), with Klüver as president, Rauschenberg as vice president, Waldhauer as secretary, and Whitman as the treasurer.

E.A.T.'s first meeting was at the end of November 1966. Klüver's wife, Julie Martin, recalled that three hundred people attended—mainly artists and engineers, all of whom "expressed huge interest. Sixty people had questions right away."

In a newsletter of June 1, 1967, Klüver and Rauschenberg stated that "the purpose of Experiments in Art and Technology Inc., is to catalyze the inevitable active involvement of industry, technology, and the arts. E.A.T. has assumed the responsibility of developing an effective collaborative relationship between artists and engineers." They stressed the symbiotic effect of such a relationship. Engineers could learn from artists about the human scale of their work, while the artist would be able to continue the "traditional involvement of the artist with the relevant forces shaping society."

On October 10, 1967, the team held a press conference at Rauschenberg's spacious three-story studio to "formally launch E.A.T. to the public and press and to ask for support from industry, labor, politics, and the technical community."

Among the speakers was John Pierce of Bell Labs, an early advocate of all that Klüver was trying to accomplish and an active participant in E.A.T. He began, "Technology is the source of the material well-being of our age. Man cannot live without art. Sometimes it seems that our culture and our lives are becoming fragmented. I am happy that dedicated and talented people are bringing art and technology together through experimentation, which is a common ground for both."

Invoking, or perhaps rediscovering, Immanuel Kant, he went on, "Surely, art cannot safely ignore or reject such a powerful force as technology. And technology should not be without art."

Among the political heavyweights who spoke were Senator Jacob Javits of New York and Herman D. Kenin, representing the powerful AFL–CIO union. In those days there were considerable misgivings that machines were going to take over the world and put millions out of work. Kenin provided the endorsement Klüver needed, making a speech extolling the benefits of computers. E.A.T.'s board of directors included Alfred H. Barr, founding director of the Museum of Modern Art; John Cage; Senator Javits and his wife, Marian; the architect Philip C. Johnson; György Kepes, the Hungarian-born artist, designer, and art theorist who had established the Center for Advanced Visual Studies, incorporating art, science, and technology, at MIT that same year; and Marie-Christophe de Menil of the Houston art-collecting family. E.A.T. was on its way.

To round off the evening, several works of art were exhibited, including Rauschenberg's *Oracle*, Warhol's *Silver Clouds*, and one of the first computer-generated works of art, *Studies in Perception I.* The last had nothing directly to do with E.A.T. but had been produced at Bell Labs by two of Klüver's colleagues, Ken Knowlton and Leon Harmon, both specialists in the then young field of computer graphics. Knowlton was already sure that his future lay in art. "The non-scientific, some say artistic, aspects of computer graphics arose for me via a sophomoric prank," Knowlton said. In 1966, he and Harmon had come up with the idea of scanning photographs with a television camera, then converting the varying halftones to small electronic symbols. When the resulting image is viewed up close it is just a mishmash of symbols, but viewed from a distance it takes on a specific shape.

Their boss, Ed David, director of Visual and Acoustic Research, happened to be away for a few days, so they decided to hang a twelve-foot-wide picture on his wall representative of their work— "of, guess what, a nude!" Knowlton gleefully recalled. As a model, they had enlisted the twenty-five-year-old dancer Deborah Hay, who was also a friend of John Cage. They paid her $50, a not inconsiderable sum at the time. Harmon and Max Mathews, a pioneer in composing music on the computer, were at the photo session, with Mathews as the photographer.

When David returned to his office he took one look at the image and was "delighted but worried." An unusual number of visitors came by David's office. The upper echelons heard about the pic-

2.2: Leon Harmon and Ken Knowlton, *Studies in Perception I*, 1966.

ture and wanted it removed immediately, but smaller versions of the photograph were in circulation already. The public relations department warned that "you may circulate this thing but be sure that you do NOT associate the name of Bell Labs with it." Bearing that warning in mind, Knowlton and Harmon brazenly approached MoMA. The curators there were not particularly excited. "Interesting but no big deal," was their response. They did, however, agree to exhibit it during a lunch hour.

A year later, Knowlton and Harmon's picture was on show in Rauschenberg's loft at the press conference to launch E.A.T. The *New York Times* published the image. "Since it appeared in the venerable *New York Times*, it was art with a capital A, instead of porn," says Knowlton. The public relations department at Bell Labs did an about-face and ordered him and Harmon to "be sure that people know that this was produced at Bell Telephone Laboratories, Inc." As for their colleague Klüver, "Billy was forever retelling the story, including pointing out that it was the first time that the *New York Times* printed a nude!"

Another of Harmon's research interests was facial recognition. For a 1973 article, "The Recognition of Faces," he used a highly pixelated image of Abraham Lincoln taken from a five-dollar bill. This intrigued Salvador Dalí, who had a keen interest in the science of perception, and he based his 1976 painting *Gala Contemplating the Mediterranean Sea* on it.

Flushed with the success of *9 Evenings: Theater and Engineering*, E.A.T. continued to bring together artists and engineers. A leaflet circulated in early 1967 announced (I leave in the misspelling to give an idea of the off-the-cuff air of the organization), "The following series of demonstrations—bull sessions—lectures by scientists and engineers is designed for artists. To stimulate the discussion and facilitate matchings, we are also suggesting the engineers attend whos field is the topic of the day." The schedule of lectures and demonstrations for February through April 1967 included computer music, television equipment and capabilities, computer films and animation (some by Knowlton), color theory, and holography. All the lectures were sold out.

The lofty aims announced at the launch of E.A.T. seemed achievable, especially considering the success of *9 Evenings: Theater and Engineering*. Sadly, this turned out not to be the case. An early attempt to raise funds failed, despite the fact that artists who contributed works to the E.A.T. show at the Leo Castelli Gallery included Walter de Maria, Marcel Duchamp, Red Grooms, Donald Judd, Sol Lewitt, Robert Rauschenberg, Larry Rivers, Cy Twombly, and Andy Warhol. Like today, despite all the publicity, only a few people actually bought art.

The Ford Foundation, IBM, and AT&T all turned down grant applications for the half a million dollars which E.A.T. needed. Requests for smaller amounts were more successful, but this did not solve financial issues concerning staff and day-to-day expenditures.

There were other problems, such as safety. One of the performances in *9 Evenings: Theater and Engineering* involved scanning volunteers' faces with a laser beam. When the Office of Radiation Control at the New York City Department of Health heard of this, they informed E.A.T. that they were to be notified in advance whenever lasers were used in performances.

E.A.T.'s most ambitious project was the Pepsi Pavilion at Expo '70 in Osaka, Japan. The planning and building involved seventy American and Japanese artists and engineers, working together. Pepsi allowed Klüver to be solely responsible for the design. For the most part all went well, but the severe cost overruns dominated the press reports. The pavilion itself was a geodesic dome 75 feet high and 165 feet in diameter, crammed with electronics and mirrors intended to transport visitors to virtual worlds. It was decades ahead of its time.

Although interesting projects were planned, E.A.T. made the mistake of trying to include too many fields. The thrust turned away from collaborations between scientists and engineers to education and art in developing countries. Its focus moved from blurring the boundaries between art and technology to those between art and life, developing instructional television films to fight illiteracy and improve dairy production around Ahmedabad, in India. They achieved some success, but they had lost sight of their original purpose. *9 Evenings: Theater and Engineering* was E.A.T.'s debut and high-water mark, with the Pepsi Pavilion in a distant second place.

Jasia Reichardt, a curator with a keen interest in the cultural importance of technology, summed up the demise of E.A.T. and the importance of Klüver: "If E.A.T. faded out, it was through a natural process of decline, precipitated by developments in technology itself. Some of the technologies became easy enough for the artist to use without much technical support. Some artists made demands that could not be met. . . . E.A.T. belonged to a decade when such ambitions could be fired by the enthusiasm of one man."

A key achievement of Klüver's *9 Evenings: Theater and Engineering* was that it inspired other collaborations and events, in particular Jasia Reichardt's own *Cybernetic Serendipity*.

## Computers and creativity: Jasia Reichardt

In 2011, when I knocked on the door of Jasia Reichardt's house in London, I had not seen her for over twenty years. Reichardt exudes ebullience, enthusiasm, and sweep of knowledge of the art scene. Her house is like an archive, housing material on the intersection of art and technology as well as on concrete poetry, in which words form images, together with a vast collection of books on art and art history and catalogues of exhibitions.

Reichardt made it clear that she was not an artist. Rather, she curates exhibitions, has been a gallery director, and writes about art, as well as about literature. She is interested in the "edges of things, of how art, literature, theater and technology intersect." In 1957, she was reading about cybernetics and became intrigued by the role technology might play in art. As she put it, "Both computer art and concrete poetry belonged to art's outer periphery."

During World War II, engineers had learned how to build sophisticated devices for gunnery platforms that, through feeding back ("feedback loops"), could make corrections in tracking a swiftly moving target such as an airplane. Immediately after the war, in 1946, the first of the Macy Foundation conferences was held in New

York, to discuss how to use such breakthroughs in peacetime. The conferences aimed to set the foundations for a general science of the human mind. One of the participants, Norbert Wiener, a professor of mathematics at MIT, had had a great deal of experience in anti-aircraft work during World War II. He used his knowledge to develop cybernetics, taking the word from the ancient Greek *kyubernētēs*, "to steer," and laid the foundations of this new field in his book *Cybernetics, or Control and Communications in the Animal and the Machine*, published in 1948.

Cybernetics is the study of how systems self-regulate through feedback and how this relates to all domains, human and mechanical. The thermostat is an example of a cybernetic system. It is made up of two parts: a heater and a sensor. Once a temperature is selected, the sensor turns the heater on when the ambient temperature drops below the chosen temperature and turns it off when the correct temperature is reached.

To artists, cybernetics opened the possibility of art as an interactive process, an open system that includes material objects (possibly interacting among themselves) as well as enabling the audience to interact with an artwork and change it.

In 1949, the Hungarian-born artist Nicolas Schöffer read Wiener's book and was stunned by the possibilities it suggested. "[Cybernetics], for the first time, makes it possible to replace man with a work of abstract art," he said. Five years later, he began producing majestic artworks, beautifully crafted, incorporating its principles. These were delicate, towering constructions made up of fine metal tubes, photoelectric cells, microphones, motors that turned rods with gears, and mirrors that caught and responded to variations in the color, temperature, and light intensity of the surroundings.

They combined complexity with beauty in movement. They were ever-changing; something new was always happening. Schöffer is now known as the father of cybernetic art.

Reichardt mulled these developments over. Then, in 1966, she heard about E.A.T. At the time, she was assistant director at the fledgling Institute of Contemporary Arts (ICA) in London. She got

2.3: Nicolas Schöffer, *CYSP 1*, 1956.

a travel grant from the US State Department and flew to New York, arriving in the midst of the hectic days leading up to the Armory show, and was immediately caught up in the excitement. She loved the way that Klüver ran meetings, adroitly dealing with "artists who wanted engineers to do the impossible—like artists wanted to fly. Some of the engineers all of a sudden wanted to be artists, too." The "meetings were enthusiastic, heartfelt, and naive." Having seen *9 Evenings: Theater and Engineering*, she was sure that part of the excitement was that things often didn't work.

Reichardt wanted to create something like E.A.T. in London. Then she recalled a comment made by the German philosopher Max Bense, whom she had met in 1965 when he visited an exhibition on concrete poetry which she had curated. He asked what plans she had for other exhibitions. None, she replied. Bense suggested, "Look into computers."

"[Bense] was adept at crossing borders from philosophy to cybernetics, aesthetics to concrete poetry, physics to the theories of chance," says Reichardt. His interests spanned the philosophy of science, logic, art, and aesthetics, and he passionately believed that the arts and the sciences should complement each other. Bense wanted to focus on what was aesthetic in computer-generated art. He called this "generative aesthetics" and tried to base it solely on objective considerations, such as mathematics and logic. He was a prime force in organizing the first exhibition of computer art in 1965 in Stuttgart, Germany. But this was overshadowed almost immediately by E.A.T. and then by Reichardt's *Cybernetic Serendipity*.

Determined to do something different from *9 Evenings: Theater and Engineering*, Reichardt planned a show whose theme would be the relationship between creativity and technology, with the emphasis on computer art. She proposed it to the co-directors and founders of the ICA, Herbert Read, the eminent poet and critic, and Roland Penrose, the artist, art historian, and friend of Picasso. At the time, the ICA was just starting up. Located on the first floor of 17 Dover Street in London, it was short of money, staff, and space. The space problem was solved in January 1968 when it moved to its present quarters on Carlton House Terrace.

Using Read's and Penrose's connections, and with interest from C. P. Snow, whom Reichardt had met at a dinner party, she managed to raise £25,000 from, among others, IBM, the Arts Council, the US Air Force Research Labs, Bell Telephone Labs, Boeing, and General Motors. Individuals who provided encouragement included Klüver's old friend Pontus Hultén; the artist György Kepes, who had founded the Center for Advanced Visual Studies at MIT; Warren McCulloch, an early champion of computers and cybernetics; Nicholas Negroponte, the architect and founder of MIT's Media Lab; the mathematician and computer scientist Seymour Papert from MIT; and John Pierce of Bell Labs. The staff worked long hours unpaid. The budget was large enough that the ICA even advertised on the London Tube.

The participants included computer artists and musicians, among them Frieder Nake of the Stuttgart Group; Ken Knowlton of Bell Labs; the Op Art painter Bridget Riley, who produced paintings resembling moiré patterns; John Cage; the Greek composer Iannis Xenakis, who used mathematics and architecture for composing; the roboticist Bruce Lacey; the media artist Nam June Paik; Jean Tinguely; and Nicholas Negroponte.

Reichardt called the show Cybernetic Serendipity "because it deals with possibilities rather than achievements . . . with exploring and demonstrating some of the relationships between technology and creativity . . . The aim is to present an area of activity which manifests artists' involvement with science, and the scientists' involvement with the arts." The show aimed for the complete eradication of boundaries between the arts and sciences. Reichardt felt that an artist's creative ability to use computers or robots was the most important factor in choosing them, rather than their artistic background. She wrote, "No visitor to the exhibition, unless he reads all the notes relating to all the works, will know whether he is looking at something made by an artist, engineer, mathematician, or architect." Some artists were critical of her approach. Stephen Willats, a computer artist with a fine arts background, complained that the show was "full of scientists masquerading as artists." But his was a minority view.

Cybernetic Serendipity was opened on October 2, 1968, by the minister of culture, Anthony Wedgwood Benn. Princess Margaret was among the guests. As Penrose emphasized at the press conference, "The ICA has always realized that the arts need to establish a very close relationship with science ... It is our custom at the ICA to explore the mysteries of technology and wherever possible to find new uses for its utilitarian inventions which will bring us enjoyment, enlightenment, and a wider consciousness."

Cybernetic Serendipity took London by storm. About 60,000 visitors came to have a look, and 325 artists and scientists participated. It was a landmark event. It was the first ever gallery show to exhibit robots, cybernetics, and computer art, and provided a technological feast of music, poetry, and prose composed on computers, together with computer-programmed choreography, computer films, computer paintings, computer-animated films, and experiments in ways to visualize data.

Radio-controlled robots roamed the halls, stopping to kiss the curious. A music computer improvised on tunes which visitors whistled into a microphone. One of the star exhibits was *Colloquy of Mobiles* by the eminent cybernetics researcher and psychologist

2.4: Gordon Pask, *Colloquy of Mobiles*, 1968.

Gordon Pask. It consisted of five rotating mobiles, three made of glass and fiberglass ("females") and two swelling aluminum forms ("males"). When a "male" emitted a beam of light, it struck the mirror inside the "female's" structure and deflected the light back to the "male," suggesting a moment of mutual satisfaction. It was a demonstration of how entities could communicate and learn about each other via feedback systems using electronic signals without human intervention.

Knowlton and Harmon's famous reclining nude, *Studies in Perception I*, now renamed *Mural*, appeared. Knowlton's pioneering computer animations were also shown, as was A. Michael Noll's study of Mondrian, Nake's line drawings, rather small and benign painting machines by Tinguely, and Schöffer's cybernetic sculptures which responded to their environment. The show was a great inspiration to young artists. One, Ernest Edmonds, then a mathematician at City of Leicester Polytechnic, recalled that it "showed a healthy lack of concern for what exactly should be seen as art, science, or technology. It demonstrated how cross-disciplinary stimulation could encourage exciting innovation."

It was the interactive pieces that most interested budding artists such as Edmonds. Besides Schöffer's cybernetic sculpture, there was Edward Ihnatowicz's *Sound Activated Mobile (SAM)*. Ihnatowicz was something of a legend in his day. Trained in sculpture at the Ruskin School of Drawing, Oxford, from 1945 to 1949, his interests included photography and electronics. He even built an oscilloscope, used for testing electronic apparatus, from spare parts. He started off at the BBC, then worked as a furniture designer. In 1962, when he was forty, he decided to devote himself to sculpture of a mechanical sort—pieces that moved.

There and then he left his comfortable house and his family for an unheated garage in Hackney, in the run-down East End of London. His son and his friends, fans of Jack Kerouac and Henry Miller, saw in his move "a great deal of existential romance, glamour and drama."

Ihnatowicz got down to work, inspired by the aesthetic shapes he saw in the car parts that he cannibalized. Using spare parts, hydrau-

lic valves from government surplus equipment, analog circuits
(circuits with wires, resistors,
capacitors, vacuum tubes, and
transistors), and servo mecha-
nisms (devices that compare
input to output signals, providing a means of correction
as in gunnery sights), he built
his *Sound Activated Mobile
(SAM)*. About three feet high, it
looked rather like a spine with
a flowerlike head consisting of
four fiberglass reflectors, each
with a microphone in front of
it. *SAM*'s circuitry enabled it to
respond to sound by bending
and trying to locate the source.
At Cybernetic Serendipity, *SAM*

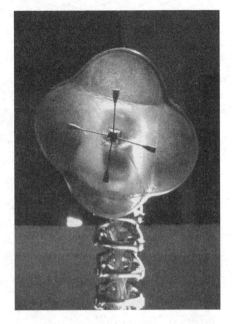

2.5: Edward Ihnatowicz, *Sound Activated Mobile (SAM)*, 1968.

leaned toward people talking as if it were listening. Visitors had the
eerie sensation that they were being watched.

The press release for the exhibition promised "a gallery full of
tame wonders which look as if they've come straight out of a science museum for the year 2000." Most of the reviews concluded
that the exhibition had indeed achieved its aim. Nigel Gosling
wrote in the *Observer*, "I think we shall look back to it one day as
a landmark." The *Evening Standard* enthused, "Where in London
could you take a hippy, a computer programmer, and a ten-year-
old schoolboy and guarantee that each would be perfectly happy
for an hour with you not having to lift a finger to entertain them?"
The *Hampstead and Highgate News* added hyperbolically, "For
breaking new ground, revealing new fields of experiment, seminal
experience, the exhibition Cybernetic Serendipity . . . is arguably
the most important exhibition in the world at the moment."

It is important to remember that Cybernetic Serendipity sprang out of a particular moment in history. The 1960s was an era of great political and social upheaval; 1968 was the height of the protests against the Vietnam War, in which technology had played an integral part. Overall, however, the mood was upbeat; England had finally emerged from the dark post–World War II days. As Catherine Mason, who writes on the history of computer art in England, notes, "Cybernetic Serendipity [was] facilitated and inspired by a postwar spirit of optimism in the (positive) power of new technologies." In her catalogue essay, Reichardt struck a more somber note: "Cybernetic Serendipity deals with possibilities rather than achievements, and in this sense it is prematurely optimistic." Cybernetic Serendipity was one of several exhibitions that year which went against the grain by overtly celebrating the computer and computer art. An exhibition in Zagreb that August looked into the social and political implications of computers; not coincidentally, the organizers found it difficult to obtain funding.

Part of the show's intent was to convince people that computers, which were new to the public at large, were not just not sinister, but cool. Reichardt noted that at first plotters—devices hooked up to computers for graphics—were used only for generating visual solutions to technical problems. Then engineers realized that they could also create pleasing images. "Thus people who would never have put pencil to paper, or brush to canvas, have started making images, both still and animated, which approximate and often look identical to what we call 'art' and put in public galleries." The press had a field day on that one. The *Sunday Telegraph* wrote, "This exhibition . . . serves to show up . . . a desolation to be seen in art generally—that we haven't the faintest idea these days what art is for or about." The *New Statesman* commented, "The winking lights, the flickering television screens and the squawks from the music machines are signaling the end of abstract art; when machines can do it, it will not be worth doing."

Catherine Mason points out a key aspect of Cybernetic Serendipity: "For computer scientists whose work was variously ignored, misunderstood or downright reviled, the importance of establishing a community of fellow practitioners and like-minded supporters was crucial to the continuance of their art."

## The Machine: As Seen at the End of the Mechanical Age

Around the same time that *9 Evenings: Theater and Engineering* and Cybernetic Serendipity were fueling an explosion of interest in technology, Pontus Hultén curated an exhibition entitled The Machine: As Seen at the End of the Mechanical Age at New York's Museum of Modern Art from November 1968 to February 1969. The catalogue was dramatically bound in tin-can steel with an enamel-painted design of the MoMA building stamped on the cover.

This was a landmark exhibition on the history of the machine as depicted in art from the Renaissance to 1968. The works included prints, paintings, sculptural and mechanical works, and a few electromechanical pieces, together with an example of laser art and a couple of examples of computer graphics. The most radical work was *Heart Beats Dust* by the artist Jean Dupuy and the engineers Ralph Mosel and Hyman Harris: a throbbing sound, light, and color installation of red dust thrust into a cone of light.

Hultén stated, "This exhibition is not intended to provide an illustrated history of the machine throughout the ages but to present a selection of works that represent artists' comments on aspects of the mechanical world." The aim was to go beyond *9 Evenings: Theater and Engineering*. "Many artists today are working closely with engineers in collaborative efforts that may have significance far beyond that of merely producing new kinds of art for our delight," Hultén went on. Technology was creating a brave new world, but it was important to bear in mind that "in planning for such a world, in helping to bring it into being, artists are more important than politicians, and even than technicians."

Jack Burnham, art critic turned sometime art theorist, took this further in the exhibition Software which he curated at the Jewish Museum, New York, between September and November 1970. By now artists were working toward eliminating nonessential forms, aiming to grasp the essence of an object in a movement they called Minimalism. This fit neatly into conceptual art, where the concept takes precedence over the object's material form. Art was becoming dematerialized.

As Sol Lewitt, one of the founders of conceptual and Minimalist art, wrote, "In conceptual art the idea or concept is the most important aspect of the work. When an artist uses a conceptual form of art, it means that all of the planning and decisions are made beforehand and the execution is a perfunctory affair. The idea becomes a machine that makes the art."

Burnham saw a connection of this with another trend, the mixing of art and technology. *9 Evenings: Theater and Engineering* had explored how systems of electronics produced art; Cybernetic Serendipity studied the role played by computers in creative acts such as art, dance, and music; and The Machine: As Seen at the End of the Mechanical Age focused on the history of mechanical technology, from Leonardo's drawings through to the 1966 E.A.T. exhibit.

Burnham sensed that the traditional media of art were being replaced by systems of electronics, sometimes guided by cybernetics. Systems analysis now stretched from policy-making to military planning all the way to Systems Art. Burnham decided to concentrate on the driver behind this: computer software, the ghost in the machine. To him, software was the essence of the new art. It was an information processing system, like the human mind. One art theorist, Edward Shanken, credits Burnham with seeing parallels between computer software and the "increasingly 'dematerialized' forms of experimental art." Burnham's insight was that it was no coincidence that conceptual art had emerged just at the moment of "intensive artistic experimentation with technology." There was a clear parallelism there.

Software, sadly, was a failure. Several exhibits did not work, due—ironically—to software problems. There were disagreements with the board of trustees of the Jewish Museum, where it was housed, and it ran way over budget. Moreover, Burnham's assumption that society was becoming increasingly centered on information and on a particular art form (conceptual art) was out of line with the times—which, in the 1970s, tended toward broader vistas such as Marxism, feminism, semiotics, and psychoanalysis. Postmodernism had arrived and immediately showed a distinct distaste for science.

On top of all these, the protests against the Vietnam War and the political climate that they engendered served to stymie any further collaboration between artists, scientists, and engineers. Students and other antiwar activists, including many artists, blamed science and technology for the development of lethal weapons such as napalm and Agent Orange. The scientific community itself was split over the issue. There was also economic recession, which meant that the funds for collaboration had pretty much dried up anyway, though the psychedelic drugs that were produced at the time no doubt played a part in inspiring later computer art.

As a result of these factors, there were few major collaborations during the 1970s and 1980s. Artists mainly worked alone on technological and scientific themes. However, interest in video art, among other experimental arts, persisted, challenging trends, particularly the growing commercialization of art as exemplified in Pop Art. Marshall McLuhan's books, examining the changes in how we communicate with one another and presaging the Internet, also helped keep the idea of collaboration between artists and scientists alive.

# 3

# The Computer Meets Art

I n the early 1950s, electronic devices—such as televisions, washing machines, and telephones—were beginning to play a growing part in everyday life. Some artists, however, saw different possibilities in them. Rather than just being useful, perhaps they might produce an entirely new form of art, one suited to the forthcoming electronic age.

The pioneer was Ben F. Laposky, an American amateur mathematician and artist. He had never studied electronics but was often described as a natural. To him, the wavy images on the television-like screens of oscilloscopes—used to test electronic equipment by observing wave signals from constantly varying voltages—were reminiscent of the art of Duchamp, Malevich, and Mondrian.

Inputting two or more signals into an oscilloscope makes the wave forms pulsate, producing intriguing, ever-changing patterns. Using long exposures, Laposky photographed the images undulating on the screen, initially in black and white and later in color, making beautiful, curved, symmetrical designs. He called these Oscillons or Electronic Abstractions and in 1953 had an exhibition of them.

These were the first "computer graphics." There were two notable points. The images were not man-made but machine-made, generated through automatic processes. The role of the artist was to work with the machine and select the most appealing images, rather than creating images himself. And there was a virtually infinite variety of possible forms.

Electronic art had begun.

## The beginnings of computer art: A. Michael Noll

In the early 1960s, computers were in their infancy—large unwieldy machines used primarily for code-breaking and number-crunching—and they were so expensive that only universities, research laboratories, and large corporations could afford them. This meant that the only people who had access to them were computer scientists and mathematicians. Nevertheless, it was not long before lateral thinkers began to be curious about what else these clumsy machines might be capable of.

A. Michael Noll, an engineer at Bell Labs, was one of the first.

In those days, computers used a plotter: a brush or pen controlled by the computer that enabled it to draw. One day, when the summer interns at Bell were using the microfilm plotter to display their data, a programming error turned one student's work into random graphical nonsense, which he jokingly called "computer art." "I can still remember him running down the hallway with his 'computer art,'" Noll writes. Inspired by the unintentional art, Noll decided to create it intentionally.

At Bell Labs, Noll had access to the latest computer technology—the IBM 7094—which was used to produce numerical solutions to complicated mathematical equations. This machine filled a huge room. Into it were fed equations and a means for solving them, in a programming language called FORTRAN, written on cards using punched holes at set positions, known as Hollerith cards. The operator would place a stack of cards in a hopper, push the on button, and the calculation would be up and running, with a whirring of tape drives in large cabinets and an array of lights flashing on and off. Such machines were essential to the Apollo space program, though today's laptops easily outpower them, with disks and hard drives replacing tapes. To analyze the numerical solution—the computer's output—engineers hooked it up to a plotter designed to make curves.

In the course of his work, Noll often did this with a microfilm plotter using 35mm film, which printed out the graphical results. After encountering the euphoric graduate student, he began to look at

these curves differently. In August 1962, he wrote a technical mem-
orandum describing them as "computer art." The Bell Labs man-
agement, however, insisted that Noll call them "patterns," because,
they argued, "only the art world could determine what was 'art.'"

Noll persisted. He was determined to produce something that
everyone would concede was computer art. But how to do it apart
from using patterns produced by solving equations? Random dots
produced by a computer and joined up by lines would not be very
interesting. But supposing he took points that appeared to be random but were not, then used a program to join them up?

So Noll wrote a program that produced dots distributed in a way
prescribed by a mathematical equation called a Gaussian distribution which generates bell-like curves. He then connected the dots
from bottom to top in a random way, producing a continuous zigzag.
The result reminded him of Picasso's *Ma Jolie*, which he had seen
many times at the Museum of Modern Art. He called his creation
*Gaussian Quadratic*.

For three years, Noll turned his attention from computer art to
experimentation with three-dimensional animations, working with
his colleague at Bell Labs, Ken Knowlton. Then, in 1965, Howard
Wise, owner of a gallery on New York's Upper West Side, invited
Noll's colleague, Béla Julesz, a thirty-seven-year-old Hungarian-
born researcher in visual and auditory perception, to show his work
there.

Wise, an art patron and a former dealer, had held exhibitions of
Pop Art and was interested in dot images, such as in Roy Lichtenstein's work. He had come across Julesz's work in an article in *Scientific American* and was intrigued by the patterns Julesz created
using pairs of random dots positioned not quite one above the other,
so that when viewed with a stereoscope they produced an illusion of
depth.

Julesz was reluctant to exhibit alone and asked Noll to join him.
Noll was delighted, though, as he recalls humorously, "Béla always
felt that his patterns were scientific with little artistic merit. I felt
that my patterns were artistic with no scientific merit." Neverthe-

less, the two young engineers were thrilled with this development. They were starting brand new artistic careers and at the very top, at the Howard Wise Gallery. They agreed to split the profits from their sales, though "in the end, not a single work was sold," remembers Noll. Noll exhibited variations on *Gaussian Quadratic*, and he and Julesz also showed joint works of 3D stereoscopic images.

Noll and Julesz thought the exhibition would be a good demonstration of the creativity of Bell Labs, but the lab directors thought otherwise. This was the year before *9 Evenings* took place, and they were still in cautious mode. They were reluctant to associate the lab with work that might be considered frivolous and generate negative publicity about the lab wasting money and resources. They put pressure on Wise to cancel the show. He responded that it had already been announced and threatened to take them to court. In the end Bell Labs backed down, but suggested to Julesz and Noll that they register the works under their own names and steer clear of journalists.

The copyright office at the Library of Congress balked at copyrighting a work as computer art, Noll recalls, claiming that something generated by a computer could not be art because a computer was merely a number-cruncher, incapable of doing anything creative. Noll replied that the computer's output was created with programs written by a human being. The bureaucrats finally relented, and *Gaussian Quadratic* became the first copyrighted piece of computer art.

The exhibition was the "very first major public showing of digital computer art (certainly in the US)," Noll asserts. Stuart Preston wrote in the *New York Times*, "the wave of the future crashes significantly at the Howard Wise Gallery . . . freed from the tedium of technique . . . the artist will simply 'create.'"

## Computer art in Stuttgart

That same year, 1965, two German artists, Frieder Nake and Georg Nees, had exhibitions of computer art in Stuttgart. Nees's exhibition was two months before Noll's and could indeed lay claim to having been the first of its kind. Whereas Noll simply made art, Nake and Nees were driven by a philosophical agenda.

Nake and Nees were inspired by the German philosopher Max Bense, who had made the critical suggestion to Jasia Reichardt, "Look into computers." Bense's work involved exploring the possibility of using computers to pin down a scientific notion of aesthetics. Part of the basis of his theory was the Harvard mathematician George Birkhoff's reduction of aesthetics to a mathematical formula: that beauty is in the inverse ratio of order to complexity, in other words, that the less complex an art object is, the more aesthetic it is. But what about intuition and subjectivity? Aesthetics was clearly more complex than this.

Bense combined Birkhoff's ideas with developments in information theory, which studies the optimal conditions for transmitting a signal so that it arrives largely intact, and Noam Chomsky's generative grammar, rules for generating language that are hardwired into the brain and therefore present from birth. How signals are transmitted through a background of random noise involves statistical analysis. Bense used statistics and uncertainty to devise a system free from the narrow mathematics of cause and effect, which had bedeviled Birkhoff's theory. He concluded that creativity did not obey any rules of logic, and neither did aesthetics.

Bense called his framework "generative aesthetics." He believed it could be applied to computers programmed with algorithms to produce almost random number sequences that generated aesthetic images which he called Generative Art.

Georg Nees, a doctoral student of Bense, put Bense's ideas into practice. In his work *Locken*, the plotter's pen traced out randomly generated arcs of circles. Nees intervened by turning the plotter off at the point where he intuitively felt that the image was complete and could be considered a work of art. Bense invited him to show his works at the Studiengalerie at Bense's institute at the Technical University in Stuttgart. When the show opened, the public was outraged. "Some guests were nervous, some hostile, and some simply left at the thought of Nees's drawings being considered art," Nake writes of the opening. Bense's response was to call Nees's work "artificial art," cleverly linking it with artificial intelligence and his own generative aesthetics.

Nake was a student of mathematics whose doctoral dissertation involved probability theory. While he was at Stuttgart, the university acquired a state-of-the-art plotter, a Zuse Z64 Graphomat. But there was no program for hooking it up to the computer, a Standard Elektronik Lorenz (ER65). The task fell to Nake. To test his programs he produced graphic representations, generating dots with pseudo (almost) random number generators and using his own intuition as to when to stop. Inspired by Bense, his aim was to write computer programs that would automatically generate drawings of an aesthetic quality with no other technical or economic purpose. Nake found the combination of intuition, creativity, and probability intriguing. He exhibited his works along with some of Nees's at the Galerie Wendelin Niedlich in Stuttgart in November 1965.

The computer art produced by Nees and Nake at Stuttgart and by Noll at Bell Labs was of its time: linear, abstract, highly geometrical. "The drawings were not very exciting. But the principle was," recalls Nake. Indeed, what the three were doing was quite revolutionary. Nake, Nees, and Noll are often referred to as the "three Ns" of computer art.

## The computer versus Mondrian

"With hindsight, the digital computer has had a big impact on art, particularly in advertising, design, and animation. And a new breed of digital artist emerged," writes Noll. With no philosophical axe to grind, Noll was freer. How about, he wondered, comparing art done by a computer, which at that time was made up of collections of lines, with art by an acknowledged master—for example, Mondrian, many of whose works also consisted of horizontal and vertical lines?

So Noll wrote a program that randomly connected randomly placed dots with horizontal and vertical bars. As in the works of Nees and Nake, the random dots were not in fact completely ran-

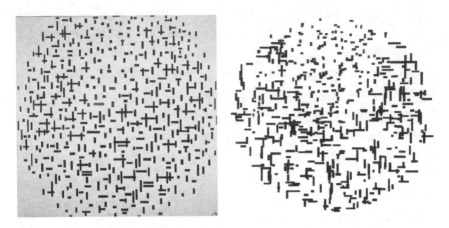

3.1: *(left)* Piet Mondrian, *Composition with Lines*, 1917. 3.2: *(right)* A. Michael Noll, *Computer Composition With Lines*, 1964.

dom. The randomness was determined by a mathematical algorithm for calculating sequences of numbers with no correlation between them. As the mathematician Robert R. Coveyou wrote, "The generation of random numbers is too important to be left to chance."

Using a plotter, Noll produced a black and white image he called *Computer Composition With Lines*. Then he showed it to a hundred people at Bell Labs alongside a black and white photograph of Mondrian's 1917 *Composition with Lines*. A mere 28 percent correctly identified the computer version, and 59 percent actually preferred it to the Mondrian. When asked, they explained that the computer image was neater, more varied, imaginative, soothing, and abstract. Non–technically-minded viewers, such as support staff, associated randomness with human creativity and incorrectly identified the computer-generated image as the Mondrian.

Here the puzzle of creativity enters, as it did for Bense, Nake, and Nees. Both pictures were actually conceived by humans but, in the case of Noll's, with the computer as an interface between the human programmer-artist and the plotter. Noll concluded that "artistic merit is not . . . something that can be determined by jury."

# 40,000 Years of Modern Art

Back in 1851, the hugely successful Great Exhibition at the Crystal Palace in London promoted both art and design. It was the brainchild of Prince Albert, Queen Victoria's consort, who had long advocated the coming together of "Raw Materials," "Machinery," "Manufactures," and "Sculptures." But it did little to counter the schism in the British art establishment between fine arts and crafts, which practitioners argued had nothing to do with each other. Crafts were generally relegated to a lower status than fine arts. And after World War II, "technology" was added to "crafts."

Then, in 1946, just after the end of World War II, the poet, anarchist, art critic, and historian Herbert Read founded the Institute of Contemporary Arts in London, with the specific aim of providing a space for avant-garde art and debate. The first exhibition was Forty Years of Modern Art, the second, in 1948, was outrageously entitled 40,000 Years of Modern Art. It was a showstopper. Curated by Read and Roland Penrose, it included Picasso's *Les Demoiselles d'Avignon*, on loan from the Museum of Modern Art in New York.

In the early 1950s, the younger artists, scientists, and designers at the ICA formed the Independent Group. The historian Catherine Mason writes, "The Group's avant-garde artistic practices, which advocated that the artist take advantage of the new technologies, contradicted the perception of a complete gulf between the sciences and the humanities in Britain." Artists and scientists were finally beginning to come together.

## Beauty in nature: Richard Hamilton

The artist Richard Hamilton, an active member of the ICA, and his scientific collaborator Kathleen Lonsdale, an eminent crystallographer at University College London, were the driving force behind the exhibition Growth and Form at the ICA in 1951. It was an attempt to bring the arts and sciences together and focused on aspects of growth, form, and structure in atoms, crystals, animals, plants, and stars. Hamilton created an environment with enlarged photos taken through microscopes and films showing the growth of crystals and sea urchins, literally engulfing viewers. Lonsdale believed that symmetries in crystals were aesthetic in an artistic as well as in a scientific sense. She was much influenced by Charles J. Biederman's *Art as the Evolution of Visual Knowledge,* in which he proposed that artists should look to science for themes rather than to the world of the emotions.

For his part, Hamilton was inspired by D'Arcy Wentworth Thompson's *On Growth and Form,* which explains the title of the exhibition. Thompson was a scientist of some note and also an accomplished classicist. In his beautifully written book, he describes how forms in nature conform to mathematics. These were the words that caught Hamilton's eye: "The world is as it is, because it must follow certain mathematical principles." If artists wanted to look for beauty in nature, they would have to turn to science.

Inspired by the exhibition, a simultaneous gathering of major artists and scientists was held, entitled Aspects of Form: A Symposium on Form in Nature and Art. Among the speakers were Rudolf Arnheim, Ernst Gombrich, and Konrad Lorenz. The Scottish financier, industrial engineer, physicist, and intellectual gadfly Lancelot Law Whyte made the comment that "There is something to gain . . . from examining the ideas of colleagues in neighboring fields." Today such a thought would seem innocuous enough. But at the time it was downright revolutionary.

Hamilton was one of the most influential artists of the twentieth century and a close friend of many avant-garde painters, including Marcel Duchamp, who often attended discussions at the ICA. His 1956 work *Just what is it that makes today's homes so different, so appealing?* is a comic book–like collage featuring a sitting room with an automobile advertisement on the wall, a woman with pneumatic breasts, expensive furniture, and a body builder holding a lollipop with the word "POP" on it. "Catch up with America and its technology" is written all over it. The work became iconic. Thus was born Pop Art, the movement that displaced American Abstract Expressionism and extolled technology as well as art. Hamilton gave his own definition of Pop Art as "Popular (designed for a mass audience), Transient (short-term-solution), Expendable (easily forgotten), Low-cost, Mass-produced, Young (aimed at youth), Witty, Sexy, Gimmicky, Glamorous, Big Business."

## Abstract art: Victor Pasmore

As early as 1947, Richard Hamilton and Victor Pasmore were teaching at the Central School of Art and Design in London following a curriculum styled after the Bauhaus School, which was founded by the German architect Walter Gropius in Weimar in 1919 and lasted twenty years until it was closed down by the Nazis. In their Basic Design Course, the Bauhaus members sought to blur distinctions between fine art and design, which they took to include technology.

Pasmore had begun his painting career as an impressionist in the fine art tradition. Then, inspired by Klee and Kandinsky, he shocked the art establishment by turning abstract. "While [abstraction] was fine in design, it was in Fine Art that it caused an uproar," he wrote. Built up of lines, with a minimalist geometry, his subsequent work had a specifically scientific and technological bent. At the same time, American Abstract Expressionism was bursting onto the European scene, exciting young artists. It too was anathema to the "fine arts."

If he was to extend his own version of the Basic Course to include abstract art, Pasmore would have to go elsewhere. His work was well known and he had no trouble obtaining a position at King's College in Newcastle-upon-Tyne in 1953. A year later, Hamilton joined him. Soon other "outlying" art schools started taking an interest in what Hamilton and Pasmore were doing. First Sunderland School of Art, and then Leeds College of Art instituted Basic Courses. By 1960, the Basic Course had become a model for first-year art studies as well as for a radical overhaul of art teaching.

## A radical new outlook: Roy Ascott

I first came across Roy Ascott at a symposium at the University of Plymouth, where he is a professor. His lecture was more a performance, presented with feet on desk, covering everything from his journey through art theory and art practice, to Zen Buddhism and consciousness studies. Short and stocky with an air of great confidence, he exudes a genuine and disarming friendliness.

Ascott is and always has been a maverick. At King's College, Newcastle, he learned painting from those two radicals who had shaken up the London scene, Hamilton and Pasmore. Jokingly he recalls having been on the verge of psychosis after studying painting under two such different artists, one Pop, the other geometrical, and who offered totally opposing viewpoints of the world in which we live.

This sparked an interest in the French artist Paul Cézanne, who exploded a single perspective point into several, drawing out from the flux of experiences that flood the eye, creating new ways of visualizing a scene. Ascott concluded that the viewer had to play an active part in a work of art; it should be a joint project. "It can be argued that the aesthetic of process and emergence in computer-mediated art starts here," with Cézanne, he writes. The key notion was process, of interaction between artist and viewer, forming a system.

This concept of process and system took further root when Ascott visited the Paris studio of Nicolas Schöffer, the father of cybernetic art, in 1957. Schöffer, he says, was the "kinetic art king; Schöffer did it big and shiny." For Schöffer, "Cybernetics was everywhere and could even lead to a new civilization." Ascott was impressed with Schöffer's dedication not only to bringing cybernetics into art but to shaping a new worldview as well.

With the core idea of cybernetics in mind—interactions within a system—he produced a series of paintings in which paint was applied in a random way to glass panels that the viewer could slide back and forth, creating new works every time. In this style of art, "gestural painting" or "action painting," the system was the glass panels plus the viewer. He called them *Change Paintings*.

Pasmore wanted to take the Basic Design Course back to London and chose Ascott, his star student, to do so. In 1961, on Pasmore's weighty recommendation, Ascott became head of Foundational Studies at the Ealing College of Art.

3.3: Roy Ascott, *Change-Painting*, 1960.

Just before leaving King's College, Ascott had come across books on cybernetics by Norbert Wiener and others. For him it was a "Eureka experience—a visionary flash of insight in which I saw something whole, complete, and entire." The mathematics was beyond him, but he happened to know someone who could help him: Gordon Pask. A distinguished researcher in cybernetics and psychology, Pask owned Systems Research Ltd in Surrey, one of whose specialties was adaptive teaching machines, designed to construct knowledge through interactions between humans and machines.

After one of his exhibitions, Ascott asked Pask to explain cybernetics to him. Back at Ascott's studio, Pask gave him the tutorial of his life, which lasted until 2 am and convinced him that "'control and communication in animal and machine' is a proper study for the artist."

At Ealing, Ascott instituted the now famous Ground Course, a radical experiment in art education. Inspired by systems thinking from cybernetics, the aim was to give students a personal involvement in their art, surroundings, and social contexts. The Ground Course was a bit like an art boot camp. One way in which Ascott put students in touch with themselves was to lock a group in a pitch-dark lecture hall and subject them to intense flashes of light. Then they were released into another room where they stumbled over a floor covered in marbles, startling passersby.

Ascott firmly believed in blurring the distinctions between art, science, and technology, and often invited scientists and technologists to lecture. Pask, the cybernetic maestro, gave a lecture on the German-born artist and political activist Gustav Metzger and his auto-destructive art. Metzger's demonstrations were a protest against the use of technology, especially computers, by the military. Yet there was a positive aspect too, as he wrote in his Third Manifesto: "Auto-destructive art and auto-creative art aim at the integration of art with the advances of science and technology."

In 1962, Metzger himself gave a lecture at Ealing. One of the students who attended was a young man named Pete Townshend. He was greatly inspired by Metzger and incorporated his attitudes into his own, now legendary, guitar-smashing performances with The Who.

Among the staff at Ealing were the Cohen brothers, Bernard and Harold. Bernard was already a famous "young Brit" artist, producing paintings that were highly geometrical and abstract. Harold was more influenced by cybernetics; in a few years he would become a computer artist, exploring whether computers could produce art indistinguishable from art made by people.

Ascott's methods of teaching, as well as his art, were too revolutionary for the majority of the staff at Ealing. In 1964, three years after he'd arrived, he moved to Ipswich Civic College, another institution at which the London art world looked down its nose. Among his students was Brian Eno, soon to be massively famous for his experimental electronic music. Eno had enrolled at Ipswich because he couldn't get in anywhere else. He was excited to find that instead of painting and sculpting, this art school was interested in concepts. It was all about "cultures being leveled. We could see that [art and science] were not different places, we felt that we could understand something about the sciences and re-digest them and do our own experiments with them."

## Systems Art: Cornock and Edmonds

In 1966, the new Labor government responded to the increasing demand for further education by merging local technical and regional colleges and forming polytechnics. This meant that expensive equipment like computers could be concentrated in a few cross-disciplinary centers, though traditional artists loathed rubbing shoulders with engineers. A few years later, the eminent writer and painter Patrick Heron published a long article in the *Guardian* complaining of the "Murder of the Art Schools."

The polytechnics may have threatened the traditional art world but they were of immense importance in the rise of computer art. They were playgrounds where artists could use equipment not originally intended for artistic purposes—and to find out how to use it they had to seek out engineers and scientists. One such artist was Ernest Edmonds, who had been impressed by the ICA's Cybernetic Serendipity in 1968.

Inspired by Cézanne and Matisse, Edmonds had wanted to be an artist. He opted, however, to study mathematics and logic, and completed a PhD in mathematical logic in 1969 at the University of Nottingham, while he was on the staff at Leicester Polytechnic. Mathematics came easily to him, which gave him free time to pursue his interest in art.

Initially Edmonds was interested in El Lissitzky's Constructivist art and in concrete poetry, in those days written with a teletype machine or drawn with plotters. He had some questions about spray-painting techniques and dropped into Leicester Art College, at Leicester Polytechnic, where he encountered Stroud Cornock, a sculptor who was adept at computers. Cornock introduced him to Systems Art, the concept of participation and interaction in art combined with rules for proceeding, based to some extent on feedback from cybernetics. The two began projects based on the theme that "creativity is not totally in the hands of the artist." The basis of this approach was the rule-governed system: the artist establishes rules and creates a work of art within set boundaries, in the same way that music, language, and science are all governed by rules. In music, there are a finite number of notes and keys, and rules for combining them. Yet composers can create an infinite number of wonderful compositions.

Edmonds realized that what he had found so appealing in Cézanne and Matisse's work was their conciseness and the "clear emphasis on composition, the power of composition and what was underlying it." He analyzed the structures in Mondrian's work and concluded that it was more "non-constrained than it appeared to be." In other words, although Mondrian's paintings seemed rigidly geometrical, as in his *Composition with Lines*, they were actually quite loose in structure.

With the rather basic computers available in the 1970s, Cornock and Edmonds constructed algorithms—rules for solving problems—and looked into how these rules could aid creativity. Perhaps, they suggested, "it may no longer be necessary to assume that an 'artist' is a specialist in art." As *9 Evenings: Theater and Engineering* had shown, the "concept of art can survive artists as we know them when their control of the situation is reduced or when they are submerged in collaboration with engineers." In other words, the concept of art will change as artists collaborate with engineers.

## Son of *SAM*

Edward Ihnatowicz, the inventor of *Sound Activated Mobile*, the flowerlike robotic creation that leaned forward and listened to people's conversations at Cybernetic Serendipity, wrote, "We live in an industrialized, technological, and commercial world and if art is to have any relevance to it, it cannot hide in the romantic, artist-in-the-garret cocoon but must be prepared to come out and join the fray." Ihnatowicz was one of the first artists to incorporate a computer in a large-scale production. For this the contacts he made in the mechanical engineering department at University College London, when he built *SAM*, were crucial. He realized he needed to learn more about computers.

Ihnatowicz was interested in doing more than just making interactive machines. Suppose it were possible to make an intelligent machine? The only way such a machine could be created would be by programming it to interact with its environment, building its intelligence from the ground up, the opposite of the "top down" approach of artificial intelligence (AI), where mental processes are taken as given and written in code. To back up this approach, he referred to the Swiss psychologist Jean Piaget's theory that children develop intelligence by interacting with their day-to-day world.

The result was the colossal *Senster*. Fifteen feet long and eight feet tall, it was a horned beast made of welded steel and looked like a cross between a giraffe and an electricity pylon or a gigantic lobster's claw. Unlike *SAM*, it responded to movement as well as sound. It was the first work of robotic sculpture to be controlled by a digital computer and was funded by Philips, the Dutch electronics giant. Ihnatowicz built it at University College London (UCL) and exhibited it between 1970 and 1974 at Evoluon, an electronics exhibition founded by Philips in Eindhoven, the birthplace of Frits Philips, the owner-founder.

James Gardner, the design engineer who created Evoluon, wrote:

> Our aim is to get people (the press) to understand the *Senster* and not just treat it as "a gimmick". Ihnatowicz is an artist (sculptor) who is interested in science (biology, electronics, stress engineering). The *Senster* combines art and science. The art bit should not be forgotten as it cannot be categorized but is far more complex than exact science is(!).

*Senster* went way beyond what Alexander Calder had accomplished with his mobiles—which, even when powered by engines, merely repeated movements. It went beyond Tinguely, whose machines were aimless, and even beyond Schöffer, whose sculptures were not computer-operated. Ihnatowicz, clearly something of a joker, liked to describe *Senster* in human terms: "In the quiet of the early morning the machine would be found with its head down, listening to the faint noise of its own hydraulic pumps. Then, if a girl walked by, the head would follow her, looking at her legs."

## The computer that paints: Harold Cohen

The Slade School of Fine Art was established in 1871, and was modeled on art schools in Europe which treated art as a means of studying nature. A hundred years later, in 1972, under the visionary leadership of Slade professor William Coldstream, the forward-thinking artist Malcolm Hughes set up a center for computer artists at the Slade called the Experimental and Computing Department, or EXP. Hughes's assistant, Chris Briscoe, was in charge of the day-to-day running of EXP. Trained as an architect, Briscoe was brilliant at electronics. He taught artists the computer language FORTRAN, built circuits, and customized plotters as well as setting up the department's Data General mainframe.

Hughes invited lecturers such as Ihnatowicz, then a research assistant in the mechanical engineering department at UCL, and Harold Cohen, then at Stanford University in California, who described the latest version of his computer program AARON, which generated paintings. The students were thrilled to be taught by such accomplished and innovative artists. One, Paul Brown, declares, "For me, Ihnatowicz and Cohen represent the first two great masters of the computational arts."

Born in Britain, Harold Cohen was an acclaimed Systems artist whose work was a highlight at the Venice Biennale of 1966. He emigrated to the US in 1968 and joined the staff at the University of California at San Diego.

There "I met my first computer," Cohen recalls. "I was simply grabbed by the experience of programming; the extraordinary excitement of finding my brain working in an entirely new way. I had the feeling that there must surely be more to computing than came through in [Cybernetic Serendipity]." Right from the start, his aim in working with computers differed from that of just about every other artist, including Edmonds. He was interested in artificial intelligence, in whether machines could learn from a set of input rules such as those that governed drawing and painting. Three years later, as a guest at Stanford University's Artificial Intelligence Laboratory, he began putting together AARON. He has continued to elaborate on it.

AARON is a highly complex series of programs based on two sorts of knowledge, declarative and procedural, declarative being knowl-

3.4: Harold Cohen, *Merging Systems*, 2013.

edge of facts and things—arms, legs, bodies, shapes, and colors—and procedural being the knowledge of how to use them. A phone number is declarative knowledge, knowing how to use it is procedural knowledge. Cohen's rules are procedural. He has taught AARON to produce paintings in endless permutations and combinations of colors, styles, and topics from figurative to abstract, though he has yet to reveal the specific AARON algorithm. AARON's paintings, such as *Merging Systems*, are pleasant and colorful though the limitation is that AARON can only produce work using whatever material Cohen inserts. Nevertheless they have been widely distributed. In effect, AARON is an artist in itself.

## Computerized special effects: Paul Brown

The late 1970s and early 1980s were a fallow time for computer art. Besides dwindling government funds due to the anti–Vietnam War movements, there was the rise of postmodernism, which favored forms less linear and rational than those produced by computers. In Britain there were also cutbacks in government spending on education in the Thatcher era, particularly on work that had no immediate practical applications.

At the Slade, a further problem was resistance among the teaching staff to the use of computers for anything outside of science and technology. In 1981, the Slade's experimental department, EXP, closed its doors, though Chris Briscoe, the computer whiz kid who had run EXP, stayed on. In partnership with the Slade, he and his former student, Paul Brown, founded Digital Pictures, the first company in England to focus on computerized special effects. They were quickly snowed under with work.

Before his time at the Slade, Brown's life had taken an unconventional path. He studied at Manchester College of Art but hated the way in which students were limited to painting, printmaking, or sculpture. After he left, he became the artistic director of a light show called *Nova Express* which worked with legendary bands such as Pink Floyd and The Who. To start off with, light shows accompanied the music and stopped when the musicians left the stage. But why not fill the break with a different sort of light show, some-

3.5: Paul Brown, *Swimming Pool*, 1997.

thing more than a play of lights? His idea was to press water and oil between thin glass plates, creating random patterns, then illuminate them from below and project them. The squashing of oil and water by the glass plates, in addition to extreme heat from the lights, causing turbulence, made endlessly changing patterns. "The audience was enraptured by the oil and water light shows." Brown himself was impressed with the "amplification of minute turbulent events and how art could be made out of random physical events."

Brown also produced a video, *Mandala*, which was shown at The Video Show at the Serpentine Gallery in London in 1974, the first retrospective of video art in the UK. But he wanted to go further, and computers offered a promising route. Then he heard about EXP at the Slade, discovered Systems Art, and set to work to take it "into the computational domain." He continued the theme of shapes existing in water, using algorithms, as in 1997's *Swimming Pool*.

## Media art as fine art: Susan Collins

Susan Collins, director of the Slade and professor of fine art, brought computer art back to the Slade fourteen years after EXP closed its doors in 1981.

Collins discovered computers during a visit to New York, where she was shown a new Mac computer with a primitive drawing program. She was greatly impressed that one could draw and save as many versions as one wished. "Now I could take enormous risks."

Back in London, in 1988, there was still considerable distrust of computers. It was difficult even to get access to one. When Collins started making computer art, she found that no one really looked at her work, just at the computer itself. Only in the 1990s did the Slade begin to pay attention again to computer art, and in 1995 Collins was asked to be the computer tutor.

Collins did not want to set up merely a "computer facility for fine art." She wanted media art to be integrated with fine art. "I had such a clear vision of what it should be like to have media art embedded in

3.6: Susan Collins, *Seascape, Folkestone, 25th October 2008 at 11:41 am.*

the fine art environment," she recalls. She was convinced that electronic media should be studio- rather than technology-driven.

Collins's work *Seascape* demonstrates her conception of "art and
technology." She positioned five webcams at various vantage points
along the south coast of England, all framing the horizon and collecting,
transmitting, and archiving images slowly, pixel by pixel. The images
were continuously updated in real time, to create a fluctuating image
on the screen. When printed out, they look like paintings of seascapes.
It is only on close inspection that the viewer can make out the pixels.

## Making art move: back to Ernest Edmonds

In 1970, Ernest Edmonds was working on Systems Art at Leicester
Polytechnic and pondering what it meant for an artist to interact
with a computer. He was sure the computer would never replace the
artist. Rather, he saw it as a control system in which the artist's job

was to figure out how it worked. The "computer as a control system," he says, "set up a landscape where participants could operate. The artist's job was to define the space of opportunity—a decision space." But at the time he could not pursue these ideas because of the limitations of the available computers and computer software (FORTRAN).

Edmonds's background in mathematics, and his interest in cognitive science, led him to consider "control issues as conceptual," and to ask questions such as, Where does software stand? What is the role of software now?

If software was art, it was closest to poetry. After all, "a poem is something more than words on a piece of paper." He concluded that the "essence of software art is not materiality." Perhaps software was like the mind controlling the hardware or brain.

Then Edmonds heard a story about a meeting between the kinetic artist Alexander Calder, famous for his mobiles, and Piet Mondrian. "Calder walked into Mondrian's studio, looked at the highly geometrical works he was producing and asked him, 'Can we see them move?'" Inspired by this, Edmonds, too, wanted to find a way to make art move. He started working out computer programs that could generate a time-based art, based on algorithms that arranged colors over time intervals; they contained rules for time, as in music, as well as rules for the spatial arrangements of form (often squares). The result was a succession of abstract artworks based on simple structures. It was computer animation in real time instead of in frames—a step forward in this art form. Eventually, Edmonds found a way to project directly from the computer. To him, the work he created was a dialogue between the computer and the artist.

Edmonds is currently exploring the use of sensors such as cameras, microphones, and telephones to trigger rules, thereby adding another layer to the system of artist, computer, audience (see Insert). He is also rethinking cybernetics and systems theory, seeking ways to extract form from chaotic situations. He tells his students, "If you want to paint with oil paint, you need to understand oil paint. If you paint with software, you really have to understand software."

## Getting connected: Roy Ascott again

"Man, you're a real artist," a colleague at San Francisco State College said to Roy Ascott after Ascott punched someone who had made derisive comments about his art at a faculty meeting.

Ascott, the maverick who revolutionized art teaching, founded the Ground Course at Ealing, and inspired Pete Townshend and Brian Eno, held a succession of posts in the United States and the UK and was sacked from most of them. He now teaches at the University of Plymouth.

Since 1980, he has been exploring telematics, the bringing together of telecommunications with informatics, i.e., information technology, though he interprets this more widely, as a "technology of thought transfer" of information through cyberspace. He felt it was important to fit art into this scheme to allay fears of a dystopian world governed by technology run amok, which was what some predicted after two highly mechanized world wars and the onset of nuclear warfare. In particular, he wanted to find a way to draw the viewer into works of computer art, which all too often seemed cold and mechanical. Ever since the 1960s, he had been convinced that cybernetics, with its feedback loops, offered a way to bring together viewer, artist, and machine in a "telematic embrace," in which there would be "love" among them. But for this to occur, it would have to be at the level of consciousness.

At Ars Electronica in 1982, the Canadian artist Robert Adrian anticipated Ascott. In his interactive piece *The World in Twenty-Four Hours*, Adrian connected artists in fifteen cities around the world by telex and telephone, enabling them to exchange ideas which were fed into a computer. The technology was primitive and the results disappointing, but this was not the point. As everyone realized, the exercise showed that the "real power of the computer was connective, rather than a device or system just to make better pictures," as Gerfried Stocker, the present artistic director of Ars Electronica, puts it.

Ascott, meanwhile, expanded telematics to include consciousness and called it "technoetics," encompassing art, technology, and the mind. He coined the term in 1994 while teaching at the University of Wales, in Cardiff, where he founded the Planetary Collegium for the study of art, science, technology, and consciousness, the idea being that the four disciplines should work together to enhance creativity.

Ascott states that "consciousness is not generated, it preexists." It cannot be reduced to matter and therefore is not open to the laws of physics and chemistry as we know them today. His point of view includes a heavy dose of speculation and nonstandard interpretations of quantum physics (regarding the effect of an observer's consciousness on measurements), topped up with shamanism, which he experienced firsthand while living with a Brazilian tribe in the Upper Amazon.

Ascott's first major retrospective took place in London in 2011. It was entitled Roy Ascott: The Syncretic Sense, in which "syncretic" refers to bringing together disciplines that seem to have no connection. His *Change Paintings* were on show, as well as other interactive works. There were few computers. The key element was the use of feedback loops in promoting syncretism.

# 4

# Computer Art Morphs into Media Art

One November evening in 1980, people passing through New York's Lincoln Center were intrigued to find a giant screen occupying a wall there. It was full of larger-than-life people, staring back at them, even talking to them. When the New Yorkers tried waving or jumping up and down or pressing up close to the screen, people on the screen responded. Finally someone asked these apparitions where they were. A figure on the screen answered, "Los Angeles!"

The passersby in New York and at the Century City Mall in LA were all participants in a giant work of media art created by two artists, Kit Galloway and Sherrie Rabinowitz, who called it *Hole in Space*. Akin to a wormhole which could bring two distant places smack up against each other, it used a high-speed satellite link, and showed that people interacted with strangers at a distance in a far more relaxed way than if they were actually face-to-face. They danced, they clapped, they shouted exuberantly (they had to, to make themselves heard above the noise), they exchanged phone numbers.

The *Hole in Space* opened up for just three days. Even by the second day, the crowds were vast. Everyone wanted to be in on it.

It was art—not in a gallery but in the street. Computer art had entered the media age.

BY THE 1970S computers were no longer a mystery, but they were still large and expensive. The first really revolutionary development was when the Apple II appeared at the end of the decade. This was a personal computer capable of color graphics. The first modems came out at almost the same moment, allowing digital signals to be transmitted over phone lines.

With the 1980s came an explosion of technological advances. A new generation of personal computers came on the market, less expensive and easier to use, with new software enabling artists to create digital images. In 1982, Adobe Systems appeared with a new line of digital-imaging software, followed by the compact disc. Larger graphics cards were developed leading to Adobe Photoshop, making it easier for artists to experiment with digital art. In 1986, Steve Jobs purchased the Lucasfilm Computer Graphics Division from George Lucas and formed Pixar Animation Studios.

By the 1990s, developments were speeding up. First came hypertext mark-up language (HTML), then Tim Berners-Lee created the World Wide Web. Browsers allowed both text and image to be viewed, and Net Art took off. Interactive art began to incorporate digital technologies.

In 1979, the first Ars Electronica festival of electronic art took place in Linz, Austria. In 1980, the architect Nicholas Negroponte founded the MIT Media Lab to foster the interplay between design and technology. Universities in the US began to offer courses in digital art. New York's School of Visual Arts, which opened in 1947 as the Cartoonists and Illustrators School, was in the vanguard, instituting in 1986 an MFA degree in computer art. The students who arrived were already adept at the latest technology. "They have computer tools already and so the focus at SVA is on art-making and aesthetics," says Bruce Wands, the current director of computer education. "Creativity is their number one focus." Today the SVA has 4,000 enthusiastic students and a teaching staff who are all practicing artists.

## Media art and music: Bruce Wands

Bruce Wands is something of a Renaissance man. He's not only an artist/technologist but also a musician. As part of his current work he records ambient sounds, chooses those that are most melodious, then superposes actual music on them. He is also working on representing geometrical shapes using the proportions inherent in classical Buddhist art, using 3D software to create abstract images that nevertheless convey a spiritual content.

When I enter Wands's office at the School of Visual Arts in New York's Midtown, the first thing I notice is his desk, piled high with reprints, preprints, papers awaiting grading, and his own work in progress. When I sit down, he almost disappears from view. Immaculately coiffed, he speaks like the entertainer he once was.

In 1975, after thirteen years on the road playing guitar and bass and taking care of various bands' multichannel electronic music, he joined the MA program at the Newhouse School of Public Communication at Syracuse University wanting to learn to operate recording studios as well as to study the more technical aspects of music. His instructor was a student of Roy Ascott, Joseph Scala, who taught him to program a powerful IBM 705 mainframe computer to generate images. In those days they still used punched cards to generate line drawings from a plotter. Primitive though this seemed, says Wands, "I saw the future at that point. Computer graphics was it— the wave of the future. You draw better with a computer." One of his works, dating from 1976, is a white-on-black image of a geometrical form like a snail shell, drawn on a plotter: *Heartline*.

From Syracuse Wands went to New York, where he created and produced computer animation for the billboard display in Times Square and developed the computer-animated opening for NBC's *Saturday Night Live*.

Wands is a great believer in social networks and in Net Art, which, he believes, has gone far to eliminate such "preconceived notions as science being logical [while] art is not. . . . A generational shift has occurred." Today's students simply don't think of making art with computers as different or unusual; they automatically use computer science in their work. They don't label a project as artsci, but see art and science coming together into something that demands a better classification.

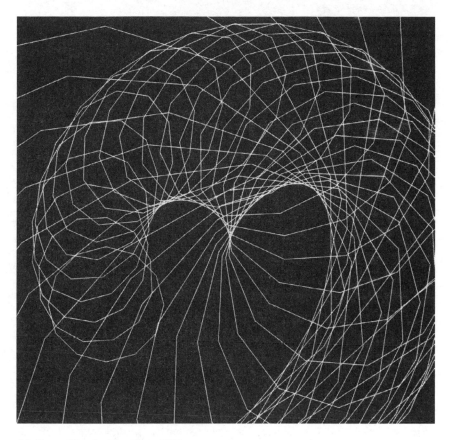

4.1: Bruce Wands, *Heartline*, 1976.

Wands believes that creativity is intuitional and experiential and emerges through experimentation. As an artist/musician, he first conceives a piece, then writes it down, then experiments. It is in improvisation, freewheeling it, where he sees creativity coming in. "To me improvisation is a non-cognitive process where 'I let the music play me,' rather than me playing the music." As a media artist, he works with algorithms capable of producing new images and sounds with every cycle. Nevertheless, "the artist has control over a wide range of programming parameters, thus giving him the ability to create images and sounds he never imagined." In traditional art, conversely, the artist has to create the permutations with no computer to help him.

The role of the artist is more to select than to create the perfect image, as in digital imaging, "where high resolution live action video can be shot and the final photograph is selected from the footage." Whatever seems intuitively to be right must be aesthetic. For Wands the creative thrust is in the process which to him is "primarily intuitive, rather than intellectual." He points out that creative mistakes often produce the greatest work. Experience primes the mind to hear the melody in, for example, a musical scale wrongly played.

In all of his work, Wands explores and pushes the boundaries of music and art with the help of computers, bringing art and technology ever closer.

## Super-real animations: Ken Perlin

In 1982 a groundbreaking science fiction film called *Tron* hit cinema screens, starring Jeff Bridges as a hacker who is trapped inside a computer and has to battle with the Master Control Program, an artificial intelligence that controls the mainframe, to get out. The film was one of the first to make extensive use of computer animation and included spectacular depictions of the bizarre world inside the computer.

The animation was extraordinarily natural, way ahead of anything that had been achieved so far, but Ken Perlin, who helped to create it, was not completely satisfied. To him, it still had too much of an unnatural machine look. So he began work at New York University's Media Research Laboratory, which he had founded, and developed a means for producing more natural textures. This works through algorithms that add random textures, known as procedural textures, to the surface, in the same way that natural textures appear random. This is akin to the random addition of signals or noise, hence the name Perlin Noise.

Procedural texture takes advantage of the natural limitations of our brains and perceptions. To create an animated image, Perlin proceeds one step at a time, building up the picture until the image is like an Impressionist painting. Blurry though it is, we can still see the scene sharply. Perlin elaborates. "Take Jane Austen. Bits of her characters are not real, but that makes no difference. She presents enough bits of characters that it will convince you they are real."

Perlin's fame spread far beyond academia when he won an Oscar in 1997. The citation read in part, "To Ken Perlin for the development of Perlin Noise, a technique used to produce natural-appearing textures on computer-generated surfaces for motion picture visual effects. The development of Perlin Noise has allowed computer graphics artists to better represent the complexity of natural phenomena in visual effects for the motion picture industry."

"Needless to say, my mom was very happy," says Perlin with a characteristically wry smile. Perlin exudes energy. Casually dressed, he is shining, despite the fact that he had been up all night editing. By the end of our interview I wonder if he ever sleeps. Bruce Wands describes him as a genius.

Perlin's research ranges over art and computer science, with the emphasis on computer graphics, animation, games, physics, and mathematics, plus an active interest in science education. To start out with he was interested in the arts, but eventually "opted for mathematics because [his] English teachers were terrible. How does anyone end up doing what they do?" After a science internship he went into computer graphics. For him this was "where it's at — inventing new things, aesthetic expression, doesn't get anywhere better than that."

"Look around," he says, "look at something in the world, and ask, 'What can we do with this? Can it lead to a computer game, a sculpture? Can its properties lead to something new mathematically?'"

Some artists say that art is art and science is science. Perlin's reply is that they start with the hypothesis that there are two sorts of people, artists and scientists. President Obama is always referred to as America's first black president, but in fact his father was black, his mother was white. The dichotomy is meaningless, just like separating artists and scientists. "Einstein was an artist. The entire universe just happened to match the one he had in his head. Categorization underplays the power of the human brain."

Perlin adds, "There is a willed ignorance by people who claim no connection with another discipline which they know little or nothing about. They have never engaged in the other discipline and are ignorant of methods and expectations there." "Great discoveries are made by people with knowledge of other disciplines."

As for him, does he consider himself artist or scientist? "I'm a researcher," he replies. For him, as for others, the labels "artist" and "scientist" are obsolete.

## Screen magic: Rick Sayre

Rick Sayre, too, is an animation maestro, a supervising technical director at Pixar Animation Studios. He has been involved in various capacities in creating the mesmerizing screen magic of films like *Toy Story*, *Monsters, Inc.*, and *The Incredibles*. In 1996, he won a Technical Achievement Award from the Academy of Motion Pictures Arts and Sciences for the Dinosaur Input Device (DID)—now known as the Digital Input Device—that helped bring the dinosaurs in *Jurassic Park* to life.

Sayre is a bit of an extraterrestrial himself. He looks every inch the sorcerer, with shoulder-length graying hair framing a long, thin face, high forehead, and chromium white complexion, always in black, adorned with earrings, painted nails, and ornate rings on every finger.

Sayre studied electrical engineering, computer science, film, and theater, a highly unusual combination, at the University of California, Davis, completing his studies at Berkeley. "For me, science and technology were only a means to an end," he says. "I was always interested in the effect it had on me, on how it created emotional states; in a way it's like the history of magic. That's what I was always interested in—magic and spectacle, performances and theater . . . particularly that technology allowed me to produce something that was unique, that had never been seen before." He was working on an animated short film with two classmates when Ed Catmull of Pixar offered him an internship. It was 1987, the year after Steve Jobs had established Pixar, and computer animation was just opening up. It was the perfect moment.

Sayre was inspired by, among others, Oskar Fischinger, the German-born nonfigurative animator. The Nazis condemned his colorful abstract animations as "degenerate art," and he saw the writing on the wall and left for Hollywood. His talent was recognized by Paramount Pictures as well as by Walt Disney, but his art was years ahead of its time, entirely different from the Mickey Mouse cartoon style of the day. Other major influences were Douglas Trumbull, a special effects pioneer who worked on Stanley Kubrick's *2001: A Space Odyssey*, and Don Coscarelli, known for the cult horror films he worked on such as *The Phantasm* and *Bubba Ho-Tep*, "doing shoestring visual effects."

The film industry operates as "hive mind," says Sayre. At a typical meeting, people "rattle off things as aesthetic touch points." Somebody might shout, "Ken Adam," suggesting that the film being discussed should have the high-gloss, futuristic look that Adam created for the classic early Bond films. Or perhaps it should have the feel of a Japanese anime.

Aesthetics can be put in and played with afterward. "Pixar can try something—test it out before a viewing audience and then alter it. Did it have the intended effect? If not, we change it."

Sayre goes on, "I don't have a precise definition of aesthetics. For us, an element of aesthetics is when there is intention behind the work and the intention is to create a certain emotional response in the viewer and that emotional response is going to be motivated by the story and is also going to be influenced by specific desires of the director. For example, we want [Mr. Incredible] to be somewhat stylized, we don't want him to feel like any particular person, we want him to have this generality and iconography of comic book characters, but not of a particular actor, a means to become, not physical texture, not detail but feeling alive."

When creating human characters for *The Incredibles*, the film-makers considered Pixar's most successful human character to that point to be Al McWhiggin in *Toy Story 2*, the greedy store owner who schemes to steal Woody. McWhiggin was depicted with skin covered in pimples and stubble. This worked in close up, but the detail was lost in medium and long shots. For *The Incredibles*, the director decided against any complicated texture. Simplicity was actually more demanding, in scientific terms. Keeping Mr. Incredible's face simple required a focus not on skin texture but on the way real human skin responds to light. "It was the most complex of any skin we had ever done—inspired by physics, medical physics, and mathematics."

Outside Pixar, Sayres collaborates with sculptors, metal fabrica-tors, musicians, dancers, and performance artists to produce inter-active artworks. In 1991 he worked with the robotic artist Chico MacMurtie to create *Tumbling Man*, for which they won the Prix Ars Electronica Distinction. *Tumbling Man* is a life-sized robot rather like a metallic skeleton, with spine, limbs, hands, feet, and a vestigial head, equipped with a computer. MacMurtie and Sayre were wired up to operate him by means of computers. When Sayre raised his arms, for example, the robot did the same thing. Problems arose when Sayre and MacMurtie made contradictory movements. Trying to obey both sets of orders, the robot ended up rocking back and forth on his back like a tortoise or doing somersaults. Neither Sayre nor MacMurtie knew who was in control. It was the robot who decided which of their instructions could best be utilized.

Sayre is interested in combining the big with the small, the abstract with the figurative. He is concerned that Hollywood, with its blockbusters, has become complacent about animated effects. The interesting developments tend to be in "peripheral areas, par-ticularly underground things like the Demoscene which are really vibrant."

He directs me toward the website pouet.net, a "wretched hive of scum and villainy." The Demoscene is a computer art subcul-ture producing computer programs which run audiovisual pre-sentations in real time, typically with no spoken words, flagging up whatever's new in that world. Members compete to show their latest demos at parties or at large meetings advertised on the website. It is alive with ideas, with creative people bursting with enthusiasm.

Sayre has always found the Demoscene exciting. One member who calls himself Gargaj reports on a gathering in Stuttgart in 2007 where "we met Rick Sayre, one of the technical leaders of Pixar, who exhibited a strange interest in the three foolish kids who were talking about demos endlessly." Sayre invited the group to visit him at Pixar. They were somewhat doubtful about whether he really meant it, but when they turned up he was true to his word. Despite their cool image, they were thrilled.

Pixar recently hired a member of the Demoscene, who played a key part in how greenery looked in the animated movie *Brave*. "Fresh blood is a great thing," says Sayre.

## A place where the future is lived: the MIT Media Lab

The MIT Media Laboratory is one of the major world players in design technology. The online course catalogue for the program in Media Arts and Sciences sets out its ambitious program as focusing "on the invention, study, and creative use of new technologies that change how we express ourselves, how we communicate with each other, how we learn, and how we perceive and interact with the world."

The Lab was the brainchild of the architect Nicholas Negroponte, together with Jerome Wiesner, ex-president of MIT, physicist, and longtime advocate of science education. It was part of the MIT School of Architecture and Planning and opened its doors in 1985. It is intended to be a "place where the future is lived, not imagined."

The futuristic new building was completed in 2010. The interior is almost entirely glass, box-shaped with a central atrium surrounded by laboratories and offices. The glass interior makes all research visible and encourages networking and collaboration, though some people I spoke to said that while this layout is good for working in groups, for creative work, which requires solitude, they tend to rent studios elsewhere. Nevertheless, the atmosphere is electric, the air alive with ideas.

The director of the Media Lab, Joichi Ito, is a Japanese American who, at forty-five, radiates boyish enthusiasm. He speaks rapid-fire, with a mind like a parallel processor; he's with you while simultaneously running through other scenarios. Occasionally he stops speaking and his fingers dart over the keyboard of his ever-present Mac.

Born in Japan in 1967, Ito moved to Canada with his family and then to Detroit when he was three. His mother raised Joi and his sister and then, at thirty-five, went to work as a secretary in Energy Conversion Devices Inc., where his father was a research scientist. She rose to vice president, then left to become an entrepreneur. Her get-up-and-go attitude was what fired him.

Ito's sister became an academic, but Ito found formal education not to his liking. "I always retained the notion of neoteny—the retention of childlike attributes in adulthood," he says. One of his many enterprises is a venture capital firm which he named Neoteny Co. Ltd. As a committed neotenist, he took the road of the nonspecialist, of continual learning. He studied computer science at Tufts University, near Boston, but dropped out "because I felt I could learn it on my own and better." He enrolled at the University of Chicago to do physics, but disliked the emphasis on problem-solving rather than understanding concepts and dropped out again.

A succession of jobs followed, which included stints as a disk jockey and the manager of a nightclub, and shuttling back and forth between the motion picture industries in Japan and Hollywood. He finally became a venture capitalist, investing in a string of Internet companies in the USA and Japan in the 1990s at a time when the world of Internet start-ups was wide open. Then, in 2011, MIT asked him to direct the Media Lab. Someone who had dropped out of college twice was a controversial choice, but as the press office wrote, "Ito, 44, is recognized as one of the world's leading thinkers and writers on innovation, global technology, and the role of the Internet in transforming society."

"The choice is radical, but brilliant," says Larry Smarr, who directs Calit2, the California Institute for Telecommunications and Information Technology, a research institution similar to the MIT Media Lab. "He can position the lab at the edge of change and propel it for a decade." Ito's understanding of finance was essential for the Media Lab as it struggled to return its funding, which had fallen radically in the downturn of 2008, back to the levels of the dotcom boom. His intellectual qualities were also outstanding. He was outspoken in support of interdisciplinarity and a focus on creativity. He is also well known as an activist, campaigning on behalf of the individual versus government.

The Lab offers a degree in media arts and science—not fine art, which is "not very collaborative," but art which is about process, the building of things. "Process, not end result, is of concern," says Ito. In the Lab, artists work with scientists and engineers on an equal footing. To be accepted as a student at the Lab requires a high degree of technological know-how. The Media Lab "is not a tech shop for artists," says Ito. It is not at the service of artists who would like to have their creations mechanized. Nor is it a place for procrastination. "People spend very little time sitting around doing PowerPoint presentations. Things are different at the Media Lab. People just do it!" Ito puts this credo even more directly. "Shut up and build it, then we'll talk about it."

## Glamour geek: Neri Oxman

Neri Oxman is an excellent example of the type of person who thrives at the MIT Media Lab. She is interested in load-bearing and has an "aesthetic fascination with forms in nature—form generation as given by nature." She is researching new uses for concrete, aiming to make possible a bold new architecture going beyond the type of building that can be made today using reinforced concrete. This encompasses engineering and art: engineering in the use of materials and art in the creation of aesthetic objects. Engineering focuses on problem-solving, but in design and art there is not always a particular problem to be solved. She is interested in "problem-seeking, as opposed to problem-solving. Problem-seeking or problem-discovery is a more difficult road to follow than trying to solve already existing problems."

After studying architecture at the Technion Israel Institute of Technology and medicine at the Hebrew University in Jerusalem, Oxman did a PhD in design computation at the Media Lab. Oxman teaches at the Lab as well as directing the Mediated Matter research group. She was featured on the cover of the high-tech magazine *Fast Company*, which described her as ushering in the "era of glamour geeks," and added, "smart is sexy."

In order to look into the way concrete shapes and supports buildings, Oxman studies how calcium shapes load-bearing bones, a process that "produces unbelievable patterns." In her research, she "tries to spec out algorithms that describe this conversation between matter and distribution of loads." This is best understood as algorithm plus context. The algorithm contains aesthetic reasoning: will calcium produce certain shapes for certain distributions and can this be applied to the distributions of concrete? The context is the conditions confronting the architect.

Oxman comes up with an illustration. Ansel Adams, the pioneering photographer, wrote books explaining his techniques. But even if you use these techniques, you won't produce photographs like those Adams took; the recipe is there, but the results will differ. The recipe—Adams's description of his technique—is a creative algorithm. Every time it is used, it produces something different. Creative algorithms are often used to power extruders to produce three-dimensional art. Algorithms are everywhere, they are in the air, they are the zeitgeist. At the turn of the twentieth century, the zeitgeist was the quest to redefine the classical, intuitive ways of understanding space and time. Einstein and Picasso both responded to this. As for the zeitgeist today, "now, with the Internet, it's about the group, the community," about collapsing distances and times. "I miss the old world," Oxman adds.

One of Oxman's best-known products is a chaise longue she calls

4.2: Neri Oxman, *Beast*, 2008–10.

*Beast*. Its single organic-seeming surface serves as the structure as well. It's a wonderful undulating creation which adapts to different bodies, i.e., different loads, varying its density and flexibility. "*Beast* lets go of boundaries such as those between structural support and comfort support. All is blended via an algorithm."

Oxman has used her studies in the distribution of materials suited to bear loads to create a variety of products including the prototype "carpal skin," inspired by the patterns on animals' coats. It is made up of organic material distributed so as to protect against carpal tunnel syndrome, a painful nerve condition in the wrist.

In her research comparing the way calcium shapes bones to the way concrete shapes buildings, Oxman works with material scientists and biologists. She studies biological specimens using CT scans (computer tomography, which uses x-rays and computers to create detailed images). The images have to be set up differently for material scientists and for biologists, who have different requirements. Eventually, education will have to be revamped to accommodate new fields of research such as hers. Biologists will have to be trained in material science and vice versa. "A da Vinci stage has to be reached. The greatest scientists of all time were artists," she says, repeating the last words emphatically.

Art has been redefined, and so has its role. Oxman mentions "iPad fever. You build a tool box and use it over and over again and it is shared by everyone." She "doesn't believe that that is the role of art, which is to question a certain reality rather than to provide entertainment." Design is art, she says. The best way to think of design is as "translation between disciplines."

## A peek into the future: Michael Bove

"The Media Lab is an idea factory," says Michael Bove. With his khaki trousers, button-down shirt, immaculately parted hair, and MIT graduation ring, Bove looks conventional enough, but behind this starchy appearance lurks a restless and creative mind. In his office every available surface is covered in books, papers, and gadgets.

Walking around his laboratory is like taking a peek into the future. Students and researchers sit around asking "what if?" questions, such as, "What if the world were literally in touch?" They've already come up with a possible way to achieve this—prototype Touch Gloves, made up of sensors containing information that is shared when you shake hands with someone else also wearing Touch Gloves.

Can you transform an object into a phone? Bove showed me the prototype for an object which looks like a bar of soap with "dimples" on it and a screen. It becomes a phone if gripped one way, a camera if gripped another way, and a recorder if gripped yet another way.

Can you use technology to teach baseball? Bove's team rigged up a baseball with sensors to teach a beginner to throw a fastball or a curveball.

Can we capture our dreams without having to jump out of bed and look for a pad of paper? Pillow Talk is a voice-activated recorder sewn into a pillow. The dreamer can store his narration on disk to be replayed later.

"Look, I don't know if this is art. I don't know," says Bove. If anything, his team's highly imaginative inventions could be called interface art, in which the performer or subject triggers a computer through movement.

Bove found his calling early. As a high school student in the late seventies, he built a personal computer, attached it to a digital camera which he had also built, and started producing computer graphics. At MIT he specialized in electrical engineering, but he was also interested in art and hung around the architecture department. Nicholas Negroponte, soon to found the Media Lab, met him and was impressed by him. Bove was the first student to do a PhD at the Lab. He now heads the Object-Based Media Group.

Bove also collaborates with artists on holographic art, holograms being two-dimensional surfaces that interact with incoming light so as to make three-dimensional shapes appear outside the plane of the surface. He is currently looking into developing a holographic television which would project images into your living room as a "cloud" which could be viewed from any direction without the need for 3D glasses. The holographic images could literally take over your living room—though this is far in the future.

Bove's chief collaborator on this project is the Australian holographic artist Paula Dawson. She has contributed to scientific papers on holography and, Bove has noted, has developed a "totally radical way of making large scale optical holograms," digital holograms which are computer-generated using a "holoprinter" device.

One of the problems in developing a holographic television is to do with questions of perception in viewing it. For a start, where exactly is the picture plane? In a painting, it is the canvas, the physical surface, but in holography it is the nonmaterial plane between the electronics producing the image and the viewer. Then there is the matter of how best to develop the image's three-dimensionality and color.

Dawson makes suggestions from her perspective as an artist about how the image should best be displayed. A major stumbling block is how to define the boundaries of the image. Looking at the image close up (in the middle of your living room), it would need to be contained in a frame, such as a task bar with decorative side columns, similar to a computer screen (see Insert).

Bove is enthusiastic about the collaboration. "Apart from the fact that Paula is creative and energetic and it's generally inspirational for my group to get together with her, she has particular interface and display requirements for her holographic drawings that have caused us to rethink certain aspects of our research and system design for interactive holographic video displays."

Dawson and Bove provide an excellent example of an artist-scientist collaboration working to benefit both.

## Digital rhythms: Joe Paradiso

Joseph Paradiso's lab at the MIT Media Lab is an electronic madhouse. Musical synthesizers, old and new, share floor space with ever more sophisticated equipment and computers. Tall, lean, and gray-haired in black jeans, a checked shirt, and sneakers, Paradiso pours out words rapid-fire. He is, to put it mildly, an electronic genius. At one point he was offered three prestigious positions at the same time—at CERN in experimental particle physics, at the Draper Lab at MIT in space science, and at the Media Lab. He chose the Media Lab because "it's all about creativity here."

Alongside his scientific research, Paradiso is a musician who builds and plays synthesizers and jams with ad hoc groups. His original massive synthesizer, probably the largest in the world, with a complex configuration of wires connecting its hundred-odd modules, has the place of honor in his lab and is slated for the MIT museum. Paradiso began work on it in the early 1970s and it took him a decade and a half to complete, starting from scratch, scrounging around electronic surplus stores for parts. Paradiso has also designed Musical Instrument Digital Interface (MIDI) systems for operating several instruments from a single control system.

He is generally involved in seven or eight projects at the same time within the Responsive Environments Group he directs. One of the group's prototype inventions is a pair of shoes housing batteries that charge as you walk, thus generating energy. Another is a pair of

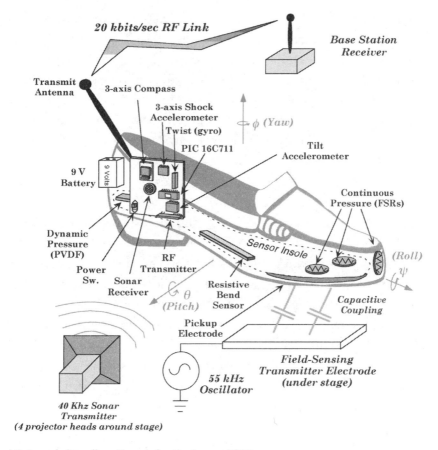

4.3: Joseph Paradiso, *Expressive Footwear*, 2000.

shoes with a sensor system that creates musical tones in response to the wearer's movements. Rhythmic movements result in pleasing sounds, encouraging the wearer to move more gracefully.

Paradiso's group has also invented a Digital Baton, which generates the sounds of different musical instruments depending on the pressure with which you hold it. It's a technological magic wand.

## At the frontier of media art: Peter Weibel

"Today art is an offspring of science and technology," says Peter Weibel. Weibel is the chairman and CEO of Zentrum für Kunst und

Medientechnologie (Center for Arts and Media Technology, known as ZKM) in Germany, a curator, an art theorist deeply versed in the classics, art history, and intellectual currents such as structuralism and psychoanalysis, and a professor who regularly gives workshops at universities across the world. From 1988 to 1995, Weibel was artistic director of Ars Electronica, a digital arts organization that produces a gigantic annual fair at Linz devoted to science- and technology-influenced art. He's also a practicing artist. In the late 1960s, he was a member of the Viennese Actionists, a group of wildly bohemian artists who practiced what they called the politics of transgression, using their bodies as canvases, covering themselves with paint and even cutting themselves with razors. He continues to be a video artist. His 2003 video work *Venus in Fur*, showing historical portrayals of women by male artists, is the only work of contemporary art at the Belvedere, the National Gallery of Art in Vienna.

Weibel speaks authoritatively on a huge spectrum of topics, ranging across art, science, technology, art theory, and theater. I meet him in Vienna at the opening of a multimedia installation showcasing the art-based research project Quantum Cinema— A Digital Vision, of which he is the director. The aim is to use digital art to explore the quantum world as well as visualizing higher-dimensional geometries. The conference takes place at the University of Applied Arts, where he is a professor.

Weibel has a huge, charismatic presence. His independence of mind springs from a turbulent youth. Born in Odessa in 1944, he was a displaced person after World War II, living in refugee camps in terrible conditions. Growing up in state institutions, he turned inward and read voraciously. He devoured Dover Publications' science books and George Boole's *An Investigation of the Laws of Thought*, which turned out to be not about thought at all but about Boolean algebra, a form of mathematical logic. Boole's book led Weibel into information theory. Then he discovered the French poets, in particular, Paul Valéry, who had studied books by the nineteenth-century English scientist Michael Faraday. Weibel did likewise. In school he enraged his science teachers by telling them that what they were teaching was wrong because they had failed to include quantum physics.

In the early 1960s, he studied medicine at the Sorbonne in Paris. Always intellectually restless, he also began reading works by the linguist Roman Jakobson, the anthropologist Claude Lévi-Strauss, and the psychoanalyst and philosopher Jacques Lacan. Medicine began to seem rather narrow once he realized how art and technology played off each other, and in 1964 he began making films. He also immersed himself in the history of the cinema.

On returning to Vienna in the mid-1960s, he found himself acknowledged as an expert on cinema. It was then that he became involved with the Viennese Actionists. It was obvious that the best way to revolt against society and the establishment was as an artist. Science was too well established. Still, Weibel found the technology involved in developing and editing film too cumbersome. Then in 1969, video cameras became available. Now he could express his "ideas about space, time, and relativity much easier," showing them in real time using simultaneous images side by side.

Weibel always works, he says, "in dialogue with scientists." In the 1970s he corresponded with Claude Shannon, a towering figure in information theory, with Marvin Minsky, the director of MIT's Artificial Intelligence Lab, with John McCarthy, who was well known within computer and cognitive science, and with René Thom, an internationally renowned French mathematician. Recently, he approached Anton Zeilinger, a physicist at the University of Vienna known for his experiments on entanglement. Once two particles on the atomic level interact, they become entangled: they forever sense what has happened to the other, like the Corsican brothers in the novel by Alexandre Dumas, separated at birth but affected ever after by the other's experiences. Einstein called this "spooky actions-at-a-distance." Zeilinger's group and others have shown that it is real. Weibel was convinced that what Zeilinger was doing "could be wonderful media art" and invited Zeilinger to speak on his experiments, complete with videos, at the 2012 Documenta Festival in Kassel.

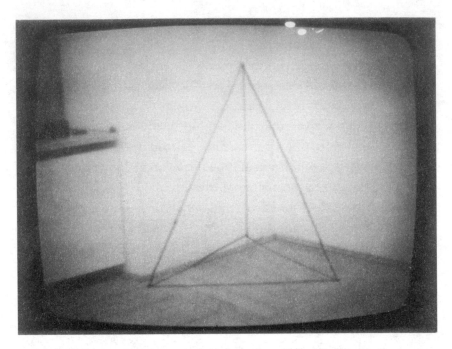

4.4: Peter Weibel, *Imaginary or Virtual Tetrahedron*. Closed-circuit video installation, 1978.

To Weibel's way of thinking, aesthetics is something our minds construct from the complex interplay of our senses. In his video *Virtual Tetrahedron*, he shows a metal frame with no particular shape, but then an actor appears and inserts a screen with a strategically placed line on it, and a tetrahedron magically appears. He also manipulates two projectors to produce a virtual tetrahedron. In neither case does the tetrahedron—one of the five Platonic solids deemed to have a degree of perfection—actually exist. It is an optical illusion, entirely constructed by our mind. "It exists only thanks to the media."

"Aesthetics is a general medium of questioning the world," he says, in this case of playing with perceptions to bring out unexpected results, hidden information. "For me aesthetics has therefore nothing to do with beauty, but with information," he continues. "Many of my works are not directly linked to reality, but to aesthetic problems."

Weibel has found that art historians and curators are often hostile to artworks involving science and technology. "It is ridiculous that art historians are not educated in science and technology because artists are much more advanced." He senses fear and resentment against ZKM, which successfully showcases science- and technology-based art.

He has crusaded for science to be included in the art history curriculum. The example he likes to give is of "how photography deeply influenced painting," arguing that many of the effects artists now use were inspired by photography, that there was "no single painting out of focus before photography." He also refers to the artworks of Andy Warhol and Gerhard Richter, created by manipulating photographs.

Like it or not, our world is driven by science and technology, which are based in theory. So, he says, "I give art a strong theoretical, political, and social urgency. And I make difficult shows." Art, he says, is "theory-dependent because it wants to be part of the modern world."

## Daring and imagination: Dunne and Raby

Fiona Raby and Tony Dunne are a dynamic globetrotting husband-and-wife team who oversee the imaginative and daring Design Interactions group at the Royal College of Art in London. According to Dunne, the philosophy behind the group is "just about anything goes. Let your mind wander." "We are interested in using design as a medium, to ask questions and provoke and stimulate people, designers, and industry," he has said.

In selecting students, they place a premium on creativity, imagination, and innovation, and look for candidates who are "highly committed to playing a significant and meaningful role in the shaping of our technological future." Incoming students need not "have a high tech savvy," says Raby. "They can be pure artists." The biggest thing they need is imagination.

The Design Interactions group is a playground for ideas, where "technological dreams" can be played out. Raby and Dunne's goal is to introduce people to technologies and "mess up all those very nicely thought-out technological systems. There is a need for the irrational, illogical, and contradictory to be considered equally."

Both Dunne and Raby lived in Tokyo for three years in the late 1980s, and were inspired by its heady mix of technology and imagination embedded in everyday life. "I don't think we ever recovered!" says Raby. Coming home, they found "no playfulness in England." They frequently return for inspiration and in 1994 established the design partnership Dunne & Raby.

"Can designers be involved in biology?" asks Raby. An early project in the field of design biology was Biojewellery by Tobie Kerridge and Nikki Stott, design researchers at the Royal College of Art, working with Ian Thompson, a bioengineer at King's College London. In the future it will be possible to grow bone tissue outside the body to be used in reconstructive surgery. But supposing it could be used to make jewelry? Supposing a couple could express their love by wearing rings made of each other's bone tissue?

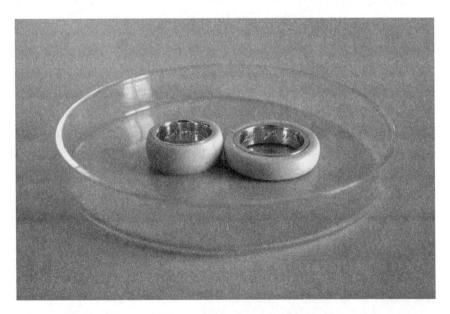

4.5: Tobie Kerridge, Nikki Stott, and Ian Thompson, *Models of Trish and Lynsey's Rings*, 2006.

To carry this out, the group used bone tissue from wisdom teeth extracted from a particular couple, mixed it with precious metals, and cultured it on a bioactive ceramic used as a scaffold for growing cells. The results—two rings of bone mounted on etched silver—were exhibited at Guy's Hospital, London, in 2007. The process involved many different disciplines: materials engineering, cell biology, oral surgery, medical imaging, computer-aided jewelry design, graphic design, interaction design, product design, fine art, media relations, journalism, science communication, sociology, and ethics.

Commercial design is not particularly speculative, in that commercial designers have to sell their products, but the work at the RCA's Design Interactions group very often is. Students look into the idea of growing hair on a host for future use, for example, or alternative uses for an Anglepoise lamp.

David Benqué, another member of the team, has a project he calls "Acoustic Botany, A Genetically Engineered Sound Garden." His starting point is that forms in nature evolve not to suit particular human needs but to satisfy irrational desires. "Beautiful flowers, mind-altering weeds, and crabs shaped like human faces all thrive on these desires, giving them an evolutionary advantage." So perhaps we could genetically modify plants for the same purpose. Supposing, for example, there could be "a controlled ecosystem of entertainment [which has an] aesthetic relationship with nature." In the future, synthetic biology might be able to actually make this possible. For now, Benqué sculpts the swelling shapes of fantastical plants and fruits wired up to produce haunting melodies reminiscent of the music of the spheres.

Projects such as these help to break down barriers between disciplines. Teams made up of biologists, engineers, mathematicians, chemists, and designers work together. "They all sit at the same table and communicate with each other, appreciate each other's skills," says Raby. "While the designers may know nothing about science, they sit with scientists, ask questions. They question dogma."

Synthetic biology explores the design and construction of entirely new biological systems. It is not biology-influenced art, but media art. Alexandra Daisy Ginsberg is a key player in this field and gives dazzling TED lectures. In 2009 she and another designer, James King, joined a team of seven undergraduates at the University of Cambridge in a project to introduce color into bacteria. Taking genes from a variety of organisms, they designed sequences of DNA known as BioBricks, which they inserted into E. coli, a bacteria common to our digestive system, treating the cell like a computer and the cell's DNA like computer code—in other words, reprogramming the cell.

The result was a new bacteria which the team called E. chromi, capable of producing colors in response to specific chemicals, which made it a potential diagnostic tool. When E. chromi is inserted into a patient's digestive system, it can change the color of the feces if certain diseases are present, thus sounding the alarm. It could also be used to check whether water is safe for drinking by turning it red if it isn't.

Ginsberg and King speculated on other uses for the new technology—dreaming dreams, as Ginsberg had learned at Design Interactions. Perhaps E. chromi might be used in food additives or personalized medicine, in antiterrorism, in organisms that excrete oil or plastic, or for analyzing or encouraging the emergence of new types of weather. The E. chromi project won the grand prize at the International Genetically Engineered Machine competition at MIT in 2009.

Synthetic biology is rife with ethical issues. Who owns the new genes? Where do we stop? Ginsberg sees part of her role in the team and in the field as "that of a provocateur—an activist." Synthetic biology "tampers with nature," making her ponder the "seductiveness of technology, blurring the boundary between the natural and unnatural, outrageous scenarios such as engineering myself."

In media art, artist and scientist are fused together. Using complex mathematics, media artists generate pleasing images. These artists are of necessity technologically savvy, and closer to the scientist than those schooled in the traditional fine arts, an area which is becoming increasingly marginalized. In this new avant-garde, the labels "artist," "engineer," and "scientist" are growing increasingly irrelevant, often replaced by "researcher." These artists distance themselves from the traditional art world, which shies away from their art, and instead take advantage of the upsurge of venues that exhibit and sell artsci, such as ZKM in Karlsruhe, Ars Electronica in Linz, GV Art in London, Le Laboratoire in Paris, Science Gallery in Dublin, and Documenta, which takes place every fifth year in Kassel. All these allow artists working on the fringes of art, science, and technology to bypass the traditional art markets. Five years ago there were only a few biennials dedicated to science. Now there are over fifty.

As to how artists can finance themselves, says Peter Weibel, "Private industry will finance them." In this way, "art will become independent, no longer dependent on markets." Property rights will become an old-fashioned notion in an open-source world where all software is on the Web and cannot be copyrighted. Ultimately, he predicts, there "will be an empire of data."

What has been emerging is an emphasis on the process rather than the end result, especially in terms of the creativity involved, which is often guided by algorithms. Aesthetics honed by experience enters as a way of selecting the proper route and product. Despite the highly mechanized methods that media artists use, they place a high premium on intuition and imagination, which resist explanation using logic. As Weibel put it, "Today art is an offspring of science and technology."

# Visualizing the Invisible

Paul Klee once wrote, "Art does not reproduce the visible; rather, it makes visible." Klee believed that we focus too much on the visible world. As if in response, today artists working alongside physicists are attempting to make the invisible visible.

The artists in this chapter include some who have studied physics and are largely scientifically self-sufficient (though they may communicate occasionally with scientists), and artists with a limited knowledge of physics who collaborate with scientists. When dealing with physics, artists are often deterred by the sheer difficulty. While artists who focus on biology may use laboratories, in physics the apparatus is often too complex or even dangerous. Artists learn what they can by reading popular science books, of high enough quality that they can begin to get to grips with underlying concepts.

## The artist of light: Paul Friedlander

"Jim Miller," of the American artists' group EyeCandy ArtWorks, wrote, "In a time when so many artists resort to bizarre and shocking gimmicks to achieve originality, I take solace in the work of Paul Friedlander and others like him. They prove that beauty still has a place in modern art."

Paul Friedlander's house, in a row of terraced houses in North London, is perfectly ordinary from the outside, exactly the same as the others. When he opens the door, I step into a cavernous space in which light sculptures twenty feet high pulse, sending out swirling patterns; he has cut a hole in the ceiling to transform the lower two floors into one. Friedlander himself looks like a friendly wizard, with a craggy face, a beard, and a warm smile.

His house is like a wizard's lair. Tools, lathes, electronics, and laptops are scattered about. The room is full of multiple varieties of light sculptures, magical spinning, whirling creations.

Friedlander was "a child of the space age," as he puts it. His mother was an artist interested in contemporary and abstract work, who frequently took him to art exhibitions, where he vividly remembers seeing kinetic sculptures as well as optical art. His father was an eminent mathematician at Cambridge, one of whose specialities was wave equations. He kept his son abreast of the latest scientific advances, thrilling him with tales of the launch of *Sputnik* in 1957 and discoveries in astronomy such as black holes. In the end, his father's enthusiasm won out and Friedlander decided to study physics, which he did at the University of Sussex, from 1973 to 1976.

Two events conspired to draw him to art. A gifted student, he attracted the attention of Tony Leggett, an extraordinary teacher and physicist, who talked him through some of the problems underlying quantum mechanics, such as the ambiguous wave/ particle nature of matter. Friedlander was spellbound but acutely aware of how hard these problems are; that ultimately contributed to his decision to quit physics. Meanwhile, he discovered the work of Nicolas Schöffer at a 1970 exhibition at the Hayward Gallery in London. Schöffer, the first to include cybernetics in his work, was to Friedlander an artist of ideas rather than techniques. Friedlander continued his studies at Sussex but started doing kinetic art as a hobby.

After graduating, Friedlander spent ten years designing stage lighting while producing works of his own. His heart was not in lighting design, he says, but he "felt squeamish about the label 'artist.'" Then in 1990, he had a breakthrough. He found a way of spinning string to produce wave forms—complex shapes revealing periodic wave behavior together with chaotic rhythmic variations—then blended these two motions together. At that time, chaos theory was an intriguing new topic and it seemed a good moment to launch himself as an artist who explicitly included science in his work.

The works Friedlander produced used a technique he had invented in 1983 called chromastrobic light. In this, colors change faster than the eye can see, blurring together into white light. When he shone this white light onto a rapidly moving object, like his whirling string, the string acted like a prism, separating out the colors. The spectacular display filling the front of his house is one of these—a vertical vibrating rope, reflecting the colors projected onto it like soap bubbles (see Insert). Spectators can interact with the sculpture, using sound and sensors that change the speed of the rope's vibrations.

Friedlander had a huge solo show, Timeless Universe, at the Sala Parpalló gallery, in Valencia, Spain, in 2006. This was a response to the English physicist Ian Barbour's contentious hypothesis that time does not exist. The Big Bang theory of the universe states that time and space exploded into existence some 13.7 billion years ago along with the universe itself. Einstein's general theory of relativity supports the belief that time had a beginning and is part of space-time, which has four dimensions, one of time and three of space. But examining space and time in smaller and smaller dimensions requires quantum theory, according to which these four dimensions coalesce into a quantum foam in which Einstein's general relativity theory eventually breaks down. This is what led Barbour to hypothesize that time actually doesn't exist.

The light sculptures that make Timeless Universe form a seemingly endless stream of wave forms colored by chromastrobic light, with an algorithm ensuring that each day there is a different variety of images. This is not an interpretation or illustration of Barbour's idea, says Friedlander, but a meditation inspired by it. His wave forms express "all possible conceivable states of the universe that govern in every detail our fates." He goes on, "I feel that I have been true to Barbour's ideas, since, for his mission to succeed, he has to show how all our sense of change and time passing—of the past and future being so different and the present moment special and unique—arises from an underlying timeless reality."

Barbour agreed to participate by writing an essay for the catalogue of Friedlander's show. However, Friedlander does not consider his interactions with Barbour as direct collaboration—rather, Barbour's ideas stimulate him to use light sculpture as a way to visualize the invisible. Thus Friedlander's physics-influenced art examines how an unorthodox view of nature might affect how we see the cosmos.

## Quantum artist: Julian Voss-Andreae

"Only as an artist am I able to do something that feels significant to me," says Julian Voss-Andreae. Voss-Andreae is a tall, youthful Austrian who radiates intensity of purpose. I first meet him in 2009 at a conference in Dortmund on Einstein and Picasso, although I am already familiar with his work.

Voss-Andreae completed his most conceptual piece, *Quantum Man*, in 2006. Eight feet tall and made up of over a hundred paral-

5.1: Julian Voss-Andreae, *Quantum Man*, 2006.

lel vertical steel sheets, when viewed from the front it looks like a man, but as you move around, the man becomes invisible. Just as in quantum physics, where if an experiment is set up to detect the wave nature of an electron, then the electron will be a wave, but if the experiment is set up to demonstrate that the electron is a particle, then it will be a particle: how you look at it, that's what it is.

Voss-Andreae understands the fundamentals of physics. At the University of Vienna, he was part of a research group under the distinguished physicist Anton Zeilinger. The group explored arcane properties such as entanglement and quantum cryptography, and called their work applied quantum philosophy. Voss-Andreae received his MSc for an experiment demonstrating that massive objects such as carbon 60 molecules (shaped like Buckminster Fuller's hexagons, and known as Bucky balls) have quantum properties. They are the heaviest particles ever found to have such properties.

Voss-Andreae's earliest artistic inspiration was German Expressionism, especially Der Brücke (The Bridge) and Der Blaue Reiter (The Blue Rider) groups. Their members included Wassily Kandinsky, and they showed works by Picasso and Klee in their exhibitions. Der Brücke was an early manifestation of Expressionism, using subjective experience to depict the world, while Der Blaue Reiter was one of the driving forces behind the development of abstraction in twentieth-century art. Voss-Andreae is particularly inspired by the poignantly expressive works of Picasso's Blue and Rose Periods, which combine Expressionism and abstraction.

His interest in science arose from a very young age through popular science magazines and chemistry and electronic sets. At twelve he got his first computer and learned code. He wanted to construct an aesthetically appealing computer game. "It was then that I realized that I needed to learn mathematics as the bridge between science and art." Five years later, he made a first attempt at computer art, writing a graphics program and modifying his needle printer to make a scanner. The result was disappointing, which he put down to the analytical manner in which he had produced it. He concluded that "probably all intellectually conjured, brain-born art is doomed to be boring and empty, a prejudice that, to this day, fuels my work: in a way all my work is an attempt to disprove this hypothesis—and to see what that secret ingredient is, beyond the purely intellectual idea, that makes the work come alive."

Voss-Andreae exudes a strong sense of independence, of going his own way, which he attributes to his friendship with the German artist Horst Janssen, a friend of his father, who lived near his family in Hamburg. Best known for his drawings, etchings, and woodcuts, Janssen "drew without the intellect interfering, [using] the influence of the consciousness on the art." Voss-Andreae's desire to work without the interference of the intellect became an essential part of his art, which represents in a free-flowing manner forms best understood using highly mathematical theory.

He is "intrigued by the time when relativity and quantum physics emerged—for example, Schrödinger, Musil, Schiele, Gödel, Kafka—from the collapsing Austro-Hungarian empire," and wonders if and how scientific and literary change related to political upheaval. He also admires van Gogh's "honest and strong dynamic," which revealed the static world as a "quantum dance. His sense of love as the primordial force of the universe always shines through his work." Listening to Voss-Andreae, one cannot help sensing the German Romantic in him.

In 2000, Voss-Andreae emigrated to the United States, and studied at the Pacific Northwest College of Art in Portland, Oregon, graduating in 2004. To Voss-Andreae aesthetics depends on satisfactory design. "To me, form and function are always a unit, and both together make a good design. Like in mathematics or engineering, I cannot separate the experience of discovering or understanding such a solution from a beautiful aesthetic experience." As both an artist and a scientist, he realized that he was in a unique position to plunge into the growing movement of science-influenced art. In fact, he had been planning *Quantum Man* since 1999.

He is currently based in Portland, where several of his works are exhibited. Others are at the Scripps Research Institute in Florida, Rockefeller University in New York, and the American Center for Physics in Maryland. Says Voss-Andreae, "I feel a strong excitement about work that really merges art and science, and I believe that this is a sign for the emergence of a new culture. But we are only at the beginning."

## Russian mysticism meets physics:
## Evelina Domnitch and Dmitry Gelfand

Although they have not formally studied physics, the husband-and-wife team of Evelina Domnitch and Dmitry Gelfand produces impressive works deeply based in the subject. Domnitch and Gelfand's goal is to stimulate the senses, stir the imagination, and even touch consciousness. They produce sonic immersion environments and ask evocative questions like, "Is it possible to create a sonic rainbow? Is it possible to render the wave behaviors of sound into those of light? Is it possible to render sound visible and allow a musician to work with the shape of sonic currents?"

One of their solutions is their *Camera Lucida*, or light chamber. They write, "The *Camera Lucida* project began as a speculative

5.2: Evelina Domnitch and Dmitry Gelfand, *Camera Lucida: Sonochemical Laboratory*, 2006.

reverie on observing sound waves with the naked eye"—observing sound waves, an extraordinary concept in itself. They took the audible sound from a musical composition and modulated it into the ultrasonic register, using a chamber containing liquid made up of over 90 percent sulphuric acid. The ultrasonic waves cause bubbles that implode violently, emitting light. This combination of sound and light is called sonoluminescence. Thus they create a "sonoluminescent universe," made up of a variety of patterns which eventually fade away. They see this work as depicting the early universe, when sound waves compressed and rarefied matter, emitting light, a process that can be seen around certain black holes.

I first met Domnitch and Gelfand when I interviewed them—for six hours!—at Ars Electronica in Linz in September 2011. Domnitch has a shaved head with nothing but an S-shaped Mohican on top, while Gelfand is fashionably dressed down in cap and T-shirt. They are a handsome couple, both of Russian origin. Domnitch grew up in Minsk, while Gelfand's family emigrated to the United States when he was a boy and he grew up in New York. They met when Domnitch was working on a PhD in philosophy at Fordham University, in the Bronx, while Gelfand was studying film at New York University.

Their partnership, they say, "came out of a series of very long philosophical discussions on the line between logic and music, between the symbolic and the nonsymbolic." I can only imagine what these discussions were like. In my own experience, conversations with them are highly non-linear, beginning along one line, then branching along others only tangentially related and sometimes not at all. They are so in touch with each other that they finish each other's sentences. Speaking to them is like speaking to one person.

Domnitch and Gelfand rejected the traditional ways of investigating fundamental issues in art and philosophy, feeling that "they offered no buzz." They were fascinated by questions such as "What is light?" They wanted to go deeper than traditional ways allowed. The clue came from John Cage. Domnitch had been familiar with his work for some years before coming to New York. She arrived in 1994, two years after his death, but she and Gelfand were able to meet close friends of his who discussed him and his work and beliefs in some depth. What particularly fired Domnitch and Gelfand was Cage's belief that "recorded music or recorded art ceases to be a real thing." It's stuck, frozen in time, never suffering the intervention of natural forces such as decay. In other words, while the world about us is transient, ever-changing, when it comes to recorded music or art "the narrative is too restrictive. Everything is forever." Gelfand's studies in film and optics offered a way to create an alternative— "live cinema," in which scenes never repeat themselves and, in fact, can never be repeated.

In their early work, Domnitch and Gelfand projected concrete images—forms painted with black ink on 35mm slides—onto clouds of water vapor. The result was spectacular. "The image was animated like a ghost that comes to life." In this way "art became a means, an arena to ask substantial questions."

One of their most recent creations, *Memory Vapor*, is a cloud chamber set up to detect cosmic rays, the elementary particles from outer space that continually bombard the earth. Coursing through the liquid in the cloud chamber, the rays make tracks like those made by peanuts dropped into a glass of beer. Domnitch and Gelfand illuminate them using a beam of white light from a laser which enhances the trail of condensation droplets, producing an almost dizzying sensation of "iridescent depth."

Other experiments encourage viewers to look within themselves. In *10,000 Peacock Feathers in Foaming Acid* they shoot laser beams into clusters of soap bubbles at just the right angles to create a projection of the molecular interactions in the bubble, giving a peacock-like display of colors as the beam splits into multiple, very fine threads. The patterns evoke the dynamics of living cells and suggest conditions in the early days of the universe. This is art as research. Exploration of the properties of soap bubbles has a long history; they are used today in modeling black holes and superstrings.

Domnitch and Gelfand have many contacts in the physics communities in St. Petersburg, Amsterdam, and Japan. Among them is Kiuchi Yasui, a leading Japanese researcher in sonoluminescence, who was struck by their *Camera Lucida* and, realizing that "there was no theoretical model for this behavior," went on to propose one. Raoul Frese, a biophysicist at Vrije University in Amsterdam who explores artificial photosynthesis, became interested in a piece called *Hydrogeny*. In this, Domnitch and Gelfand evoke the ways in which the earliest forms of photosynthesis might have occurred in oceans. "Raoul claimed that his encounter with us and our approach to physics led him to upscale his experiments," they say.

Domnitch recalls a physicist friend who was moved to tears by *Memory Vapor*. He perfectly understood the technical aspects of a cloud chamber, but in *Memory Vapor* he saw something beyond what one usually sees in cloud chamber events. "Aesthetics is to be emotionally moved," Domnitch concludes.

Aesthetics is not something they put in. Rather, it has to emerge. "We always hope the unexpected will appear; this is a most important ingredient—especially in an artwork."

## Between dimensions: Nathan Cohen

"I'm an optimist," says Nathan Cohen. "I want to see just over the horizon, try to capture the world in which we live."

When I met Nathan Cohen in the 1990s, he was experimenting with slats of wood arranged at different angles, carefully planned in meticulously drawn sketches like scientific diagrams. I've always believed that an artist's preparatory sketches, delineating his thought processes, can stand alone. Cohen's were exemplary. In order to achieve symmetry in his constructions, he worked on graph paper, orienting each slat in relation to fixed horizontal and vertical lines—in mathematical terms, their positions relative to the x and y axes, their x- and y-coordinates.

One construction reached its final form when the coordinates of the slats were adjusted one after another in a specific way. Cohen had rediscovered a set of interchanges known in mathematics as a permutation group. Thus does an artist working in a scientific way chance upon mathematical symmetries.

For Cohen, art runs in the family. His father, Bernard Cohen, is an eminent British abstract artist and former Slade professor at University College London, while his uncle, Harold Cohen, is a pioneer of computer art. From an early age Nathan loved to draw and often went to art museums, such as the Pitt Rivers Museum in Oxford. There were frequent trips to New Mexico, an area that his father found particularly inspiring. There he became entranced with rocks and minerals. "I almost became a geologist," he recalls. In the end he opted for art, but he always looked for ways to include science in his artistic research.

Cohen studied at the Slade School of Fine Art, where he sought out scientists for conversations and developed a keen interest in collaboration. Why not, he thought, "look for ways to share a common understanding?" Over the years he has found that artist and scientist "need to be sensitive to the needs of the other." He describes this relationship as "ongoing and organic." Process is of the essence.

In 2006 Cohen started working with Japanese scientists at the Graduate School of Media Design at Keio University, in Tokyo. He works with a team of researchers seeking ways to blur the distinction between the real and virtual worlds, trying to represent "that space which exists between two and three dimensions," examining the conundrum "Where does the picture end and the real world begin?"

He plays with light, shining it on geometrical constructions or placing a light in the construction itself, or projecting digitally onto three-dimensional constructions or onto the floor or walls. The constructions are beautifully made, so precisely machined that sometimes they seem to float. Cohen wants the viewer to see them as "part of the real world, the viewer's world." When a person walks between the projector and the image, they're picked up by a sensing device, giving a second level of imagery and apparently creating extra dimensions, as in his *Interactive RPT Wall Installation*.

5.3 (a): Nathan Cohen, *Interactive RPT Wall Installation* (2008), Ars Electronica, Linz, Austria (without projection).

5.3 (b): Nathan Cohen, *Interactive RPT Wall Installation* (2008), Ars Electronica, Linz, Austria (detail with projection).

Cohen's constructions express his concept of aesthetics: "finding patterns in what we see, enjoying or being disturbed by (im)balance and (dis)harmony that can be evinced by symmetrical and asymmetrical arrangements of elements, and the construction of forms where the reading of the elements that compose them are spatially ambiguous and yet clearly defined."

He believes that his collaboration with Japanese scientists "has made possible new ways of creating artworks which challenge the way we perceive our environment." The lab's director, Susumu Tachi, spoke of his interest in bringing together "technology and fine art," which will require "a new integration of advanced technology and art." Both artist and scientist clearly benefited from this collaboration.

In 2012 Cohen initiated an MA course in art and science at Central Saint Martins College of Arts and Design, in London, the first such program in the UK. The aim of the program is to teach students to build a common ground and learn to communicate with scientists, to "make a journey toward understanding where the other person comes from." To do this, Cohen takes students into laboratories. He believes that this is important in fostering art and science collaborations. "There is nothing new here," he stresses. "The history of Renaissance art was primed by mathematics and physiology. There should necessarily be a science/art continuum."

## Drawing with light from the stars: Tim Roth

Tim Roth and Robert (Bob) Fosbury's spectacular art installation *From the Distant Past* used data from the Hubble Space Telescope to evoke the "heartbeat of the primordial universe."

Roth is a light artist who is interested in what light and color can tell us about the universe. In an exhibition at the Kunstfassade in Munich, in 2004, he used light panels on a building and arranged for them to be set off by light captured by telescopes on the earth and in space, as well as triggered by events from particle accelerator facilities. The matrix, driven by the macro- and microcosms, blinked on and off in dazzling colors.

Roth is an imposing figure, broad-shouldered and tall, whose loose-fitting suits make him seem even bigger, and who wears his hair swept back in a braid. He is German but speaks excellent English, liberally sprinkled with "cool." His route to art was indirect. Although he always had an affinity for the sciences, he studied philosophy and politics at the university in Tübingen and had an avid interest in photography. He obtained a degree in visual communication from the University of Art and Design in Kassel.

In 2009, Roth was a guest artist at the headquarters of the European Southern Observatory in Garching, near Munich. There he met the astrophysicist Bob Fosbury, at the time head of the European coordinating facility of the Hubble Space Telescope. Fosbury was eager to present astronomical data in an artistic way but had never had the opportunity until he met Roth.

The Hubble Space Telescope is akin to a time machine. It tells us what the universe was like ten billion years ago, when it was a mere three billion years old. Instruments on Hubble separate the light it receives from objects in the universe into their constituent colors, each of which carries characteristics of certain chemical elements. In this way Hubble puts together the fingerprint of the stars that are being observed, revealing what they are made of.

In 2010, Roth and Fosbury made their presentation in Venice, projecting the jagged curve of these data—similar to a chart of the day's trading on Wall Street—in a line of bright green laser light onto the façade of the Palazzo Cavalli-Franchetti. They called it *From the Distant Past*. As Fosbury puts it, the squiggly green line, which looked uncannily like a cardiogram, held "symbolic meaning—it was the 'heartbeat of the universe.'"

Fosbury was delighted at this eye-catching way to represent scientific data while avoiding technical details. People who saw *From the Distant Past* realized that they were seeing light from billions of years ago as well as evidence of what these hugely distant, very ancient stars, most likely long since blown to smithereens, were

5.4: Tim Roth in collaboration with Bob Fosbury, *From the Distant Past*, 2010.

made of. Although his collaboration with Roth did not affect his day-to-day work, it made Fosbury aware of ways to present knowledge about the universe that would spark people's imagination.

Since 2011 Roth has also collaborated on *From the Distant Past* with Antonella Nota of the European Space Agency, who is the Hubble Space Telescope project scientist and mission manager at Space Telescope Science Institute (STScI), as well as with Ken Sembach who is mission head from STScI. Nota bewails the lack of awe today in those dealing with astronomical data. In the past, she says, astronomers used to sit alone at their telescopes, "observing the night sky on top of a mountain [and] would breathe amazement. Now the data arrives directly to your computer screen." Collaborating with Roth allowed her "to reestablish an equilibrium I had partially lost, to appreciate in a deeper way the beauty of the data I am working on. It has reinforced that my role, as a scientist, is not only to find the answer, but to communicate the same sense of wonder to all people."

## Glimpsing the universe: Vanessa Harden

Vanessa Harden is a woman with a sense of humor. Her multifarious projects include guerrilla gardening, which involves her and her coconspirators planting flowers secreted in a false-bottomed briefcase into unadorned earth in traffic islands and other neglected public places. Another of her design experiments is a GPS chastity belt—the ultimate antirape device—which activates an electric circuit in the garment and locks it when the wearer enters an area she has previously classified as dangerous. It is, of course, more conceptual than practicable.

In person Harden is ebullient and always flamboyantly dressed. She is a fashion designer, among much else. Harden, who is Canadian, studied at Ontario College of Art and Design, then at the MIT Media Lab and the Royal College of Art in the Design Interactions group, which is where she realized that she wanted to "interpret science with design." At the RCA she had the chance to work with the synthetic biology group, but opted for a different sort of design.

Harden's knowledge of physics derives almost exclusively from conversations with physicist collaborators, to which she responds by creating works of art. Her work bears seriously on science, however; it is not just free creation. In 2009, she was approached by a group of astrophysicists based at Imperial College London, to participate in a project designed to put across certain astrophysical concepts and phenomena. Harden came up with a name for the project: Urban Sputnik. The idea was to use the scientists' knowledge of physics and Harden's design experience to devise something that would connect people in a sensory way with "distant cosmological phenomena that cannot otherwise be directly perceived nor experienced on a human scale."

Andrew Jaffe was one of the astrophysicists involved. The scientists' role, he says, "was crucial throughout the project, from the original ideas for topics to the final design, as well as in writing the explanatory notes." The scientists saw Harden's role as translating their ideas into art, a venture requiring hours of labor. The col-

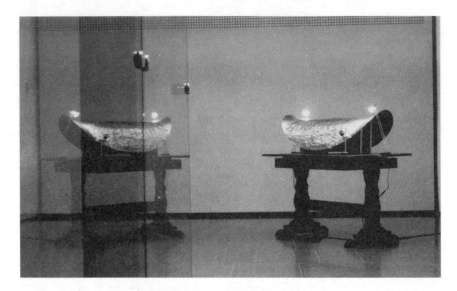

5.5: Vanessa Harden, *Shape of the Universe*, 2011.

laboration, says Jaffe, "enabled/forced me to think about some of my ideas more visually than previously, as well as being forced to describe them to Vanessa." It has certainly impacted on his teaching and his public lectures though, as far as his own research goes, "I think the effect has been minimal."

The result of the collaboration was an exhibition, Urban Sputnik: Interactive Cosmology, which excited a lot of interest, with events at both arts and science venues, including the Royal Astronomical Society and the Royal Institution in London. One of the exhibits— *Shape of the Universe*—is a "saddle" of beaten copper which represents the curved structure of four-dimensional space-time. Viewers can actually touch it and peer through a magnifying glass to see our solar system, a minute dot in the immensity of space-time.

## Cosmology is my theme: Josiah McElheny

Josiah McElheny's spectacular glass and chrome installations represent nothing less than the entire universe. When *An End to Modernity* was shown at the Andrea Rosen Gallery in Chelsea, New York, in

5.6: Josiah McElheny, *An End to Modernity*, 2005.

2005, it generated acres of admiring coverage. The *New York Times* described it as "The Entire Universe on a Dimmer Switch."

The sculpture, a gigantic starburst of 1,000 blown-glass orbs and some 5,000 rods, mostly of chromed aluminum at least 3 feet long, is some 14 feet across and 12 feet high. It represents the explosion of space 13.7 billion years ago. The length of the rods is proportional to the time duration since the Big Bang, while the glass spheres represent galaxies at the corresponding moment with lamps representing quasars. It hangs at eye level, with the bottom of the work just six inches above the floor. It's majestic and also incredibly fragile. It looks as if the slightest touch could shatter it.

McElheny is a glassblower by training. He studied at the Rhode Island School of Design, then with a master glassblower in Sweden, and won a MacArthur "genius" grant. He is intrigued by the concept of reflections and was inspired by a series of conversations in 1929 between the sculptor Isamu Noguchi and the designer and thinker Buckminster Fuller, in which they discussed reflective sculptures in a reflective space, in which no shadows would be cast, making it totally self-enclosed, a perfectly formed utopian environment. McElheny's response was to explore how "the act of looking at a reflective object could be connected to the mental act of reflecting on an idea."

Then in 2000, he noticed the striking spiky modernist cut-glass chandeliers at the Metropolitan Opera House in New York. It occurred to him that they had been made around 1965, when Penzias and Wilson at Bell Labs found astounding evidence to support the hypothesis that our universe came into being with a Big Bang, and struck him as the perfect way to depict it. But he didn't just want to illustrate it. He wanted to get the science right.

Everything fell into place in 2004 when he approached the cosmologist David Weinberg, a professor of astronomy at Ohio State University and author of the "Dark Matter rap," a rap which Weinberg recorded, all about the wonder and strangeness of dark matter, the mysterious substance which physicists hypothesize constitutes 85 percent of the total matter of the universe.

The two immediately bonded. They met frequently and Weinberg gave tutorials to McElheny. McElheny recalls that "David's view of cosmology is closer to an artist's view"—that is, highly visual. It was a "two-way dialogue in which Weinberg learned about issues facing the artist."

The result was *An End to Modernity*, followed by a succession of other enormously successful works on similar themes. *Island Universe* was exhibited in the White Cube Gallery in Hoxton, in London's arty East End, and in prestigious galleries around the world.

It is a collection of five glass and chrome starburst sculptures, each representing a different model of the cosmos and other potential universes. As in *An End to Modernity*, McElheny collaborated with Weinberg and was scrupulously careful to create a scientifically accurate model.

Says Weinberg, "More people saw *Island Universe* in one day in Madrid than have ever read my *Astrophysical Journal* articles." I met Weinberg some years ago when I gave a lecture at the astrophysics department at Ohio State. I was struck by his interest in the humanities, particularly in art, which is unusual for a physicist. Though he says that the effect of his collaboration with McElheny on his research is small, since the "big gap in technical level makes it hard for the art to affect the science," he found that it affected his day-to-day activities, such as teaching, lecturing, and helping him to see his scientific work "in a broader cultural perspective, as part of a long intellectual tradition." The collaboration spurred him to read writings on the cosmos written many centuries ago, as well as current studies relating this work to modern theories such as the multiverse. The works he collaborates on with McElheny draw on ideas from historical figures with whom he was not previously familiar. Thus, intellectually speaking, their collaboration has been mutually beneficial.

Despite the hard work and time he has put into his collaboration with McElheny, a time equal to the length of time he has spent on his own research, Weinberg is not credited as the co-creator of the sculptures. This is a delicate as well as a contentious area. Scientists expect to have their name on works resulting from collaborations. After all, if there were no science, there would be no art. But this is not as straightforward as it seems. Weinberg knew from the start that only McElheny's name would appear on the finished work. "This was the deal," he says, "not that we ever discussed it—and I decided it was okay. In the end, I got quite a bit more recognition out of it than I initially expected." However, he continues, "my colleagues are sometimes miffed on my behalf that I am not listed as the co-creator of the sculptures."

## Collaboration: a personal perspective

In 2008 I gave a lecture at the Wellcome Collection on the mural Picasso painted in 1950 in London—known as *Bernal's Picasso*—which they had recently acquired. At the dinner afterward, I met the artist Fiorella Lavado. She was looking for a scientist to collaborate with. She had no scientific background but was an avid reader of science articles in newspapers and periodicals such as *New Scientist*, and also read serious popularizations of science.

In her mid-thirties, Lavado was born in Lima, Peru. She studied audiovisual communication, then worked in advertising and documentaries and eventually focused on graphic design. She also studied with the Spanish performance and video artist José Luis Arbulú and the German neo-Expressionist painter, printmaker, and sculptor Ralf Winkler, alias A. R. Penck—a bête noire of the East German regime—which lent an edge to her work. She had lived in Germany, then moved to London at around the time we met.

Lavado is striking, with springy black hair, a sparkling personality, and the ability to get things done. When we met, her work consisted mainly of weaving, using ultrathin stainless steel wire in layers to create objects of various shapes and sizes, often spherical. One of her creations was like a representation of a black hole, going beyond the usual scientific visualizations drawn by artists from mathematical models. It is seven centimeters across. Lavado wove it from the inside out using a single strand of wire, spun into many layers with an indentation in the middle.

We decided to collaborate on a joint project which we entitled Weaving the Universe: From Atoms to Stars. We already had one completed object: the black hole. The problem was how to transform it into something that could evoke the frightening grandeur and poetry behind an object in the heavens of such incredible size—trillions of miles across—and awesome power. We found a way to communicate all this using Adobe Photoshop. In the resulting image, the wire sphere is transformed into a massive engine, spinning and spewing out the light generated by captured gas molecules. The wire texture adds to the mystery as it hovers in the darkness of deep space.

Our first exhibition was at the Benaki Museum in Athens in September 2009, in a program celebrating the international year of astronomy. There we showed the original black hole and other depictions of astronomical objects: a wormhole and a planetary nebula. We also gave a talk explaining how our collaboration worked.

We exhibited the same works at the Royal Astronomical Society in London in February 2010 and again gave a talk. For the first time that anyone could remember, the monthly RAS meeting was attended by artists. Other invitations to lecture arrived and we always took along the miniature black hole.

Shortly afterward, we were invited to exhibit at the famous Schering Gallery in Berlin in October 2010. Previously we had shown only half of our proposed project; the full concept included atoms as well as stars. For this show Lavado wanted to find a way to depict the wave/particle duality of light and matter, where an electron can be both a wave and a particle at the same time. After much discussion, she created a floor-to-ceiling sculpture consisting of many strands of stainless steel wire, each of which is alternately made up of wavy sections and solid clumps, which, when viewed together, form a tapestry evoking waves and particles blending into each other. We called the work *Anschaulichkeit* (Visualizability), a term coined by the eighteenth-century philosopher Immanuel Kant, who pondered the way in which we form images of the world and what these images have to do with science, in this case the quantum world.

These were not conceptual artworks capable of many interpretations, so I thought we needed explanatory text to make their meanings clear. With quantum physics text is indispensable, because it's so hard to imagine or understand. Most unlikely of all is the wave/particle duality. The text for *Anschaulichkeit*, however, was not as helpful as it could have been because the gallery placed restrictions on the amount of text that could be displayed so that what I had originally written was severely edited—an example of how galleries may have to rethink their rules when it comes to artsci.

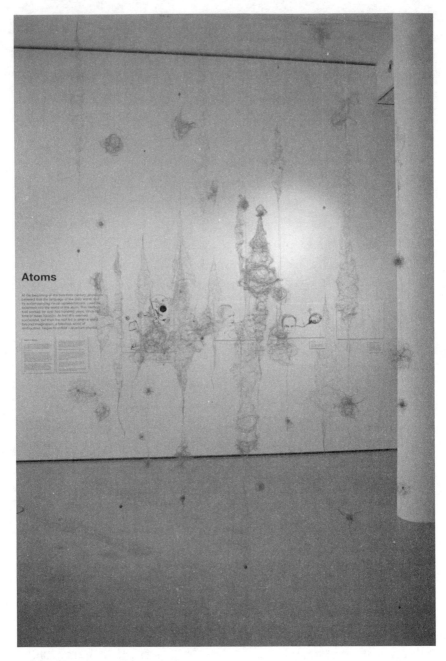

5.7: Fiorella Lavado, *Anschaulichkeit*, 2010.

The exhibition at the Schering Gallery marked the completion of our project. Sadly, our collaboration had already begun to break down in ways that seem to be common when artists and scientists work together. A number of issues arose, such as the independence of the artwork from the science. I have always felt that without the science there can be no artwork. Sometimes, however, the artist begins to believe that the scientist has simply given them a gift of information and knowledge.

The tipping point was the same realization that struck David Weinberg when he collaborated with Josiah McElheny. My name appeared only once, beneath the title of the exhibition. Only Lavado's name was on the many artworks, even though the work could not have been created without my scientific input. I have discussed this with other artists and scientists. We concluded that an agreement should be drawn up beforehand stating that the project is a total collaboration and that both names should appear on each work.

I can't imagine how our collaboration could influence scientific research, but nevertheless it was an interesting attempt to visualize highly counterintuitive phenomena in a way that is aesthetic and could also be inspirational. The effects worked both ways. I became more sensitive in dealing with concepts of aesthetics and beauty, and the collaboration opened new vistas for Lavado. It was also a helpful learning experience, enabling me to look at artsci from the point of view of a practitioner as well as of a writer.

## Art at CERN

The Conseil Européen pour la Recherche Nucléaire (CERN) was founded in 1954 as a United Nations for science, another way to unite Europe in the wake of World War II. It has been a phenomenal success. Originally there were twelve member states. There are now eighteen, all from the European Union. At present there are some 10,000 visiting scientists from over 600 universities, and over 100 different nationalities working there. In addition to CERN's scientific achievements, there have been momentous technological spin-offs, such as the creation of the World Wide Web.

A highlight of the CERN day is lunch at the cafeteria, where scientists huddle, discussing the nature of matter, the number of dimensions in the universe, or why we are here, oblivious to the stunning backdrop of the snowcapped Alps. In recent years the public, too, has become engrossed in news from CERN. In pubs and restaurants, people discuss reports of neutrinos found to travel faster than light, shedding doubt on Einstein's relativity theory, though in fact those particular data turned out to be incorrect. Then came the buzz over CERN's discovery of the Higgs boson: what it is and what it means.

There has also been a huge amount of interest in CERN from the art community. As Renilde Vanden Broeck, CERN's senior press officer, tells me, CERN "always welcomed artists [but] the thing was unorganized until Ken McMullen," the distinguished independent filmmaker, arrived on the scene. McMullen's movies are grounded in history, psychoanalysis, and literature, and have been described as painterly and theatrical. *1867*, based around Manet's painting of the execution of the Emperor Maximilian in Mexico, is a deconstruction in time, space, and memory of the eleven months Manet labored over the work and the events surrounding the execution, while *1871* pivots on the violence of the 1871 commune and the writings of Émile Zola. *Cinema plus Psychoanalysis = the Science of Ghosts*, based on the philosopher Jacques Derrida's suggestion of a link between cinema and psychoanalysis, is a cinema verité in which Derrida plays himself conversing with the actress Pascale Ogier. In the 1970s, McMullen made a film featuring the German artist Joseph Beuys, who uses science in his work, and this inspired his interest in science.

Trained in painting at Liverpool University and then at the Slade in London, McMullen decided to swap brush for camera while keeping his fine art background in mind. There was no difference between making a film and painting a picture, he tells me. In the late 1990s he was living in Paris and dating the actress Irène Jacob. Her father, Maurice, was a theoretical physicist at CERN and invited him to visit. Many of the physicists there had seen his films. Then in 1997 McMullen became a professor at the London Institute, an umbrella organization covering the major London art schools. The rector asked him to find a scheme to boost the rating of the institute so that it would receive more government funding. McMullen contacted Jacob, who suggested he approach the Particle Physics and Astronomy Research Council in London. McMullen's idea was to take a group of art students to CERN to make a film that would be "not merely art as illustration; it would embody physics concepts into the art world." The London Institute suggested that instead of students, McMullen should look for artists with international reputations.

McMullen invited the artists Roger Ackling, Jerôme Basserode, Sylvie Blocher, Richard Deacon, Patrick Hughes, Paola Pivi, Tim O'Riley, Monica Sand, and Bartolomeu dos Santos, and the scientists John March-Russell, Stephan Russenschuck, Michael Doser, John Ellis, Helenka Przysieniak, and Hans Drevermann. Along with Maurice Jacob, they all met at CERN in December 1999. They spent the morning touring the complex, wandering through the tunnels, riding the monorails, visiting experiments, and learning about the CERN workshops. In the afternoon, each gave a brief talk. "The effect is spellbinding and the thing is begun," Michael Benson, director of communications at the London Institute, wrote in his diary. He told the *CERN Courier*, "Artists today are beginning to realize that science provides fertile territory for the imagination."

During the following year, the artists returned to CERN from time to time to consult with their scientist collaborators. The result was an exhibition entitled Signatures of the Invisible, which opened in London's Atlantis Gallery on March 1, 2001. The press release described its overall aim:

> The laws of physics are not going to go away, although these laws are no longer intuitive—relativity, antimatter, quantum mechanics are how Nature works. Sooner or later the artists will have to confront the challenge of representing and commenting on these foundations of human life. Never has there

been a better opportunity as the new media of video and computer graphics, again direct spinoffs of physics research, give new possibilities for presenting apparently abstruse scientific concepts.

Luciano Maiani, director general of CERN, wrote, "Signatures of the Invisible is a groundbreaking initiative. For the first time researchers at the leading edge of current work in two apparently different, unrelated fields—contemporary art and high energy physics—have worked together." Emmanuel Grandjean wrote in *Tribune de Genève*, "There is fusion in the air at CERN. [CERN] attempts the marriage of art and science."

McMullen emphasized the unique collaborative aspect of the exhibition. It was, he says, "not a straightforward illustrative sci-art project but a free flow of ideas." Many spectacular works were produced, including engravings depicting particle collisions, installations using nonradioactive lead from the Roman era, metal sculpture created in collaboration with workshop technicians, computer art, printmaking, fashion drawings, and video art.

The sculptor Richard Deacon, famous for his abstract works created using everyday materials, was struck by the vast size of the Large Hadron Collider (LHC) compared to the microscopic nature of the Higgs boson, which it was designed to catch—rather like a Tibetan ghost trap, a wooden frame crisscrossed with threads, trying to capture an invisible entity. He built a "particle accelerator" out of plaster and Sainsbury's plastic bags, describing it as "a kind of archetype."

Patrick Hughes creates three-dimensional illusions which seem to move as you walk in front of them. He has been described as a painter of paradoxes rooted in perceptions. He was struck by the array of instruments at CERN that seem to open doors onto nature. To represent them he painted instruments on six panels which open to reveal a view of the Alps as the viewer walks by.

Monica Sand was interested in particle physics, the trapping and absence of light. Physics represents light as a continuous glow, a field of light like a field of grass. Sand called her work *Maxwell's Field*, in reference to the great nineteenth-century physicist James Clerk Maxwell, who discovered the equations for electricity, magnetism, and light. To represent the field of light she used an array of boxes, into which she slid panels made of material from particle detectors at CERN, and "black glass" to stop light passing through. "The mathematical, the simple, and the complex are hidden in the world you are invited to contemplate," she writes.

The late Bartolomeu dos Santos was a larger-than-life Portuguese artist. Of considerable girth, he exuded good humor and deep thoughts on art, as I recall him on the many occasions we met at University College London. He worked in the printed image, and at CERN was fascinated by the whiteboards covered in equations, like hieroglyphics, which he saw as maps to the unknown. Once an equation has been written, it exists forever, even if it is erased. He wrote on one of his works the words, "*Ne pas effacer.*" (Do not erase.)

The French sculptor and installation artist Jerôme Basserode was struck by the way physics incorporates past, present, and future, and the relative nature of time as one turns into the next. To represent this he used several huge spinning tops, machined in the CERN workshops, inscribed with the words "past," "present," and "future." Visitors see the tops slowly rotating past, taking time along with them.

McMullen produced seven works which, along with dos Santos's piece, were less illustrative of physics and more to do with "signatures of the invisible," in line with the exhibition's aim. In *Skin Without Skin*, McMullen considered what would happen if you took a piece of paper, drew parallel lines on it, then crumpled it up into a three-dimensional shape, like a relief map of mountains and valleys. He did this several times. Each crumpled paper came out different even though he applied the same amount of force. The theoretical physicist John March-Russell commented that the uniqueness of each crumpled paper was analogous to the assertion, in quantum theory, that the act of measuring alters the system which is being measured so that we never know the system itself, only the system plus the measuring device. "There can never be a 'quantum Xerox machine,'" March-Russell writes. McMullen realized that the magic of the crumpled papers was that we perceived them differently depending on how they were illuminated; the crumpled paper (the system) had no meaning until it was illuminated by light (the measuring device). By now he had taken to referring to each piece of lined, crumpled paper as a "crumple theory." With the help of Ian Sexton, a CERN craftsman, he scanned a crumpled paper in the super high-tech laser scanner, scaled it up to nine feet long by three feet wide, then cast it in three stainless steel segments. Using lasers, Sexton sliced the stainless steel segment into thin plates that slotted into a horizontal base. The result was like a mountain range put through an egg slicer. There were different shadows and textures depending on the light; slices seemed to disappear in the same way that measurements in quantum physics occur. *Skin Without Skin* resulted from a collaboration between an artist, a physicist, and a technician. In 2001, the Chinese government awarded it a prize for outstanding work in art and science.

Among other work which McMullen showed in Signatures of the

5.8: Ken McMullen, *Skin Without Skin/Shadows and Errors*, 2000–01.

5.9: Ken McMullen, *?Lumen de Lumine*, 2000–01.

Invisible is *?Lumen de Lumine*, an eight-minute video of a woman in a cadmium red dress standing in a steel circle—a cross-section of an old accelerator at CERN. She swings a cord with a lit 60-watt bulb at the end over her head, slowly letting out the cord. At first she is in control of the lightbulb, but as the cord becomes longer, it becomes more difficult to control. Soon it is controlling her and she sways rhythmically to keep her balance. Her difficulty in controlling the bulb is a metaphor for the collisions and near misses as millions of particles whip around the accelerator at speeds near that of light. McMullen then reversed the image and ran it next to the original, suggesting the beams of particles flying in opposite directions around the circular accelerator and further emphasizing near misses. In some versions the woman recites Hamlet's famous soliloquy, "To be or not to be," in German, the language of Einstein, Heisenberg, and Schrödinger, pioneers of the quantum revolution. *?Lumen de Lumine* is also a meditation on the solitude deep underground in an abandoned tunnel. The light emanating from the whirling lightbulb, a metaphor for the whirling particles, is "light emanating from light"—*Lumen de Lumine*—an ancient alchemical notion referring to the divine light that reveals the mysteries of nature.

On February 9, 2006, *?Lumen de Lumine* was projected in a six-minute loop onto Torness nuclear power station in Scotland, a major landmark on the East Lothian coast. It was the largest motion picture installation ever seen in Europe, and was viewed by over 20,000 people.

Another McMullen piece, *Signatures: Commemorationem*, was a thirty-minute film loop of a conversation between McMullen, the physicist John March-Russell, Michael Doser, an experimental physicist at CERN known for his experiments with antimatter, and the critic John Berger, with whom McMullen had previously worked. Berger started the proceedings, stating, "To find something that was previously hidden, you first have to be lost." The film was made in a natural setting without complex cameras or lighting equipment. Berger insisted that there should be no preparation, just free-flowing conversation, so that "what we see is somebody thinking and not necessarily responding." The aim of the film, McMullen adds, was "so that anyone can get the drift as to what goes on in particle physics." The conversation roamed across aesthetics, ethics, and what quantum physics has to say about causality and determinism.

The reviews of Signatures of the Invisible were excellent, although several journalists were mystified as to what they were seeing. But they all got the key point. As Stephen Pile of the *Daily Telegraph* wrote, "What would C. P. Snow make of all this? It is 42 years since he observed that art and science are separate cultures, but suddenly they are getting along just fine. Science-inspired art events are opening all over the country." Some 15,000 people saw the exhibition over a period of three weeks.

The CERN physicists Michael Doser and John March-Russell were on hand and gave explanations which were helpful and piqued viewers' interest. They were extensively quoted, especially March-Russell, whose explanations contained much theoretical physics, right up to high-dimensional membranes and parallel worlds. As Rachel Campbell-Johnston of *The Times* wrote, "A conversation with a theoretical physicist is like having a pair of jump leads clipped to your brain. Sparks of inspiration fly."

Signatures of the Invisible was shown in Stockholm, Lisbon, Paris, Strasbourg, Brussels, Tokyo, Australia, and at MoMA PS1 in New York, where 250,000 people came to see it, with 10,000 attending the private viewing. For the PS1 exhibition, the Stanford Linear Accelerator Center shipped over particle detectors. The concrete scientific equipment stood alongside the abstract depictions of science.

OVER THE NEXT few years, CERN continued to grab headlines. In 2009, Dan Brown's blockbuster novel *Angels and Demons*, bristling with secret societies, symbolism, and the unique addition of an antimatter bomb manufactured at CERN, was filmed there. On its website, CERN emphasized that both the story and the bomb were pure fiction. More recently, Robert Harris based his thriller *The Fear Index* on CERN and the research going on there.

That same year, 2009, the artist Josef Kristofoletti painted an image of the Atlas Detector, part of CERN's Large Hadron Collider, on the huge building housing it. But in terms of advancing art at CERN, the most important event that year was the arrival of Ariane Koek to start an arts residency program there. She called her new initiative Collide@CERN.

Koek is a dynamo. In the course of a sentence, her voice can shift over an octave. She radiates enthusiasm and sees herself as a cultural specialist. She was formerly at the BBC, where she produced and directed cultural programs including science documentaries. The Clore Leadership Program awarded her a fellowship to pursue cultural projects, and she opted for one outside her comfort zone: art and physics. "For me it was a no-brainer," she says. "I immediately chose CERN. I've always been fascinated by physics, the way it combines an intense logic with an intense imagination." Collide@ CERN came into being in 2011, with an initial plan for it to run for three years, funded in part by Ars Electronica, the digital art center in Linz, Austria.

Koek instigated a formal competition in which the winning artist would spend two months in residence at CERN and a month at Ars Electronica to develop a work inspired by CERN. There was also a second strand of dance and performance inspired by physics, sponsored by the city and canton of Geneva. Koek dubbed this new cultural policy for engaging with the arts "great arts for great science." At CERN the chosen artist would seek an "inspiration partner" by "speed dating," the idea being to meet as many scientists as possible, each of whom would explain their work.

Koek felt it was important to keep the artist's stay at CERN short so that artist and scientist "don't get too close, so that the artist doesn't get involved in details. Artists shouldn't have to prove themselves to the scientist." The pair would meet regularly, present an onstage conversation at the end of the artist's residency, and hopefully produce a joint work.

Koek was adamant that Collide@CERN should not be an outreach program, explaining CERN's activities. Rather, it should aim to put art and science on an equal footing.

## CERN's first artist in residence: Julius von Bismarck

Koek's first artist in residence was Julius von Bismarck. To call von Bismarck avant-garde is an understatement; he is an *artiste provocateur*. His works are extraordinary and often set up deliberately to provoke. In 2008 Ars Electronica awarded him the Golden Nica for *Image Fulgurator*, a camera equipped with a telephoto lens and pistol grip, electronically primed to manipulate photographs as they are being taken—for example, inserting a dove in front of the huge image of Mao Tse Tung in Tiananmen Square. Von Bismarck handles his camera like a pistol, looking appropriately sinister with his long black beard, black stocking cap, and tight-fitting black sweater and trousers.

Behind the dramatic image is a delightful personality and a wry sense of humor, which can be seen in his recent video, *Punishing Nature*, shot at the Statue of Liberty by a colleague standing unobtrusively in the crowd. First von Bismarck is seen bull-whipping a large rock, then the scene shifts to him whipping the base of the statue. Three burly US Park Police jump on him and handcuff him. One version concludes with von Bismarck lying handcuffed on the deck of the boat that takes tourists back to Manhattan, surrounded by police. The video was made, as he puts it, "in collaboration with the local police."

Von Bismarck arrived at CERN in 2012 and chose the theoretical physicist James Wells as his "inspiration partner." He insists that he did not collaborate as such with a scientist at CERN, but was greatly inspired by them all. "I was more interested in an intellectual exchange than in working with their media," he says. He and Wells hit it off, as their joint presentations demonstrate. Their joint publication has yet to materialize, however, and neither will reveal what it will be, though von Bismarck hints that "international galleries are interested already."

Von Bismarck's role was not just to learn but to impart some knowledge of art. He did this in a typically provocative fashion. Some of the physicists had given him a crash course in physics. In exchange, he gave them a crash course in art: five minutes of art history, after which he asked each person to come up with an idea for an artwork. Then they all had to critique each other's work, a rite of passage in art schools. Finally, he led them out to the parking lot and told them to run around like elementary particles. For another "intervention," he took physicists into a dark room and asked them to describe things they could not see.

Von Bismarck was impressed by the seriousness of the conversations in the CERN cafeteria. Recalling how, after the war, captured German physicists were interned at Farm Hall, near Cambridge, which was bugged to record their conversations, he decided to record conversations at as many tables at CERN as possible. His plan was to play them back over different channels, making a sound installation scanning the soundscape of the cafeteria.

5.10: Julius von Bismarck, *Versuch unter Kreisen*, 2012.

In fact, the only work he completed at CERN was one he had begun before coming. *Versuch unter Kreisen (Research under Circles)* consists of four lights hung by cables of adjustable lengths with three motors in each unit. As the lights swing back and forth, a laptop controls the lengths and frequencies of the movements until, after a certain number of random swings, all four move in unison.

It is, says von Bismarck, to do with numbers and thus combines art and science.

What did James Wells get out of the collaboration with von Bismarck? In a panel discussion at CERN in September 2012, he gave his impressions. His previous conception of how artists work had been rather naïve, he confessed. He had thought they just painted and that the first version was generally enough—a Hollywood view of the artist's life. From von Bismarck he learned that artists work as hard as physicists and need the same dedication. He saw von Bismarck as a positive role model for young physicists, who sometimes emerge from their PhD studies confused, with their creativity undermined. Von Bismarck, said Wells, seemed to be a man who asked himself, "What do I want to do? Here's how I'll do it." Wells added that "von Bismarck brings humor to physics. People at CERN think he is great."

In fact, Wells's enthusiasm was not universally shared. A number of physicists at CERN complained that there was little transparency in the program, in that almost nothing was discussed or explained to the group as a whole. Many of the physicists who interacted with von Bismarck got little out of it. As a result few chose to participate, which ran counter to the aim of Collide@CERN, which was to put the artist in close touch with physicists. Von Bismarck was chosen as artist in residence at CERN because of his superstar status as an artist, underscored by the fact that he had won a Golden Nica at Ars Electronica, despite the paucity of his knowledge of physics. Perhaps a person with greater involvement with physics might have left a more enduring footprint.

Another comment was that Collide@CERN was too structured, and there was too much control over the artist. A two months' residency at CERN is perhaps too long. It might have been better if the artist could have come and gone over a period of time, conversed with physicists, imbibed the atmosphere, then returned to his studio to produce sketches and prototypes. Von Bismarck agrees. "After three months of brain work [two at CERN and one at Ars Electronica], it was so nice to work with my hands again."

Joe Paradiso of the MIT Media Lab, who made the dancing sneakers that created music in response to the wearer's movements, did extensive work at CERN and was offered a position there. Regarding artists today, he says, "It's all about credentials. The modern artist is an artist, technician, and manager." His advice to Collide@CERN is, "Get a great artist," someone who "will use the data." He's referring to data visualization as a medium. "Use the data, live in it, live in it. That'll be astounding."

Luis Álvarez-Gaumé shares these sentiments. What CERN needs is "plenty of space for installations," he says. "Its laboratories should be filled with inspired art."

A former head of the theory group at CERN, Álvarez-Gaumé is a renowned physicist and an outspoken advocate of the need for an interplay between physics and art. An aficionado of literature, painting, and music, he often introduces lectures by Koek on Collide@CERN. He is committed to developing an arts dimension at CERN and suggested introducing a 3 percent arts tax on all income there, to be set aside for the arts.

Álvarez-Gaumé suggests letting artists loose, with no constraints such as an "inspiration partner" and no limited residency— no residency at all, for that matter. Let them wander around, then go back to their studios and encourage them to return. CERN, he continues, has the cachet to attract top-rate artists, who should be invited and paid for their services, and "hopefully they will leave something behind." Since research at CERN explores the invisible, he suggests that "artists with abstract ideas" should be preferred.

When I ask whether he has any particular artist in mind, he names Keith Tyson, who had recently visited CERN.

## Abstract physics/abstract art: Keith Tyson

In 2002 Keith Tyson won the Turner Prize for *Artmachine*, an arcane system which he developed using computer programs, flow charts, and books to generate chance combinations of words and ideas which he then turns into artworks. He calls it "the fusion of mathematics and reality through a computer." Some of the works he developed from *Artmachine* and put on show for the Turner Prize included sci-fi, Heath Robinson–type devices such as *The Thinker (After Rodin)—Technical/Notes, 2001*, a hexagonal monolithic black structure with computers humming inside which he called "a comatose god running its own universe" (see Insert), and two paintings entitled *Two Discrete Molecules of Simultaneity.*

Tyson is often referred to as the "mad professor," and his work regularly touches on and incorporates science. He is a man with big ideas that emerge from a well-thought-out theme based in science: the emergence of simplicity from complexity. Self-taught, he knows a lot about complexity theory and sees the "world as information, as emergence out of nothing."

Born in the north of England, Tyson is proud of his working-class roots. He was interested in science from an early age, building computers and designing games with code which he taught himself. His artistic talent was encouraged by his primary school teacher. Tyson left school at fifteen and spent five years working as a welder in a shipyard, building nuclear submarines and learning engineering in the process; then, when he was twenty, he went to art school.

Tyson thinks deeply and carefully about what he does. "Richness is always at the frontiers," he says. "Stuff happens at the edge of a cloud, at boundaries between countries. But reality is beyond all that—reality doesn't care about boundaries." He sees no "dichotomy between science and art [but is] interested in modeling the world." Art and science are not, for him, exclusive categories. "They are ways of exploring the same things, one culture with two fascinating abilities. One should talk about exploring the already existing fusion between the two. Society suffers from reduction and compartmentalization, leading to 'social autism.'"

Armed with such thoughts, Tyson enjoyed the visit he made to CERN in 2009. He recalls it as an "extraordinary place where curiosity is still important." He was impressed that "like good artists, good scientists are simple in their explanations, open-minded, humbled by experience. The discussions were fascinating."

## Seeking a visual language for scientific art: Steve Miller

Steve Miller's work ranges from x-ray portraits and Rorschach tests to DNA art. We meet outside his spacious studio on Manhattan's Lower East Side. He has a second studio in Sagaponack, in the Hamptons, in a converted potato barn once owned by the artist Frank Stella. A lot of Miller's savvy comes from jobs he took early on to support his art—bartender, ski instructor, studio assistant, construction worker, and commodities trader, all jobs that require personality, confidence, and verve, which he has in abundance. Nowadays, as well as being a practicing artist, Miller teaches at the School of Visual Arts in New York. His interests include biology art and physics art.

Born in Buffalo, New York, he was taken to museums by his parents so often as a child that eventually he came to appreciate them, especially the Albright-Knox Art Gallery, where he encountered Jackson Pollock's spectacular *Convergence*, a massive work nearly eight by thirteen feet. It was a major event for a twelve-year-old. "It, like, knocked me out," he says. Miller's family offered no direct support for his artistic aspirations but "artistic encouragement was in the genes." Miller's grandfather was the glass designer who created the iconic Coca-Cola bottle, while his great-grandfather was a portrait painter of some renown as well as an agent for Tsar Nicholas II charged with acquiring paintings, particularly Impressionists.

Miller did some painting in high school and went on to Middlebury College, a small liberal arts college in Vermont, where he majored in fine arts, specializing in sculpture. He then spent two years as a fellow at the Fine Arts Work Center in Provincetown, Massachusetts, then by degrees relocated to New York and the Hamptons.

In New York in the early 1970s, Miller continued to sculpt, then turned to Abstract Expressionist painting. Soon disenchantment set in: "The habitual gestures of making paintings had become frustrating and were feeling meaningless." He started making movies and working in commercial film. But this still wasn't immediate enough for him. He longed to put brush to canvas, pencil to paper once again.

Looking into Cubism he became "totally enchanted with that epoch" and its new visual language, developed in response to innovations in technology. This was the sort of art he wanted to do.

Perhaps, like Picasso at the turn of the twentieth century, he was in at the beginning of a new avant-garde. He took in the dramatic changes happening in computers, technology, and science. This was how it happened in Cubism, he reasoned. After all, "changes in technology and changes in science allow for changes in consciousness." "Man, I wanted to be in on the next epistemological break," he says, "epistemological break" being a term from the French philosopher Michel Foucault meaning a rupture from accepted knowledge that is felt immediately. Miller was sure that the new technologies could be used "to reinvent new genres, to literally get inside, to reinvent portraiture."

Miller has always believed that the job of the artist should be to reveal truths. The 1980s was an era fraught with corruption, greed, and the dominance of the military-industrial complex, with AIDS running rampant. There seemed to be "a breakdown in culture, society, and the human body." Why not "use science to look at pathology as a metaphor" for this breakdown?

He was looking for a way to do this when he came across Rorschach tests, in which subjects interpret ink blots. Psychologists claimed to be able to deduce a person's personality and emotional state from their interpretations. If a prisoner up for parole gives a sinister interpretation of a Rorschach blot, he would be less likely to be freed than if he gave a more benign one. This social aspect fascinated Miller; he also liked the fact that the blots were "somebody else's piece of paint." He liked the randomness of the process, the fact that he was not responsible for whatever shapes emerged. So he took the blots and made them into art. He scanned them, photographed them from the computer screen, then printed them onto silk screens and touched up the result with acrylics. Now, "a blob of paint had a scientific meaning." And each person still saw them in a different way.

"Technology allows you to penetrate, to get inside," he says, referring to x-rays, MRI and CAT scans that make the invisible visible. To him, the "beauty of these images is that they are the biological, technological, scientific equivalents of the Rorschach blots." They tell us about pathology, about the state of the body. To the untrained eye, an x-ray or an MRI scan looks like a surreal jumble, but a scientist looking at the same scan can make out a virus or a cancer cell. In the same way that the Rorschach test can be a litmus test of a culture—instead of seeing a butterfly, people might see "Hiroshima or bloody fetuses on the pavement"—scans are used to reveal diseases of the body.

Reasoning in this way, Miller turned to medical pathology as a metaphor for cultural diseases. First he took images from medical texts. Then he tried using medical imaging as a new way of making a portrait. "It was a new way to take on a historically dead genre, to reinvent a genre that had been marginalized by photography." Miller's aim was to use "new technologies to reinvent new genres, literally to get inside, to reinvent portraiture"—to produce an internal portraiture. "All of a sudden, I realized there was this whole other world that couldn't be seen by the eye but could be visualized through the new technology." Imaging enabled him to move away from traditional portraiture, in which the eyes communicate the subject's inner being, to a new view of the subject, a new way to understand what a human being is. "We really have windows into the human body."

Through friends in medical research, Miller was able to use x-ray machines, sonogram apparatus, MRI and CAT instruments, as well as electron microscopes. He was like a child in a toy store. In 1993, as a gift for two patrons, Jacques and Véronique Mauguin, Miller combined sonograms of two fetuses in their mother's womb which he silk-screened and sandwiched beneath a radiograph of the father's hip. The finished work is entitled *Portrait of Jacques and Véronique Mauguin*.

In *Portrait of Pierre Restany*, also produced in 1993, in Paris, he depicts the French art critic and philosopher Pierre Restany as an x-ray profile, complete with glasses and cigar, with another x-ray image of his hands (see Insert). The entire canvas is silk-screened and bleached by acrylic, with a graduated scale like those used in word-processing software along the side.

Since the 1980s, computers had been part and parcel of his art. But as late as 1993 the art world still considered computers a gimmick, not part of serious art, "whereas today all artists use computers." In Paris, the mood was different; France had embraced the new technology. Everyone had a Minitel, like a small computer, which you could use with your telephone line to make purchases or reserve seats at a restaurant, or—famously—to flirt. Handheld devices for paying with a credit card also started in France. The graduated markings in the portrait of Pierre Restany emphasize Miller's respect for computers as well as his interest in cognitive science, a quantitative study of the mind based on seeing it as an information processing system, somewhat like a computer. Indeed, Miller considers his portraiture a new way of looking into the mind.

Another 1993 work, *Portrait of Dr. William Frosch*, his psychiatrist, superimposes slices of MRI scans of the doctor's head with a Rorschach image, once again a silk screen on canvas, bleached with acrylic to produce a surreal juxtaposition. Besides being a portrait of Dr. Frosch, it is also a view into the mind of the person looking at it. One of Miller's self-portraits, *Self Portrait Yellow*, completed earlier, in 1992, is an MRI of his own spine superimposed with EKG signals together with an MRI of his scrotum. Another self-portrait consists of a CAT scan of his head and spine, with EKG signals superimposed to evoke his bodily rhythms.

He has also made forays into DNA art, producing a portrait by mixing the nuclei of some of the subject's blood cells into a bean culture. Using an electron microscope, he photographed the nuclei after they had undergone differentiation (mitosis). He then photographed the chromosomes under an electron microscope as they divided. Then he scanned the images into a computer to produce a genetic portrait, an extraordinary and beautiful image that looks like a cloud of bubbles.

Miller began x-raying and scanning anything he could lay his hands on—women's high-heeled shoes, his mother's handbag. In 2007 he made a powerful x-ray image of a Glock pistol being loaded by a skeletal hand and called it *Glock*. Scientists were amazed and envious of what he got away with. He "brought into radiological labs all sorts of things, like snakes, guitars, violins. Scientists never get to play like that." This, in his view, is what an artist can bring to a medical laboratory.

Back in 2000, the director of the Brookhaven National Lab (BNL) in Upton, Long Island, had invited a group of artists to meet the scientists who work there. For Miller the big attraction was the BNL's Relativistic Heavy Ion Collider (RHIC), 2.5 miles across, which was used to study the debris from collisions of protons hitting protons, and of heavy nuclei such as gold and lead smashing into each other at speeds close to that of light. At the time it was the only accelerator studying heavy nuclei. Now the Large Hadron Collider at CERN devotes a month each year to this research.

The aim of smashing heavy nuclei together was to re-create conditions at the moment of the Big Bang, the creation of the universe, when temperatures were about a million trillion degrees Centigrade. At Brookhaven, they actually did it. For a few billionths of a second, the protons and neutrons that made up the heavy nuclei melted. Then their fundamental constituents, quarks—the building blocks of matter—and gluons—the particles that glue them together—burst free to form the quark–gluon plasma. It was as if they had moved back in time. The existence of the quark–gluon plasma was first shown at Brookhaven.

The quark–gluon plasma is like a sort of hot primordial soup. Studying it could give us further information about what conditions were like in the very early universe, before quarks bound together to form protons and neutrons and stopped existing as free particles. To Miller the RHIC was like an imaging machine that looked into the body, except that it looked into nature instead.

In the course of that visit, Miller met Steve Adler, who worked on one of the giant detectors, PHENIX (Pioneering High Energy Nuclear Interaction eXperiment), whose task was to actually detect the quark–gluon plasma. "They invited me to do whatever I wanted," he tells me. It turned out to be an embarrassment of riches. Adler was an engineer in information technology. His job was to write the code instructing PHENIX how to detect the quark–gluon plasma. "Steve showed up at the site of the experiment in Brookhaven," Adler recalls, "and I showed him around, giving him everything he wanted—items like physics doodles on the back of paper scraps, software source code printouts, and all the photos he wished to take. From that he started creating his art which I found inspiring."

Miller's starting point was a discovery he had made the previous year in Singapore, when he was teaching art there. He came across stores selling Neolithic pottery, dating from 5000 BC, for almost nothing. The pots struck him as not only beautiful but an early "investigation of matter, taking dirt and doing this." His studio is now full of ancient pottery. His realization was "how to tell the story." The RHIC enabled him to tell it as a "time line of human development, from mud pies to the [discovery of the] quark–gluon plasma." Among the works he produced is *Neolithic Quark* (2001), a series of paintings overlaying images of Chinese Neolithic pots with software codes and physics equations, providing a timeline from one of the earliest experiments in transforming matter right up to the present. *Untitled 91101* is from that series (see next page).

Miller also spent time at the National Synchrotron Light Source (NSLS), also at Brookhaven. This massive machine accelerates electrons so that they emit light and is used to study the structure of crystals. One of the scientists there introduced Miller to Roderick MacKinnon, based at Rockefeller University, who was using the light emitted from electrons to study the structure of proteins. MacKinnon was interested in how certain proteins—ion channels—permit ions such as calcium, sodium, potassium, and chloride to enter cells. These ions cross cell membranes at up to a million ions per second. This is an electrical process. Our bodies, of course, are electrochemical engines. MacKinnon took snapshots of ion channels using the NSLS which helped him discover their spatial structure and how they can pass through cellular walls, for which he won the Nobel Prize in 2003.

5.11: Steve Miller, *Untitled 91101*, 2001.

The two men hit it off, and MacKinnon invited Miller to Rockefeller and gave him access to his notebooks. In 2003, Miller produced a series of works, among them *Protein #330*. In this rather magical work he uses a photograph of a potassium ion channel, identified using light from the NSLS, silk-screened onto a canvas, with a photograph of equations which MacKinnon had written on his blackboard superimposed on top of it. The equations also describe the process by which ions diffuse through cell walls.

Miller's work is sometimes misunderstood. In November 2007, in a publication called *Big Red & Shiny*, the art critic Katie Hargrave wrote a review of his exhibition Spiralling Inward at the Rose Art

Museum at Brandeis University in Waltham, Massachusetts. This is what she wrote of *Protein #330*: "Silk-screened on canvas, Dr. MacKinnon's notebook page is obstructed by this confusing protein image, with only the title words and a mess of partial equations showing through. What qualities do we need, Mr. Miller? The work creates a conversation between the neurobiologists who understand these equations, but in their obstruction, he denies the usability of the science. Instead, viewers are left with undeniably beautiful and poetic images. If Miller is disallowing the scientists to be able to understand this work, then who is it for?"

At least she saw the totality as beautiful and poetic. "If you read this review through the lens of an art critic, I fared pretty well on many fronts," says Miller. Cate McQuaid, a more scientifically literate art critic, described the exhibition as "an affirmation of life."

In avant-garde circles, Miller says, "beauty is, for many, a forbidden word." Beauty is not something he considers before beginning a work, rather "it's the result of a process" which encompasses past experience, thought, and the making of the work itself. As artists often do, he emphasizes the process over the end result. Yet in the end he chooses not to make any definite statement, simply noting that beauty is, like a Rorschach test, in the eye of the beholder. Perhaps more explanation might have helped in the exhibition at Brandeis. These works are not responses to emotional situations or landscapes. Nevertheless, like the works of old masters, works that bring together science and art deserve close study and much thought.

Miller does not collaborate with scientists. Rather, "scientists cooperate with me." What he brings to his relationship with scientists is a thorough knowledge of art and its history, gleaned from years of study. But if he produced a work in very close cooperation with a scientist, he would certainly add their name to it, he says. What is important to Miller is the "cross-pollination that can loosen everybody up. Art gives scientists permission to play."

Too often, he says, scientists "appreciate art but know nothing about it." Once, when he gave a lecture at CERN, a scientist, thinking of classical art, asked how Miller could consider his own work to be art. Artists' visits to labs such as Brookhaven and CERN give scientists some idea as to what the present world of art is all about. Certainly art can be useful "to present complicated information to an audience, as for example, in Michael Frayn's play *Copenhagen*, which portrays a drama about the bomb. It's not about science, however. I would like to do that," says Miller.

Steve Adler, who works on PHENIX searching for the quark-gluon plasma, says that Miller's art had no direct effect on his work. But, he adds, "Steve's influence on my work at that time would have been like a breeze blowing cross-bow on an oil tanker. But I did feel the breeze." Keenly aware of the art world, Adler has always appreciated and "been aware of the creative intellect we share with artists." Both seek to understand the "world out there"—artists with canvas, pen, and paper, scientists with mathematics. Adler's contact with Miller reinforced his belief "that both art and science play equal roles in our quest to understand the world we live in."

Adler is also well aware of the schism between art and science that came with the Industrial Revolution, raising the scientist's status to that of "new magicians, with artists relegated to more of an intellectual pastime status—an unfortunate asymmetry." With regard to his work as a scientist, "I see my motivation in a larger context, one which Steve [Miller] helped me to understand."

"Expertise demands discipline and focus," says Miller. "There is so much to learn that it's very difficult to do both [art and science]." Nevertheless, his work with scientists has produced some fascinating, profound, and—dare I say—beautiful works of art.

## Beyond space-time and matter: Antony Gormley

Most of Antony Gormley's sculptures are based on the human body, often his own. One of the best known is *Event Horizon*, thirty-one statues of naked men cast from his own body, which appeared dotted around streets and the tops of buildings in London in 2007. Eight years earlier, in 1999, Gormley created *Quantum Cloud* for a site next to London's Millennium Dome. This is a massive piece some 90 feet high, made up of 5-foot-long steel sections assembled according to a "random walk" procedure, an algorithm according to which successive pieces can have no correlation with what went before. Although the steel sections are arranged entirely at random, the base of the sculpture is an enlarged image of Gormley's body, of which a shadowy form can still be seen. Gormley has said that the inspiration for *Quantum Cloud* came from conversations he had with the physicist Basil Hiley on what might have been the mathematical structure of the universe before space-time and matter came into existence.

Born in Yorkshire in 1950, as a child Gormley loved Ernst Gombrich's *The Story of Art*. He was equally interested in science, particularly biology. He had a crystal garden, his "own laboratory in the garden shed where I would make explosions and precipitate crystals." Books by writers such as David Bohm and Fritjof Capra, who explored consciousness and the connections between quantum physics and Buddhism, suggested that his twin interests of art and science could be brought together.

He went on to study archaeology, anthropology, and the history of art at Trinity College, Cambridge, then traveled to India and Sri Lanka, to explore his interest in Buddhism, and finally went to Central Saint Martins College of Art in London, then on to Goldsmiths, and then to a postgraduate course in sculpture at the Slade. The works that followed included the celebrated *Angel of the North*, completed in 1998. A gigantic human figure 66 feet high, cast in steel with two massive outstretched wings 177 feet across, standing on a hillside near Gateshead keeping watch over the main road and railways headed north, it's a well-loved landmark for all who pass by.

In 2006, Gormley, by now a famous sculptor, had a major exhibition at Galerie Ropac in Paris. Among the exhibits was an installation called *Breathing Room*. It was like a three-dimensional drawing, a framework of aluminum tubes forming the outline of rooms nesting one inside the other, which the viewer could walk into and through. Covered in phosphorescent paint, it glowed at night. Gormley called it a "three-dimensional drawing in space." Making the installation led him to think about how space was contained within the structures and how they changed their nature depending on the light. He suspected that these were the sorts of questions considered by scientists at leading scientific establishments such as CERN. So he convened a "little conference," inviting Michael Doser from CERN, as well as other scientists. Doser in turn invited Gormley to visit CERN in 2008, and in the course of the visit enquired as to who they should approach for a commission. Gormley replied that he didn't need a commission. "I would be very honored to be able to donate them a work and I gave them *Feeling Material XXXIV*."

The work, a swirl of iron spirals in which the viewer can just about make out a human form, Gormley's preferred theme, was installed above the central staircase in the main administration building.

5.12: Antony Gormley, *Feeling Material XXXIV*, 2008. 5mm square section mild steel bar, 155 x 244 x 153 cm. Collection of CERN, Geneva, Switzerland. Photograph by Stephen White, London. © the artist.

Gormley describes it as "an attempt to materialize the place at the other side of appearance where we all live." He was, he says, inspired by the dramatic discoveries at CERN, which have "led to a reassessment of the standard view of atomic structure, and I felt that, maybe, *Feeling Material XXXIV* evoked that." As in *Breathing Room*, Gormley continues to explore the meaning of the space inside his wire sculptures.

CERN physicists have described *Feeling Material XXXIV* as representing the wave/particle duality of elementary particles, or a particle's trajectory as it moves through the liquid in a bubble chamber. Whatever it represents, it succeeds in sparking conversations about art in a scientific laboratory.

Another of Gormley's works that involved science was *Blind Light*, an installation which formed part of the first major exhibition of his work at the Hayward Gallery in London in 2007. *Blind Light* is deceptively simple: a glass-enclosed room filled with clouds of vapor. Visitors stepping inside are immediately disoriented. They find themselves stumbling about in a dense fog in which they can hear other people but not see them until they lurch into them, a little like being at the bottom of the ocean. Viewers outside the room can see the emerging silhouettes of those inside. To create this pea soup, Gormley consulted scientists at Imperial College who specialize in cloud science, who advised him on the necessary conditions regarding water vapor and temperature, helping him develop "the highest density cloud composed of the smallest possible water particles."

Gormley has also spoken with the British mathematician Roger Penrose about how to construct geometrical figures out of foam. Penrose contributed an essay to the Hayward's exhibition catalogue on how Gormley's objects relate to topology, the mathematics of surfaces.

"Collaboration is alive and well in the scientific community. Artists could learn a lot from them," Gormley says. "I don't think the two cultures should ever have parted company. Anybody with a curious mind (and I think all human beings have curiosity) has the potential to mix intuition and empirical analysis, and I think that's what art is." All the scientists he has met, he says, "in particular Roger Penrose, have been the most curious and open-minded about the world around them." He sums up: "Art does a similar thing: it proposes a model through which we can look at the world around us."

## The buck stops here: Rolf-Dieter Heuer

Having risen through the ranks of German physics, Rolf-Dieter Heuer has been director general of CERN since 2009, a position akin to that of CEO of a huge multinational corporation. An imposing man with a mane of white hair and a full goatee, he focuses fully on the person to whom he is speaking at the time. I am well aware that fifteen minutes after I begin the interview it will be over. Both he and I have to be super-efficient.

His favorite quote, he says, is the famous words of Paul Klee which form the motif of this book: "Art does not reproduce the visible; rather, it makes visible." What this means to him, he says, is that art makes the invisible visible. This is also his own personal quest as a physicist, as he studies the invisible world of elementary particles, the building blocks of nature. He has clear views on CERN's artist in residence program, Collide@CERN. "I hope that scientists realize the importance of art and will connect with society. We have to show society what we are worth and what we are doing for society. To transmit that through art, partially at least, in my mind is important and it opens horizons." He is sensitive to critics of CERN, who ask what exactly it contributes to society given the large amount of money it absorbs. The artistic dimension could be at least a part of a response to those criticisms, he says.

Heuer does not believe that art and science will ever come together. The "questions are so deep today that they cry out for specialization." He points out the "difficulty to act as an individual today. All is teamwork, to function as a creative group." I ask whether this holds for theoretical physicists as well as for experimentalists,

5.13: Steve Miller, *Cables*, 2012.

who work in groups numbering in the hundreds. Even in theoretical physics, Heuer answers, "The solitary worker is gone."

Theoretical physicists have a well-defined sense of beauty—as in, for example, a beautiful equation. Heuer is an experimental physicist who uses instruments that dwarf people, made up of thousands of wires connecting magnets and complex electronic circuits. What he seeks, he says, is "functionality rather than beauty." To him functionality is aesthetic, "but this goes along with beauty as in the alignment of cables." At CERN, the alignment of cables—wires laid out in parallel with each other—are works of art in themselves. There are no unsightly tangled masses here. Some laboratories have row upon row of cabinets with transparent plastic doors enclosing shelves, each holding a myriad of color-coded circuits, which might well be works of minimalist art.

"If it functions well, it has to be beautiful," Heuer concludes.

# Intermezzo: How Science Helped Resolve the World's Greatest Art Scandal

At a creativity meeting in Leiden in 2011, I met an American physicist named Richard Taylor who told me that he was actually an artist as well as a scientist. When I interviewed him, he mentioned his fascination with fractals, and we turned to the gripping story of the role he had played in the "world's greatest art scandal." This story is all about the importance of science in authenticating works of art, supplementing the methods of art historians, who use their training to ascertain a painting's chain of ownership, or provenance, and connoisseurs who apply their intuition, honed by expertise in art history and a great deal of experience.

What role can science play? In using x-rays, infrared analysis, high-resolution multispectral cameras, analysis of digital images, carbon dating, paint pigment analysis, and so on to authenticate works of art, science and technology are tools, nothing more. Taylor's case was something else. It employed mathematics to explore the authenticity of a work as well as to analyze why a work of art is aesthetic and how it was created by the artist. It offers a glimpse into the artist's mind.

Born in the UK in 1961, Taylor was attracted to both art and science. A turning point occurred when he was nine, when he came across the catalogue for the Pollock exhibition at New York's Museum of Modern Art in 1967, written by Francis V. O'Connor, the preeminent Pollock scholar. Taylor immediately became hooked on Pollock's patterns.

When the time came to choose a career, Taylor chose his other passion, science. He found, however, that he could pursue his interest in patterns while studying processes such as the way electricity courses through electronic devices, like a river splitting again and again into tributaries. Then he came across fractals.

In 1977 the Polish-born mathematician, Benoît Mandelbrot, put together decades of research on patterns that repeat themselves no matter how many times they are magnified—a property mathematicians call self-similarity—and are built up of shapes of incredible complexity. Zooming in on one small part reveals that it is identical to the larger segments. He called these fractals because their dimensions need not be whole numbers. They can be fractions.

Fractals occur in nature. In a mature tree, the branches sprout smaller branches which end in even smaller branches, and so on. Although the self-similarity in a tree and its branches is not exact, it is statistically present—that is, close enough on average, as are the fractal properties of clouds, river networks, mountain ranges, coastlines, and electricity surging through electronic devices. Complex systems such as the nervous system and the blood and lung vasculature can be modeled, and better understood, as fractals.

Taylor studied fractals in the physics department at the University of New South Wales in Sydney, and meanwhile continued to paint. Then in 1994, the urge to pursue his love of art became irresistible and he took a sabbatical for a year to do a foundation course at Manchester School of Art, where he focused on art and photography.

At one point his group was sent to the Yorkshire moors for a week to draw. The weather turned stormy, however, which made it impossible to sketch the landscape. Then Taylor remembered a story about the French painter Yves Klein. In the late 1950s, he was ensconced in a café in Paris with his agent during a rainstorm when he remembered he was supposed to deliver a painting to a gallery in Toulouse. But he had no painting to deliver. He asked his agent to remind him of the subject. "Patterns in nature," his agent supposedly replied. "No problem," said Klein. A pioneer in performance art and minimalism, he tied a partially painted canvas to the top of his car and drove through the storm. Nature completed the painting and it sold for $10,000.

So Taylor and his fellow students assembled a contraption out of fallen branches, part of which blew back and forth in the wind like a pendulum. This moved another part which held paint cans, which dripped a pattern determined by the wind's direction onto a canvas placed underneath it on the ground. When the storm forced them indoors, they left their apparatus to paint all night, driven by nature. The irregular pendulum movement produced patterns "similar to Pollock's work," Taylor recalls.

"Suddenly the secrets of Jackson Pollock seemed to fall into place for me. He must have adopted nature's rhythms when he painted." The next step was to use science to find out whether he could identify tangible traces of those rhythms in Pollock's artwork.

Pollock had a major alcohol problem and in 1938, when he was twenty-six, started seeing a Jungian analyst who encouraged him to express his unconscious through his painting. Meanwhile he was also studying Surrealism, inspired by its forays into the fantastic, and the works of Picasso, where he was impressed by the structure offered by Cubism. He settled into a style in which he applied daubs of paint, one layer at a time, using brushes or sticks.

Pollock's great idea was to involve his body in his painting. First he put the canvas on the floor, which gave him greater freedom of movement. Then he had an epiphany: why not pour paint directly from the can? Sometimes he used as many as fifteen clusters of pourings. The resulting "drip style," begun in the 1940s, was the embodiment of Action Painting and of the new art movement, Abstract Expressionism. Imitators popped up everywhere. But Pollock's paintings were not simply random splashes of paint. Even a cursory glance shows a periodicity, a rhythm, in their patterns. Beneath the rhythm there was something else that separated Pollock's paintings from even the most scrupulous imitators: fractals.

Soon after returning to Australia in 1995, Taylor established himself as an expert in the fractal analysis of systems ranging from nanoelectronics to the retina and solar cells. He also completed his master's degree in art history. His thesis was on Pollock, of course. The more he looked at Pollock's paintings, the more the paint splatterings seemed to resemble electricity flowing through electronic devices, with the telltale fractal property of self-similarity. Was this Pollock's fingerprint?

Taylor scanned Pollock's paintings into a computer, and then fitted a computer-generated mesh over them. No matter how much he decreased the mesh size—corresponding to higher magnifications—the patterns remained fractal, statistically self-similar, like the branches of a tree. Pollock had built up these fractal patterns unknowingly, of course, beginning with small islands of paint which he then connected. Taylor studied fourteen authenticated Pollock paintings and they all showed fractal patterns.

He also looked into the fractal dimensions of Pollock's paintings—the fractal dimension being a measure of how patterns fill space. A straight line has one dimension, while the completely filled space of a flat canvas has two. Depending on how filled the canvas is, the fractal dimension can take on values between one and two. The less sparsely filled the canvas, the lower the dimension. Pollock's early paintings have a fractal dimension of around 1.2, and later works vary from 1.3 to 1.5, giving smooth, somewhat sparse images, as in his 1950 painting *Autumn*. Taylor found that when Pollock altered the sequence in which he introduced colors, it changed the fractal dimension.

Some of Pollock's later paintings had a higher fractal dimension and were more complex, perhaps because he wished to "keep the viewer alert by engaging their eyes in a constant search through the dense structure of a high D [dimensional] pattern," says Taylor. As one of Pollock's friends, Reuben Kadish, noted, "I think that one of the most important things about Pollock's work is that it isn't so much what you're looking at but what is happening to you," which seems to back up Taylor's analysis.

Taylor—who is now professor of the extraordinary combination of physics, psychology, and art at the University of Oregon—believes that the human eye is a remarkable fractal detector, a reasonable assumption given that we make our way through the world largely by recognizing patterns, such as when individual components of a face—eyes, nose, mouth—instantaneously snap into place and we realize it's someone we know, or when a master chess player sees not a mass of individual pieces on the board but the overall pattern. It also explains why there is close agreement between the eye of the art expert and Taylor's computer analysis.

Why do people find Pollock's paintings beautiful—that is, aesthetic? Taylor used skin conductance tests—measuring the conductivity of the skin using EEG (electroencephalography) and fMRI (functional magnetic resonance imaging)—to check neurological responses to psychological and physiological stimuli. He found that people were more at ease and also more attentive when looking at mid-range fractal patterns. So perhaps fractal analysis could reveal a regularity hidden within apparent randomness. This was what made Pollock's paintings so pleasing.

Pollock once said, "There was a reviewer a while back who wrote that my pictures didn't have any beginning or any end. He didn't mean it as a compliment, but it was. It was a fine compliment." He died in 1956, before either chaos theory or fractals were known.

So how did Pollock create fractal patterns? Taylor explored this question with a pendulum that deposited paint on a canvas as it swung, a sleeker version of the device he had used on the Yorkshire moors. This pendulum did not swing smoothly. It was driven by motors and swung chaotically this way and that, suddenly changing direction. Taylor studied the patterns it produced. "A striking visual similarity exists between the drip patterns of Pollock and those generated by a chaotic drip system," he wrote excitedly.

A famous photograph shows Pollock at work, leaning way over a huge canvas with his wife, Lee Krasner, in the background. He looks in excellent physical shape, a graceful man. But in fact his alcoholism made him shaky and left him often trying to keep his balance, especially while leaning. His medical records attest to his problems with balance. Researchers have found that when people try to keep their balance—tightrope walkers, for example—their hands trace out fractals. Taylor found that children aged five, just perfecting their balance, produced Pollock-like paintings. "To fake a Pollock one needs the same physiology as Pollock. It's not easy to paint a Pollock," Taylor concluded.

His scientific investigation pointed to a way to authenticate Pollocks. But perhaps it went even further. "Perhaps it may even be able to throw a narrow beam of light into those dim corners of the mind where great paintings exert their power." Taylor had uncovered a far-reaching result of the interplay between art and science, one that could lead to a better understanding of creativity: we are born with structures in our mind capable of generating practically anything, including fractal patterns and complex mathematics; and our minds are constructed to see beauty in these, to appreciate and respond to them with pleasure.

IN 2006, POLLOCK'S *No. 5, 1948* sold at auction for $140 million, at that time the highest price ever paid for a painting. It had better have been the real thing. Not surprisingly, threats of lawsuits to prevent experts publishing contrary opinions loomed.

The controversy that involved Taylor began to unfold in late 2002 when Alex Matter, a New York filmmaker, made what seemed to be one of the most exciting discoveries in recent art history. He found a cache of thirty-two paintings in Pollock's familiar drip style that his father, Herbert Matter, had stored away. Herbert, a photographer and graphic designer, and his wife Mercedes, an artist, had lived near East Hampton, where Pollock held court from 1945 until he died in a car crash in 1956. They had been close friends of his. Two years after Pollock's death, in 1958, Herbert had wrapped the paintings in brown paper and placed labels on them stating that they were produced in the 1940s and were acquired by "gift + purchase." He did not, however, state that they were Pollocks. Some were on boards and all were much smaller than Pollock's usual huge canvases. In 1978 he put them in a storage locker in Wainscott, New York, near East Hampton. Herbert died in 1984 and Mercedes in 2001.

Alex Matter contacted the Manhattan art dealer Mark Borghi. Borghi had been the victim of fraud before, but sensed that in this case the Pollocks were genuine. Their provenance—Matter's father's direct connection with Pollock—seemed gold-plated. Nevertheless, he wanted a second opinion. In late summer 2004, he consulted Ellen Landau, a professor of art history at Case Western University in Cleveland, Ohio, who had written several authoritative books on Pollock. She dropped everything and flew straight to New York to view the cache. "I was completely blown away—the scholarly thrill of a lifetime," she says. The style, the boards, the initials "J" and "P"—"there are too many things about them that are pure Jackson." Her authentication seemed conclusive.

The trouble began when another Pollock expert, Eugene V. Thaw, a veteran art dealer, disagreed, arguing that Pollock did not use the sort of boards in the Matter cache and that it would have been out of character for him to borrow materials from Mercedes Matter when he worked in the Matters' New York studio. Besides, they just didn't look right. He proposed that they had been done by Mercedes and her art students in order to study Pollock's technique. Francis V. O'Connor, the guru of Pollock studies, whose book had inspired Taylor as a boy, agreed with Thaw. O'Connor and Thaw had coedited the Pollock catalogue raisonné, the list of all known Pollock works.

Back in 1985, Lee Krasner, Pollock's wife, had created the Pollock-Krasner Foundation to fund promising artists. In those days there were Pollocks popping up everywhere and, in 1990, the foundation established the Pollock-Krasner Authentication Board to sort the wheat from the chaff. However, after the supplemental volume to the Pollock catalogue raisonné was completed in 1996, the foundation decided to disband the authentication committee. They had just successfully defended two lawsuits from owners who charged "restraint of trade" when their canvases were not authenticated, claiming that nonauthentication adversely affected their value. Besides, they were fed up with the huge numbers of fake Pollocks, as well as with the people who bought them and then reapplied for authentication. The committee had investigated over seven hundred paintings and were exhausted.

Landau, Thaw, and O'Connor had been members of the authentication committee and had worked together in harmony. Now the situation changed dramatically. In light of the number of works in the Matter cache and the money involved—if authenticated, each painting would be worth upwards of a million dollars—Ronald Spencer, an attorney for the foundation, announced that the foundation would "rethink its involvement in authorship questions." The problem was that the authentication board could not be reconvened because the members were at odds over the authentication issue at hand, which by now had gone public.

Sparks flew. Thaw said that Landau should have included him from the start and that if she had done so the whole dispute would have been avoided. "If Ellen Landau's opinion prevails," Thaw told the *New York Times*, "people will happily buy [the paintings] and they'll go into museums and books, but not the ones that I have anything to do with."

Landau replied that she had not been paid for her advice, though she would receive a fee for organizing a show to celebrate the fiftieth anniversary of Pollock's death, which would feature the Matter paintings. She declined to reveal how much the fee would be. In a slap at Thaw and O'Connor, she insisted that scholarship and objectivity were her only goals. "Unlike the authors of the Pollock catalogue raisonné," who bought and sold Pollocks, she said, "I am an art historian with an impeccable reputation, not an art dealer."

Aware of Taylor's fractal analysis of Pollock's work, O'Connor urged the Pollock-Krasner Foundation to contact him discreetly and ask him to assess the cache. Taylor agreed and in June 2005 received scans of six paintings selected by O'Connor. Three had appeared on Alex Matter's website and so were widely known, while the others were representative of the style in the entire cache. The

Pollock-Krasner Foundation, Taylor recalled, "made me sign a confidentiality agreement that I would not disclose the findings." In fact, he concluded that the works did not exhibit the fractal structure typical of known Pollock works, thus casting serious doubt on their authenticity. Anticipating accusations of bribery, he was adamant that the foundation had paid only his research expenses.

The foundation called a meeting at their offices in New York, and Taylor met his intellectual hero, Francis O'Connor. All the participants acknowledged "the unprecedented nature of the meeting [in which] for the first time computers were playing a significant role in determining the fate of artworks." All agreed that every clue as to authenticity had to be followed up. The foundation even hired detectives to check the history behind the discovery and whether it squared with Pollock's life.

Then, in November 2005, Mark Borghi, the Manhattan art dealer contacted by Alex Matter, approached Taylor. He also wanted him to do a fractal analysis. This put Taylor in a bind. He couldn't tell Borghi that he had already done one or what his results had been. In fact, by commissioning him to do another, Borghi's request bypassed the confidentiality agreement Taylor had signed. Although the foundation agreed that he could do another analysis, Taylor felt it was unfair and requested them to release his original findings.

While the foundation ruminated and continued work on a comprehensive report, rumors surfaced that Ellen Landau's team had new evidence of the cache's veracity—specifically, a photograph of Pollock standing in front of one of the paintings that Taylor had said did not display a fractal signature. If so, it would prove that the painting was genuine. "Thus the PKF [Pollock-Krasner Foundation] decided it was 'put up or shut up time,'" says Taylor, "and so announced my findings" to the *New York Times*, which published them on February 9, 2006.

Landau learned of Taylor's results when she was interviewed by the reporter Randy Kennedy, who was writing the story. Livid, Landau criticized the foundation for not revealing them to her or to Alex Matter. "Secrecy," she said, "impeded scholarly debate and consensus." She added that fractal analysis was a "very new and contested field in art authentication." Yet just three months earlier, Landau had been happy for Borghi to request Taylor to do just such a fractal analysis, which suggested that she was unhappy with the results rather than unsure of the method. She stated that a more exhaustive investigation would appear in "a full-scale catalogue as soon as it [was] completed."

Francis O'Connor commented that Taylor's results had "reinforced his initial doubts after examining the paintings." The tide was beginning to turn, although the foundation still said it was awaiting further research and the moment when all the experts reached a consensus.

There were criticisms of Taylor's work. One Pollock scholar claimed that Taylor's analysis was inconclusive since he had examined only six paintings. Taylor agreed. "But this was not an ideal world. The story was unfolding in the international press at a rapid speed. I was asked to analyze six paintings in three weeks. My team had to work around the clock (literally)." Reputations were at stake—Taylor's and, even more, Landau's. But when Taylor phoned the foundation, a lawyer there broke the news that some of the cache had already been sold to a New York art dealer and "you've just lost them $40 million." The foundation warned Taylor to expect an all-out assault.

The only expert that Landau and her group could come up with was a physics graduate student from her own university. In an interview with the *New York Times*, the student, Katherine Jones-Smith, stated categorically that Pollock's paintings were not fractal. Taylor recalls at this point picking up the phone in his office to hear Mandelbrot's voice. "Oh my God. What have I done wrong?" was his first thought. In fact, Mandelbrot had called to offer his support. He was delighted to find another case in which "mathematics had an incredible link with physicality."

Jones-Smith, along with a senior colleague, Harsh Mathur, published their evidence in the journal *Nature* on November 30, 2006. Landau was triumphant. "I am pleased they have successfully refuted Richard Taylor's thesis. . . . Irrespective of whatever determination is ultimately made on the authenticity of the recently found Matter paintings, fractal analysis should not be considered a foolproof technique for authenticating works by Pollock. The fact that Taylor has refused to fully share his testing criteria casts further doubt on the credibility of his claims."

In fact, Taylor had published his response alongside the article by Jones-Smith and Mathur. Jones-Smith and Mathur claimed that Taylor had not used a small enough mesh to prove conclusively that he was dealing with fractals. Not so, he replied. He had employed the guidelines used by all fractal researchers to examine physical systems—which, unlike mathematical curves, are not fractal "all the way down"—and had worked according to the concept of limited-range fractals, that is, of a system displaying fractal patterns statistically, like a tree. If Jones-Smith and Mathur's demands were met, he added, half the published papers on fractals would have to be dismissed.

Jones-Smith and Mathur offered a childlike doodle of stars and claimed that it showed the same fractal patterns that Taylor had found in Pollock paintings. But Taylor's analysis of similar star patterns found them to be, in fact, not fractal. As another fractal expert, Lazaros Gallos, put it, "What [Jones-Smith and Mathur] have done is a simple trick. This is bad science about fractals." In addition, Taylor says, his fractal analysis was purpose-built specifically for Pollock paintings, not star patterns. "This is like taking an analysis of elephant ears and applying it mindlessly to a giraffe!"

So why was Jones-Smith and Mathur's simple-minded analysis published in the first place? Referees had recommended against publication, says Taylor, but *Nature* liked controversy." As the physicist Wolfgang Pauli would have said of their work, "Why, that's not even wrong."

An exhibition was being planned for the fiftieth anniversary of Pollock's death in his hometown of East Hampton in 2006, to feature the Matter cache. Matter insisted that the paintings be displayed as Pollocks, which would thereby authenticate them. Aware of the controversy, Ruth Appelhof, the director of Guild Hall, where the show was to be held, refused and the exhibition was canceled.

The air was thick with innuendos, promises of new evidence in support of the cache, and rumors of lawsuits. All this came to light in a series of emails published in the *Cleveland Plain Dealer*. The emails were between Landau, Borghi, Albert Albano, the director of the Intermuseum Conservation Association in Cleveland and a former conservator at the Museum of Modern Art, Robin Zucker, Matter's publicist, and James Martin, a forensic scientist, and had been sent between November 2005 and January 2007. Steven Litt, the paper's art critic, had been following the Matter story closely. To begin with, he had intended to write about the controversy from the viewpoint of Ellen Landau, who, being a professor at Case Western University, was a local hero. Litt had no reason to doubt her assessment, given her high position in the art world. But then Thaw's opposition surfaced, followed by O'Connor's, and he "began aggressively pursuing the story."

In November 2005, not long after declaring the Pollocks genuine and unbeknownst to the foundation or to Taylor, Landau had asked Albano to recommend an expert to cross-check a pigment analysis of three of the paintings which was already under way at the Harvard University Art Museums, for no charge. Albano recommended James Martin, founder of Orion Analytical in Williamstown, Massachusetts. Orion's website states: "Orion uses microscopy, spectroscopy, and scientific imaging to investigate the structure and chemical composition of materials found in more than 4000 years of cultural property, forensic evidence, and manufactured goods—from ancient Egyptian artifacts to printed circuit boards. We consult on materials at issue in authenticity studies and insurance claims, and serve as consulting or testifying expert in civil and criminal proceedings." Mark Borghi, the gallery owner, hired Martin in December 2005, unaware of Taylor's results. Martin got down to work. After a month, 350 extensive tests revealed that twenty-three of the purported Pollocks contained pigments and resins not available in Pollock's lifetime. Furthermore, in some of the paintings Pollock's initials, "JP," appeared on top of the modern pigments.

Matter admitted that the paintings had been in poor condition and that in 2003 and 2004 he had had them heavily restored, rather than having them documented immediately by an art museum or materials analyst. It was a decision he now regretted. But the restoration merely complicated Martin's analysis. He was able to use sophisticated state-of-the-art microscopy to search beneath the surface to the original pigments.

Albano tried to think of ways that Pollock could have come to use such materials. One of his suggestions concerned some words Herbert Matter had written on the brown paper wrapping: "Robi paints." Robi was the nickname of Robert Rebetez, a Basel art store owner and Herbert Matter's brother-in-law. There were a lot of chemical firms in Basel, so perhaps he had stocked unusual pigments, not yet patented, that Pollock might have used. In February 2006, Albano reported his thoughts to Landau with the comment, "Hang in there. No one's dead yet!" Landau took them on board, continuing to give high-profile interviews, insisting unequivocally that she was convinced "these are Pollocks." But a month later Albano gave up his hypothesis. He was now convinced that attribution could not be maintained.

At the time, Landau was planning another exhibition of the Matter cache at the McMullen Museum of Art at Boston College, to open in September 2007. In an email dated September 5, 2006, the gallery owner Borghi informed Albano that "Jamie's [James Martin's] research [was] going to lead to the fact that the works were produced after Pollock's death." Borghi recommended canceling the exhibition as "a strategy for a graceful exit by Ellen [Landau]." Robin Zucker, the publicist working for Matter on the planned exhibition at the McMullen Museum, disagreed. In an email to Landau, he claimed new evidence had come to light that contradicted Martin's early findings.

Landau was delighted. She told Zucker to ignore her statements about giving up. "Despite the paint analysis," she wrote to Zucker, there was a great deal of "documentary and circumstantial evidence" that linked "these paintings to Pollock's relationship with the Matters."

In reply, Zucker emailed that Alex Matter was "troubled by both Martin's initial findings and Taylor's subsequent fractal support." This was not a problem, however, because he had "learned that fractal analysis is invalid, and that Martin's report still leaves questions." In fact, of course, Taylor had already dealt with the fractal criticism and Martin's results were not "initial" but completed. As circumstantial evidence for the cache's veracity, Matter pointed to the "personal and artistic integrity of his parents," meaning they would not have attempted fraud.

Jeremy Epstein, the lawyer whom Matter had hired after the cancellation of the 2006 show in East Hampton, had also claimed that fractal analysis was "dubious and unproven as a way to analyze paintings," and added that soon-to-be-released "circumstantial evidence" was strongly on the side of Landau and her backers.

In October 2006, Martin presented his report and all the supporting data to Borghi and Landau but remained silent about what he had found, fearing a lawsuit by Matter. Epstein denied threatening Martin. But in an interview in February 2007, Matter mentioned that the possibility of a lawsuit was "very negotiable." He told the *Cleveland Plain Dealer* that he would release Martin's report when the exhibition at the McMullen Museum of Art opened, because it was as yet unfinished, though he left unanswered the question of why it was as yet unfinished.

As for the bombshells that were supposed to vindicate Landau and her team, neither the photograph of Pollock standing next to one of Matter's cache nor the "circumstantial evidence" mentioned by Zucker and Epstein ever materialized. In January 2007, the team at Harvard came to the same conclusion as Martin had—that the pigments were modern.

Landau soldiered on and assembled a show with the title "Pollock Matters" at the McMullen Museum of Art to run from September 1, 2007, through December 9, 2007. Landau's new strategy was to focus on the relationship between Pollock and Herbert Matter rather than on the newly discovered works. The disputed paintings were exhibited separately without attribution and the publicity blurb made no mention of any controversy. The catalogue mentioned the Harvard University analysis, but only in passing. In reply to the Harvard team, Landau claimed that more extensive investigation might be able to prove that Pollock could have used materials not yet patented or commercially available. She alluded to "still other avenues of exploration," referring to the "Robi paints" that Pollock may have obtained from Robert Rebetez. In fact, the Harvard team had learned from one of Robi's daughters that "neither she nor her sister recalled paints being sent to relatives in the U.S. and their father's store stocked only standard brands of artist's paints." In any case, Landau stated, the science of dating pigments "is not as hard and fast as is often assumed." Eyeballing or "cold canvassing" sufficed for her.

Landau also mentioned a fingerprint found on a paint can in Pollock's studio which matched one on a painting in the Matter cache. This had been mentioned in an article by Paul Biro, a forensic art expert who examines paintings for the artist's fingerprints. He had suggested that this could be evidence for the authenticity of the Matter cache.

Martin did not contribute to the catalogue even though he had been invited by the McMullen to do so and had done the most extensive analysis, and despite the fact that a press release by the McMullen's director, Nancy Netzer, stated that the exhibition would make available all known evidence regarding the attribution of the Matter cache. In an email to the *Cleveland Plain Dealer*, Netzer explained that Martin needed permission from Borghi and Matter to publish his results, as it was they who had hired him to do the analysis. But along with their permission came a heavy-handed agreement, as Martin's attorney, Stanley Parese, explained to the paper. The agreement drawn up by Epstein, Matter's lawyer, prohibited Martin from speaking about his findings before or after the publication of the catalogue. According to Martin's original contract he was free to go public with his analysis, but he did not do so for fear of a lawsuit. What Epstein proposed would have relegated Martin's work to obscurity.

Parese continued, "As a scientist and a scholar, Martin was not willing to have the owners of the paintings [including the New York art dealer who had recently purchased several of them] dictate the terms under which he would participate in a scholarly publication." One of Martin's collaborators on the authentication work told the *Cleveland Plain Dealer* that he was appalled that the McMullen would stand by and permit a scientist to be silenced. Richard Newman, head of scientific research at the Museum of Fine Arts, Boston, and one of the contributors to the catalogue, refused to be silenced and in his article mentioned Martin's work "which has not been published," citing Martin's report filed at Orion which scholars could consult and study. He did not, however, mention Martin's conclusion.

Landau mentioned Richard Taylor's fractal analysis and then dismissed it, saying it had been shown by Joncs-Smith to be "demonstrably flawed." There was no mention of Taylor's rebuttal.

At this point, Jones-Smith and Mathur teamed up with Jones-Smith's PhD supervisor, the astrophysicist Lawrence Krauss, to produce what they claimed to be a more precise critique of Taylor's fractal analysis. In the October 2007 issue, *Scientific American* triumphantly announced their results, which was strange as their paper had merely been submitted for publication to the prestigious physics journal *Physical Review Letters* and not yet refereed. Despite this, they had gone ahead and released their results to *Scientific American*. This time, instead of dealing with star pattern doodles, they analyzed three authenticated Pollocks and claimed to find no evidence of fractal patterns, contradicting Taylor's hypothesis that fractal patterns were Pollock's signature.

Hany Farid, a computer scientist at Dartmouth College who had been following the controversy, told *Scientific American* there were flaws in their results. "I think they took a fairly simplistic way of separating those colors," he said, which skewed their results away from the verification of fractal patterns in the Pollocks they analyzed.

Taylor explains that their color separation technique, separating out Pollock's layers of poured paint, was simply too primitive to detect fractals, while his own was highly sophisticated and complex, far exceeding the techniques that the Case Western scientists had at their disposal. In the end, their detailed article was rejected by *Physical Review Letters*. Both the author of the *Scientific American* article and the magazine itself were "rightly embarrassed," Taylor recalls.

The article in *Scientific American* includes an image of a drip painting with a fractal signature that was not Pollock's. This suggested, wrote the author, that fractal patterns are "no reliable way to distinguish a Pollock." In fact, Taylor had always insisted that fractal analysis had to be supplemented with other analyses, such as of paint pigments and style, to identify Pollock's hand. In the end, the concluding factor had to be that the picture looked like a Pollock.

The situation had reached flash point. On September 28, 2007, a symposium was convened to decide the fate of the Matter cache. Sponsored by the International Foundation for Art Research, whose brief was to look into issues of attribution and authenticity of works of art, it was held at the National Academy of Design in New York. The symposium's title was simply, "Are They Pollocks?" The subtitle is significant—"What Science Tells Us About the Matter Paintings"—because in the end it was science that decided the issue.

Francis O'Connor recalls, "The entire art world was present at the meeting. I have never before or since seen such a gathering of everybody who matters." Three major players were absent: Matter, Richard Taylor, who was in New Zealand, and Landau, who "graciously declined" the invitation. The three speakers were Pepe Karmel, chair of the art history department at New York University and a co-curator of the 1998 Pollock retrospective at the Museum of Modern Art, Richard Newman, head of scientific research at the Museum of Fine Arts, Boston, and James Martin of Orion, who had carried out the pigment analysis.

Karmel gave an in-depth formal analysis, comparing the Pollock cache with authenticated Pollocks in terms of style and paint application. He concluded that the works were probably experimental paintings by Herbert or Mercedes Matter done in the style of Pollock, as many artists were doing in the 1950s. This showed how great Pollock's influence had been and how it had pointed to new possibilities. "These pictures are not a new bunch of Pollocks," he concluded.

Newman's stance differed from the position he had taken in his article for the catalogue for the McMullen Museum exhibition. He emphasized the presence in the Matter cache of paint pigments that were not available during Pollock's lifetime and reminded his listeners that even the appearance of material available to Pollock on paintings in the Matter cache "does not mean that Pollock created the painting." Connoisseurship—the application of aesthetic principles and the expert eye—was also needed.

Martin, no longer relegated to a footnote, gave a virtuoso performance, stating firmly that several of the pigments on the painting had not been produced until 1977, that the boards were manufactured in the 1970s, and that the initials "JP" appear on top of paint that was definitely not available to Pollock. Martin concluded that these findings, in conjunction with the statements of the other two speakers, "are patently *inconsistent* with the claimed attributes of the Matter Paintings as objects that were created by 1949 or, indeed, as objects that were created prior to Jackson Pollock's death in 1956."

Martin had also had the fingerprints on some of the paintings investigated by two qualified law enforcement fingerprint examiners, who concluded that there was insufficient detail for an identification. He had even contacted officials at the federal, state, and local level and found no known samples of Pollock's fingerprints. It seems that Pollock had never been fingerprinted. Martin concluded that perhaps the fingerprints on the paintings were Pollock's, or perhaps they belonged to others in his studio, such as Krasner or Herbert Matter.

For Karmel it was pigment analysis that had decided the issue, not fractals, which he dismissed as a "red herring." (In this he misunderstood fractals, defining a fractal pattern as possessing self-similarity at any magnification, all the way down. But this refers only to mathematical patterns, not patterns in nature such as trees, craggy mountain ranges, and Pollock's paintings.) What if the pigment analyses had turned out to be inconclusive? Surely in that case fractals would have been given added weight by art historians. They weren't such a red herring after all. Taylor continues to be called upon for his advice on Pollocks and reports that his "fractal analysis technique has a 100% record."

O'Connor tells me that he "can still see the expression on the face of the dealer who was revealed to be their owner, surrounded by lawyer types, fuming as he left. There never was any litigation, the evidence was overwhelming." The dealer had bought the paintings from the Mark Borghi Gallery.

Epstein, Matter's lawyer, insisted on having the final word. In an interview in the *East Hampton Star*, he questioned Martin's professional reputation, claiming that Martin had not completed his research, had destroyed data, and demanded extra payment for redoing work he had destroyed. Martin's response was that Epstein's comments were "patently wrong," but that he welcomed scholarly debate. He continued, "given the statements attributed to Mr. Epstein, it is no mystery why very few experts are willing to speak publicly on matters related to authenticity of fine art and other cultural property."

It was Epstein's remarks that prompted Albano of the Intermuseum Conservation Association in Cleveland to release the whole sequence of emails to the *Cleveland Plain Dealer*. "My concern is that the credibility of a highly regarded and intelligent colleague is being impugned," he wrote. He went on to assert that Martin's results were "irrefutable," had been corroborated by other laboratories, and had to be taken with the "utmost seriousness." The *Cleveland Plain Dealer* reported a survey of pigment scientists and paint industry executives all stating that the possibility that Pollock could have used the materials in the paintings "ranges from unlikely to virtually impossible." Litt concluded that the emails Albano passed to him "underscore the central importance of scientific evidence" in rendering it "extremely improbable" that the Matter cache were Pollocks.

In 2010, Ellen Landau removed the paragraph claiming she had authenticated the Matter cache from her biography on the Case Western University website.

Some mysteries remain. If Herbert Matter wrapped the paintings up in 1958, how can it be that some of the paint was not produced until 1977, as Martin said? Why did Herbert Matter place the cache in a storage locker in 1978, twenty years after he had wrapped them up in brown paper? 1978 was the year that the Pollock catalogue raisonné appeared, edited by Francis O'Connor and Eugene Thaw. Why didn't Herbert Matter contact the editors so that they could add the paintings to the catalogue, instead of putting them into storage? Did someone open the locker before 2002 and tamper with the contents, and if so, who? Who put the initials "JP" on the paintings? And were they done by Mercedes Matter and her students in the style of Pollock, which remains the most likely explanation?

The Matter paintings are currently in limbo, whereabouts unknown.

Obviously Pollock didn't discover fractals. Rather, he created patterns that he instinctively found to be beautiful, and Taylor later established that this was for the very reason that they were fractals. Fractals are all around us and the eye is an excellent detector, all the more so in that our daily movements frequently involve analyzing patterns. As Taylor has noted, this was behind O'Connor's uncanny intuition as to what was a real Pollock. In this way fractal analysis goes way beyond matters of authentication and is far from being merely a red herring.

In the backyard of Pollock's house in East Hampton there are trees with branches sprouting more branches, like a river dividing into rivulets. Perhaps Pollock stared at them, wondering whether they contained a deeper structure, which would come to be called self-similarity. Pollock attempted to represent these deeper structures in the most aesthetic way possible by unknowingly adjusting the fractal dimension (it is around 1.3) and cropping his painting by removing outer regions where fractal quality deteriorated.

The Matter matter was a monumental clash between art, science, and the art world. It was precipitated when the eyeball judgment of one eminent art historian was called into question by connoisseurs whose own eyeball judgment differed. Two of the "legs" of artistic assessment were at odds: art history and connoisseurship; art historians with their focus on provenance couldn't tell a fake Pollock from a real one. The third leg, science, entered in the form of pigment analysis and, for the first time, fractal analysis. In the end pigment analysis decided the issue, but fractal analysis played its part in "identifying the artist's 'hand' rather than simply the materials they used," as Taylor put it. It also emphasizes the inherent distrust of the art world for scientific methods that might undermine the monopoly of connoisseurs and art historians.

An important spin-off, one that goes far beyond the authentication of Pollocks, is the realization that, as Richard Taylor puts it so well, the study of fractals and even science itself could perhaps "throw a narrow beam of light into those dim corners of the mind where great paintings exert their power."

# 7

# Imagining and Designing Life

As Marta de Menezes, a biology-influenced artist who creates new forms of butterflies, puts it evocatively, "We are witnessing the birth of a new form of art: art created in test tubes, using laboratories as art studios."

Biology-influenced art was the next major artsci movement to emerge after computer art, kick-started with the help of a great deal of funding from the Wellcome Trust. It has really hit its stride in the twenty-first century. Biology is the science of life. Of all the sciences, biology touches us all most immediately. While other science-influenced art movements can expand our horizons and inspire us, biology-influenced art has the potential actually to affect our lives.

Biology-inspired art dates back some 32,000 years, ever since humans first picked up a piece of charcoal and drew an animal on a cave wall, and reappears in the works of Leonardo. It's old, yet new, for the modern-day art movement involves collaborations between artists and scientists, making up-to-date research tools available to the artist.

Unraveling the structure of DNA in 1953 was the greatest scientific discovery of the twentieth century, with its potential to cure disease and alter life forms through genetic engineering. Metaphors abound to make the new biology understandable: that blood flows through vessels like a liquid through pipes, that the heart acts like a mechanical pump, that ions flow through cell walls like particles through a mesh, and that the mind is a digital computer. And biology is, of course, of more immediate relevance than exotic subjects such as supernovas or Schrödinger's cat, simultaneously dead and alive. Moreover, artists have been welcomed into biology laboratories to work alongside scientists.

Biology-influenced artists have at their disposal objects like organs, that can actually be glimpsed by cutting open the body or using functional magnetic resonance imaging ($f$MRI), or like cells that can be seen with microscopes. In contrast, physics depends on visual images generated by mathematical models to provide a glimpse of unimaginable objects such as black holes, while observing the heavens with the naked eye reveals only dots of light twinkling on a vast canopy. In biology-influenced art, the object of study can actually be altered by artists exploring the boundaries between humans, animals, and robots, working alongside scientists.

Artists can even genetically engineer new forms of life in sterile conditions, in vitro, or in bioreactors (devices that maintain the optimum conditions for growth). Their startling experiments on objects that straddle the border between living and nonliving remind us that in the long course of our evolutionary history we come from— and still have within us—organisms other than human.

Science is changing our world and our lives at an ever-increasing rate. Today artists are bringing science out of the laboratory.

## Art & Science: merging art and science to make a revolutionary new art movement

In 2011 I co-curated a show at GV Art in London entitled Art & Science: Merging Art & Science to Make a Revolutionary New Art Movement. My co-curator was Robert Devcic, the dynamic owner of GV Art, which at the time specialized in biology-influenced art. Instead of the usual minimalist title for an exhibition, we chose a longer, more challenging one to reflect the debates swirling around this new art movement. Most of the artists in the show collaborate with scientists and the benefits sometimes run in both directions. Their works explore the meaning of life and our bodies and of systems on the boundary between living and nonliving.

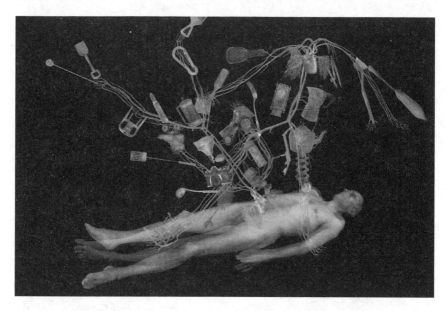

7.1: Andrew Carnie, *In Out*, 2008.

Artists in the laboratory often produce works of considerable interest to scientists. Andrew Carnie plays with the notion of how our view of the human body is mediated by science. *In Out* is an exploration of how our sense of self is disrupted when a foreign body such as a heart is inserted into us. How does awareness of such an invasive event affect our perception of who we are? Carnie claims that "art is too important to be left to artists—science too important to be left to scientists."

David Marron has a different take on biology-inspired art. As a paramedic he ponders the body in death, whether from violence, accident, or natural causes, sometimes in the loneliest of circumstances. For our exhibition he created an extraordinary shamanic figure that he calls *The Physician*, with a birdlike head and proboscis-like nose that he says was inspired by a stethoscope, a chest that doubles as a medicine cabinet, and a long blue hospital robe. It's oddly dignified and imposing and certainly unforgettable. "Each work is approached differently but a generalized underlying subject is humanity, scratching at our fragility and durability, violence

7.2: David Marron, *The Physician*, 2011.

and emotion—our habits," he explains.

These artists explore, interpret, and reinterpret forms in nature, attempting to identify forms that are successful and find out what makes them so. Davide Angheleddu describes his investigations thus: "My artistic production gets inspiration from nature, particularly from nature sublimely described in the book *Kunstformen der Natur* (*Art Forms in Nature*) by the German philosopher and biologist Ernst Haeckel." To investigate the essence of natural forms, he produces crisp, perfect, organic shapes using digital technologies to sculpt.

Katharine Dowson is inspired by the way that science and technology can inform us about the hidden world of the human body, beyond even what we can see in anatomy museums. She uses laser technology to work with transparent materials, particularly glass, because it is "a major component in scientific discovery, from test tubes to lenses, revealing the microcosmic and macrocosmic universe and their visual similarities." She makes amazingly fine sculptures of the brain and the vascular system. In Dowson's *Memory of a Brain Malformation*, the blood vessels from a cerebral angiogram of her cousin's brain have been etched by laser into glass. It depicts a tumor that was later removed by laser, which was also the means for creating the work.

Ken and Julia Yonetani look to the environment for inspiration and use salt or sugar mixed with water to make extraordinary sculptures of anything from a still life as perfect as something in an old

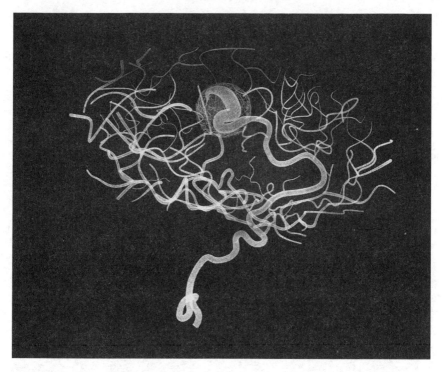

7.3: Katharine Dowson, *Memory of a Brain Malformation*, 2006.

master painting to a salt mountain, an image of human veins, or a map of the world. "Our work tries to retrace lines of connections that have been broken or lost, particularly between acts of consumption and the environment," they write.

Susan Aldworth works on the border between philosophy of mind and neurophysiology. She studies the relationship of the self to the physical brain, "how to define one's personality and whether it can be physically located." Among her tools is *f*MRI. For the piece in our show she took twenty prints of a scan of her own brain, then scratched sketches on them and added text, so that the "scan might look like it could show what was going on in my imagination as well as the physical structure and function of my brain." She called the work *Cogito Ergo Sum 3*. While the scan reveals the brain's physical structure and function, it does not tell you anything about the self, she points out. "Where am I in those pictures?" she asks. "You will look into my brain but you will never find me."

Annie Cattrell too looks into the self. She contributed *Pleasure*

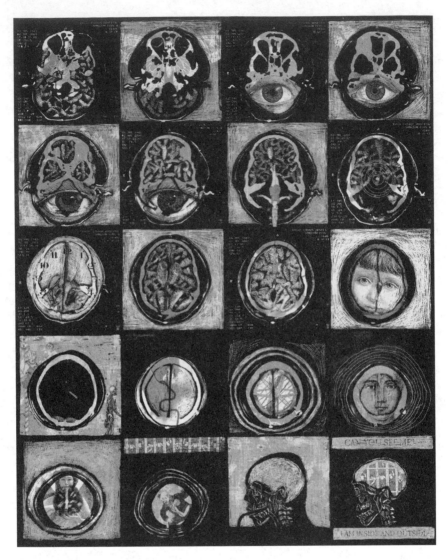

7.4: Susan Aldworth, *Cogito Ergo Sum 3*, 2006.

*and Pain*, created in collaboration with Morton Kringelbach, an eminent neuroscientist. They used *f*MRI to study the region in the brain stem that responds to pleasure and pain, then, using rapid prototyping, a technique akin to a three-dimensional print, Cattrell made a model of the brain stem to show its oscillations. The resulting sculpture, made of resin, is beautiful as well as interesting.

7.5: Annie Cattrell, *Pleasure and Pain*, 2010.

As Martin Kemp, the art historian, put it in a review in *Nature*, it "invites analogies—an incredibly complicated vertebra perhaps, or a wondrous fungal growth, or even a blossom flowering posthumously in the scalding water of a teapot."

What we see is a very precise depiction of how the mind actually responds to pleasure and pain. Together Cattrell and Kringelbach try to understand human consciousness and unconscious processing from a specifically neuroscientific viewpoint. The question is, however, whether this could be interpreted as reducing mental events to mere chemical processes. Kringelbach says that Cattrell's work, which turns "ephemeral, oscillating brain activity into concrete sculptures, has helped me rethink what I thought I knew"—high praise indeed from a neuroscientist of Kringelbach's stature.

Nina Sellars showed *Lumen*, a beautiful kinetic installation that was a scan of a fictional interior made using the play of light.

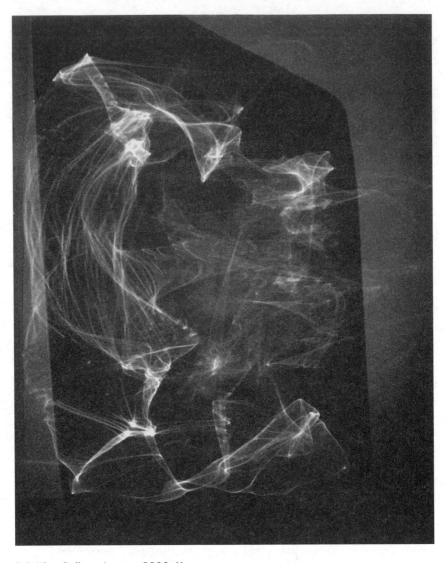

7.6: Nina Sellars, *Lumen*, 2008–11.

The image was produced by directing a small but intensely bright light source at an imperfect glass lens turning on a platform, with the resulting ghostly image projected onto the gallery wall. The image appeared both mechanical and organic, revealing a vision only attainable through technology, with light not only illuminating surfaces but also defining volumes without mass. Exhibited in a darkened room, the resulting images were highly poetic. It was one of the most popular exhibits and the most abstract piece in the show, making visible an invisible world and thereby touching on physics-influenced art. In fact, to make it Sellars collaborated with scientists at the Laser Physics Centre at the Australian National University, Canberra.

Helen Pynor's work is equally controversial in its own way. Organs, the all-too-real components of human and animal bodies, are something that we tend to prefer to look away from or see only in sanitized form, wrapped in plastic in a butcher's shop. Pynor uses them to make beautiful and thought-provoking works of art. For our show she created *Liquid Ground 6,* inspired by drownings in the Thames, depicting floating garments evoking human bodies, with organs washing about in the middle of them.

7.7: Helen Pynor, *Liquid Ground 6,* 2010.

Each piece was a huge collage using photographs made into prints and mounted on glass to give the effect of floating. Her aim, she says, was to make a connection between the body as a cultural thing and the visceral self and to do so in a sensitive way that "avoids either the sensationalism of gore or the clinical neutrality" of medical practice.

The fact that individual organs can be kept alive outside the body means that their integrity must be considered, especially if one believes that consciousness is spread throughout the body. To understand organs fully, Pynor believes that one must take into account the question of the philosophical status of living versus nonliving organisms and the various historical and cultural narratives that surround organisms. She writes: "I'm fascinated by the mystery of our status as biological beings whose bodies are the repository for experience, language, and consciousness in and beyond the central nervous system."

One of the highlights of the exhibition was Oron Catts and Ionat Zurr's *Pig Wings*. Catts is director of SymbioticA, the controversial laboratory established in 2000 at the University of Western Australia, Perth. One of its specialties is tissue engineering, creating entities on the borderline between living and nonliving—"semi-living sculptures." A husband-and-wife team, Catts and Zurr work on the cutting edge of biology-influenced art, growing things rather than sculpting them. Usually to be found in lab coats, they have replaced their easels with petri dishes to grow different types of tissue from the cells of dead animals. Once the skin has grown they dye it, photograph it, and use it to create extraordinary objects. They are, they say, interested in direct engagement with living tissue sculptures, rather than using traditional representational techniques.

Their *Pig Wings* project is a meditation on the shape a pig's wings would take if pigs could fly (see Insert). To make it, the two used stem cells from a pig's bone marrow, grown into pig tissue on a biodegradable polymer frame. Coated with gold, the tiny wings (4 cm x 2 cm x .5 cm) were exhibited in jewelry boxes. The first two times the pig wings were exhibited, there was also a living version of the wings growing in the gallery. They write, "Wet biology art practices are engaged in manipulation of living systems. [We are] exploring the manipulation of living tissues as a medium for artistic expression." "Wet biology" means the use of living tissues. In growing cells in order to explore the shape of wings on pigs, they explore the aesthetic as well, following the rule that in nature forms possessing pleasing properties are naturally preferred because they are the most efficient, like beautiful theories in physics.

Catts was much inspired by the Vacanti brothers, who in 1995 engineered the Vacanti mouse, a mouse which looked as if it had a human ear growing out of its back. The "ear" was actually a structure formed by planting cells from cow cartilage in biodegradable mold in the shape of an ear and implanting it under the mouse's skin. The Vacantis' aim was to create human organs in the laboratory from the patient's own tissue so there would be little chance of rejection—groundbreaking research. Catts and Zurr actually worked under Joseph Vacanti and produced *Pig Wings* at his laboratory in Harvard in 2001.

Catts and Zurr explore "what it is to be human, a quest in which art is going to play a major role." Their aim is to provide a platform to study ethical issues around life and the incompatibility of the way that science and society deal with it. "The function of art," says Catts, "is to expose areas of life that we don't have the proper language to describe," such as his semiliving forms: what to call them, how to deal with their dignity. He seeks out "areas of incompatibility, zones of discomfort."

Stelarc was the undisputed star of our exhibition at GV Art, famous for his spectacular performances dating back some forty years. To him, "what connects artists and scientists is technology." One of his two contributions to our show was a left ear growing on his left arm. To make it, some of his stem cells were seeded into a biodegradable polymer frame shaped like an ear and implanted on his arm in a series of operations, begun in 2007, which are still going on. Eventually a microphone will be inserted into the ear, connected to a Bluetooth system, so that the ear can "hear." There will also be a receiver in Stelarc's mouth so that when he opens it a person near him will hear the voice of someone else, who is telephoning him.

Besides the ear, the operation as performance was shown at our exhibition in the form of six stunning color photographs by Nina Sellars, entitled *Oblique: Images from Stelarc's Extra Ear Surgery* (see Insert). Sellars's artwork utilizes drawing, photography, installation, and up-to-date technologies. "In the 21st century we have become captivated by technologies for realms that exist beyond what is normally visible," she writes. She uses techniques such as medical imaging technology to create exquisite light works. Her giant photographs of Stelarc undergoing surgery ask the question: Who is the artist—the surgeon, the patient, or the photographer?

The photographs—especially vivid in that they were not framed—show the initial operation, in which a scaffold in the shape of an ear is inserted into a hole in Stelarc's arm. The arm seems almost disembodied, with its gaping hole waiting to receive the microphone. The harsh lighting focused on the arm with everything else fading into darkness is reminiscent of paintings of surgical postmortems hundreds of years ago, lit by candles inserted into the cadaver's opened body cavity. We staged debates in which two of the photographs were directly behind the speakers. Some of the audience had to look away now and then.

Stelarc's other piece was *Stretched Skin*, a huge digital portrait of his own face flattened and floating, like a Mercator projection of the world. Printed on three separate panels, it was exhibited not on the wall but horizontally, 40 cm off the floor, illuminated from above. Many viewers found it one of the most shocking pieces there. For Stelarc, art is a way to explore "what it means to be a body."

Stelarc pushes the limits of the body, and much of his work is space-age stuff. He has been hung up by hooks through his skin, sported a robotic third arm, and operated a vast metallic spiderlike six-legged walking machine, the Muscle Machine, which was acti-

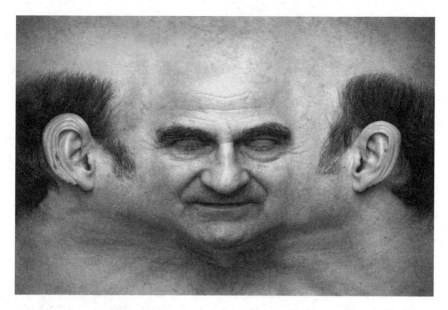

7.8: Stelarc, *Stretched Skin*, 2010.

vated by responding to his movements. Bodily extensions of this sort, he asserts, are destined eventually to actually become part of the body. After all, we are only part human. Part of us is already inorganic in that our physiology contains viruses from billions of years ago, made up of inorganic matter.

This should not surprise us. We live in an age in which organs await transplantation and mechanical parts such as hearts can be inserted into us—even animal hearts. Human eggs can be frozen and then fertilized by sperm that may have also been frozen. Blood transfusions can make my blood into yours, corpses can be cryogenically frozen to be reanimated when a cure for a particular disease is found, and comatose bodies can be kept alive on life support machines. Soon, he says, there will "be no more biological death." Instead, it will be a question of being disconnected from whatever machine is keeping us alive.

All of these artists are typical of the extremes of science-influenced art. Like practitioners of extreme sports, they push themselves and their art to and beyond limits.

A GOOD NUMBER of people—some 2,800—visited our exhibition on "merging art and science to make a revolutionary new art movement." The reviews were generally positive. Sally Carter wrote in the *British Medical Journal*, "This exhibit is an adventure." Matthew Reisz commented in the *Times Higher Education*, "Wherever science is leading us, these artists are following and finding powerful ways to address important issues of progress, identity, and what it means to be human." Katia Sowels of the *Independent* described the exhibition as "intriguing, because at first glance art and science are generally considered opposites." Helen Lewis, in the *New Statesman*, identified a problem that the new movement of science-influenced art has yet to deal with: "Scientists are leading the way—several of these pieces have already been sold to museums or professional bodies—but getting art critics interested is proving harder." This is apposite. When curators at major museums and galleries agree that science-influenced art is important and meaningful, it will have arrived.

During the exhibition I chaired three debates. The intense give-and-take paints a telling picture of the complex issues underlying this newly emerging art movement: how it should proceed, the role that science plays in the work of artists, how and whether the coming together of art and science should or should not proceed, and what it actually means for artists and scientists to work together. Tiffany O'Callaghan wrote in the *New Scientist*, "Miller spoke about an emerging third avenue of art in the 21st century, a true convergence of art and science, but the artists were at pains to dissemble his vision," and indeed on several occasions I found myself the minority voice. She went on, "Perhaps as artists increasingly collaborate with scientists, and head into labs for their technology, the distinction between tools and inspiration will blur."

The first debate took place on June 8, 2011. It was a warm evening and over a hundred people were crammed into a small room, overflowing into the garden. This debate had a sharp critical edge and spirited discussion, not surprising as the panel consisted of

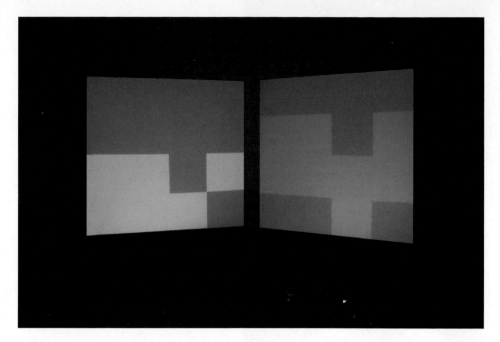

Ernest Edmonds, *Shaping Space One*, 2012.

Michael Bove and Paula Dawson, *Fig. 3 Phantombrush*, 2009.

Paul Friedlander, *String Theory II*, 2010.

Keith Tyson, *Studio Wall Drawing: 2001, The Thinker (After Rodin)– Technical/Notes*, 2001.

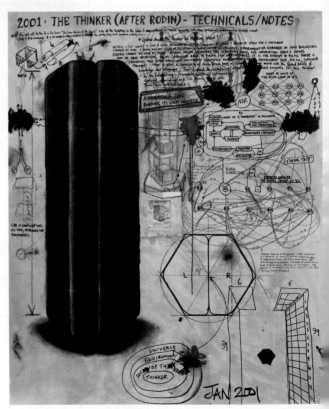

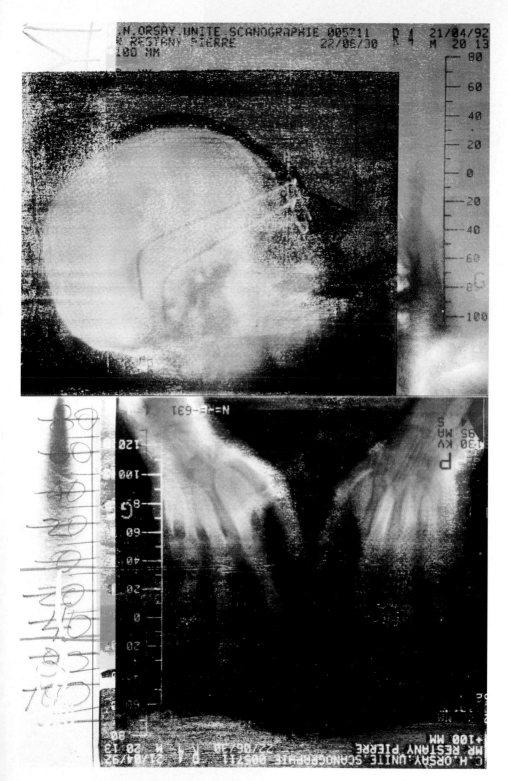

Steve Miller, *Portrait of Pierre Restany*, 1993.

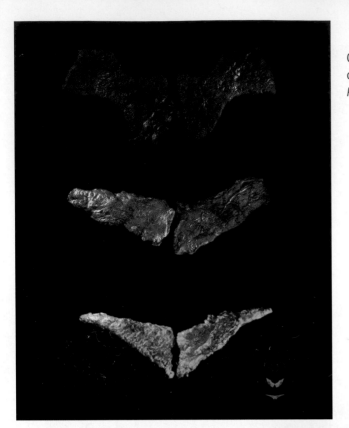

Oron Catts and Ionat Zurr in collaboration with Guy Ben-Ary, *Pig Wings*, 2000–02.

Nina Sellars, *Oblique: Stelarc's Extra Ear Surgery*, 2008.

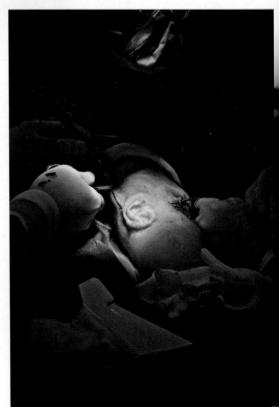

Eduardo Kac, *GFP Bunny*, 2000.

Brandon Ballengée, *DFA186: Hades*, 2012.

Erik Guzman, *Weather Beacon*, 2008–09.

Scott Draves, *Electric Sheep*, 1999.

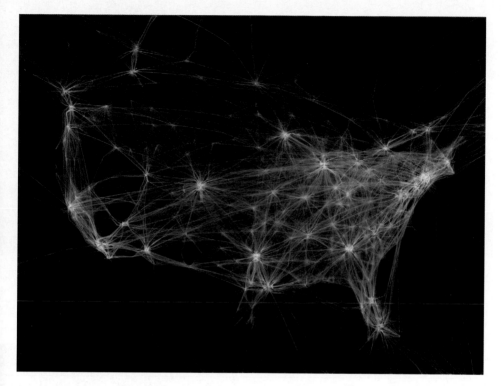

Aaron Koblin, *Flight Patterns*, 2005.

W. Bradford Paley, Kevin Boyack, and Dick Klavans, *Relationships among Scientific Paradigms*, 2006 (first version).

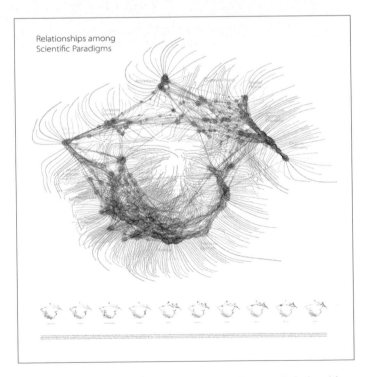

W. Bradford Paley, Kevin Boyack, and Dick Klavans, *Relationships among Scientific Paradigms*, 2006–10 (second version).

Oron Catts, Nina Sellars, and Stelarc. To start the discussion, I asked whether science-influenced art might prove to be at the forefront of a whole new culture in the twenty-first century. To my surprise, the panelists attacked me for using the words "science-influenced," which they interpreted as suggesting a hierarchy of disciplines—that science was above art. I replied that artists have always taken their inspiration from outside themselves. In this day and age, science pervades our culture, so it makes sense that artists should use it as a source of inspiration. The panelists were disturbed by what they saw as a lack of symmetry in my remarks, the suggestion that science affects art but not vice versa. I replied that while science-influenced art certainly exists, there are very few examples of art-influenced science, although I am sure this will evolve.

The artists also claimed that they were not so much influenced by science but simply used it like paint in a can. They finally conceded that they were challenged by science. But technology was different.

Oron Catts spoke of the evils of technology and the importance of separating it from science. This is an old refrain dating from the post–atomic bomb days into the 1960s and 1970s—that technology, not science, is responsible for nuclear weapons and the evils of society. Today this stance seems archaic, in light of the blurring of the line between science and technology.

To the audience's surprise, the panel also insisted on a "dignity of disciplines," arguing that erasing the line between art and science could lead to scientists producing bad art and artists producing bad science. In other words, art is art and science is science—a position that seems increasingly parochial. The panel went on to argue that scientists think less creatively than artists because of the restrictions on their work imposed by the need to apply for grants. This may apply in the case of scientists who are experimentalists, whose research grants are for a very specific project, but not to scientists who specialize in theory, whose research can range wider than their grant specifies. The panelists seemed to know little of what went on in the world of science outside biology. This held for most of the audience too, the vast majority of whom were artists.

The panelists complained that while they visit scientists in their laboratories and offices, the scientists never visit them. Scientists tend to get lost in their own work and not have time to look outside it. The educational process too often does not give scientists the urge to expand their horizons. The importance of looking into the arts is critical for wide-ranging research.

Art and music can certainly inspire great science. A Bach sonata can help recharge one's energy to return to what seems an intractable problem. But this is different from being *influenced* by an artist as a result of the collaboration to such an extent that the scientist changes his view or looks again at his day-to-day research. There is a difference between science-inspired art and science-influenced art, as there is between art-inspired science and art-influenced science. I'm interested in artists influenced by science, not artists inspired by science, "influenced" being stronger than "inspired."

## The body as canvas: ORLAN

Like Stelarc, ORLAN's canvas is her body. In fact, if anything, the experiments she carries out on it are even more extreme.

Before I even meet ORLAN, her assistant reminds me by email that her name is spelled in capitals. She is a brand. The underlying theme of ORLAN's work has always been an attack on traditional concepts of female beauty—invented, she says, by men for their own pleasure—and on Christianity, which treats the female body as unclean.

Always controversial, ORLAN had her great epiphany in 1978, when she was thirty-one. She was about to address a symposium on video and performance art when she collapsed and had to be taken to the hospital for emergency surgery for what turned out to be an ectopic pregnancy. She took a video crew to film the operation and insisted on remaining conscious throughout. Looking up from the operating table, she saw light beaming down as if from heaven, and the surgeon standing over her like a priest officiating at a Mass with his assistants around him like fellow celebrants. There and then she realized that surgery, sculpting her own body, was the way to express herself.

ORLAN was always a rebel. At twenty-four she flew in the face of her bourgeois upbringing and changed her name from Mireille Suzanne Francette Porte to what she calls her nom de guerre. At art school she dutifully reproduced sculptures and copied paintings of the masters. "Very boring," she recalls. She soon turned to avant-garde art and filmmaking.

For her, feminist issues were always paramount, especially in the revolutionary 1960s. She launched herself into performance art early on. In 1964, when she was seventeen, she had herself photographed naked, "giving birth" to an androgynous mannequin. Later she entitled the photograph *ORLAN Giving Birth to Her Beloved Self.* Like the mannequin, the name ORLAN is neither masculine nor feminine. "I am both man and woman" is her message.

In 1971 she transformed herself into a Catholic icon, exhibiting photographs, videos, and sculptures of herself as "Saint ORLAN," swathed in swirling fabrics, both living doll and living sculpture. Like the nursing Virgin Mary and Amazons in battle, she has one breast exposed, thus pointing up the hypocrisy of society's traditional depiction of women as either Madonna or whore.

In 1990 she began a new project, *The Reincarnation of Saint ORLAN*, involving a series of nine operations to transform her into the precise image of female beauty as depicted by artists, invariably male. The operations are conceived as performances, with the surgeons dressed in outrageous garments designed by Issey Miyake and other top designers. ORLAN is awake throughout, listening to poetry and music, and the entire procedure is filmed and often streamed live. Taking the operating room as her studio, ORLAN directs, sometimes requesting that parts of the procedure be repeated because the surgeon's gesture is not just as she wants it or because her face is hidden, or requesting that the camera angle be changed. Sometimes the surgeons want to direct. To ORLAN, these operations are artist–scientist collaborations, but with herself in charge. "I must have control over my own body."

In one operation ORLAN had "beef implants," bones from a cow, inserted at her temples to suggest the Mona Lisa's protruding brow. She decorates the implants with colored makeup and glitter.

When I meet her in Paris, ORLAN beams friendliness and warmth combined with a formidable intellect. Her hair, à la Cruella de Vil, is black on one side, white on the other, and she wears gray-rimmed round glasses with yellow earpieces. The implants on the sides of her face accentuate her expressions as she describes her life and work. Her combination studio, living quarters, and office is immense, filling two floors, with most of the ceiling of the ground floor removed. The upper floor has a ramp for accessing books and files, while the ground floor is lined with bookcases filled with books by or about ORLAN and artworks depicting her.

Our conversation quickly turns to her present interest: cells. She is struck by the immortality of cancer cells, which live on even when the body in which they are growing dies. "Cells expand in time and space, like the body," she says.

She started her recent project, *Harlequin Coat*, at SymbioticA in 2007. Harlequin in his diamond-patterned suit, the nimble, irrepressible trickster of commedia dell'arte, has always intrigued artists. In his Rose Period, Picasso was entranced by him. Shrouded in mystery, supposedly possessing the secrets of alchemy, Harlequin has the power to become invisible or take other forms. According to some versions, Harlequin can be male or female.

ORLAN's harlequin coat is a patchwork of diamonds of different skins. ORLAN uses her own skin cells, mixed with those of others—cells from a twelve-year-old female fetus of African origin collected in a biopsy, muscle cells from a marsupial left over from scientific research—together with human blood cells, mouse connective tissue, goldfish neurons, human brain cells, menstrual endometrium, umbilical cord and vaginal cells, together forming an assemblage of skins of different colors, ages, races, and origins. The cells are cultivated in petri dishes set inside diamond-shaped Perspex of different colors. During her surgical procedures ORLAN wears her harlequin's coat. A bioreactor forms the harlequin's head.

7.9: ORLAN, *Harlequin Coat*, 2007.

As more of ORLAN's cells are co-cultured with other cells in the bioreactor, so the installation evolves. Gradually this living material replaces the dead cells in the petri dishes. But this drama, as ORLAN calls it, is invisible, played out inside the bioreactor, in contrast to ORLAN's other works, in which her body is opened up, revealing everything. "This particular work is caught between the folly of wanting to see and the impossibility of seeing," she explains.

But how can this be art if the drama is invisible? It differs from visual art in that it is strongly performance-based, but it is art nonetheless. ORLAN gives Jackson Pollock as an example. His art is as much his performance as the painting he makes—the act of painting as a form of visual art.

In *Harlequin's Coat*, says ORLAN, "biotechnology is taken out of the laboratory and turned into a spectacle through the commedia dell'arte character of Harlequin and ORLAN's cells which play an actor's role, in both the theatrical and the linguistic sense of the word."

## The fluorescent rabbit

The American artist Eduardo Kac works on the borderline between humans, animals, and robots. He calls the products of genetic engineering "transgenic art." In 1997 he was one of the first to have a microchip implanted in his body, in his case in his ankle, the place where slaves were most often branded.

Kac's most famous creation is Alba the rabbit, a white rabbit implanted with a green fluorescent protein gene from a type of jellyfish. Like the jellyfish, Alba shone green when exposed to blue light (see Insert).

Roy Ascott, a pioneer of computer art, supervised Kac's PhD thesis. He is cynical about Kac's motivation. "Kac's point of view was that he saw the rabbit as Duchamp's urinal." He intended it to shock, to ensure that bioart would never be the same.

The director of the laboratory in France where Alba was born in 2000 was responsible for the experiment and refused to allow Alba to leave. So Kac launched a "Free Alba" campaign, with T-shirts and flags, claiming that his campaign highlighted the complexities of how an artwork is received and the pressures placed on the artist. He set up performances involving himself, the media, and the public "to *enact* critical views, to make present in the physical world invented new entities (artworks that include transgenic organisms) which seek to open a new space for both emotional and intellectual aesthetic experience."

A recent work of his, *Cypher* (2009), includes a DIY transgenic kit, giving everyone the chance to breathe life into their own piece of biological art.

## Beauty in asymmetry: Marta de Menezes

Marta de Menezes also interferes with nature, though her works are far more benign. De Menezes works at the intersection between art and biology. In her best-known work, *Nature?*, she tampers with butterflies' wings to create asymmetries. The asymmetrical butterflies flutter around a greenhouse.

I met de Menezes in 2009 at a symposium entitled "Einstein, Picasso" in Dortmund and was struck by her seriousness of purpose and expertise in biology, a subject she largely taught herself. De Menezes is Portuguese. She studied fine arts at the University of Lisbon, intending to become a high school art teacher. Then she went to Oxford. She found university life highly stimulating and began collaborating with scientists.

The artists who have most influenced her, she says, are Joseph Beuys, Gerhard Richter, and John Cage, all of whom defined their own paths, breaking away from the norm.

In 1998, de Menezes came across a paper in the journal *Nature* about scientists at the University of Leiden who were making studies of butterflies. To explore the formation of wing patterns, they altered the factors affecting wing appearance by interfering with normal butterfly development, but without modifying any genes. De Menezes was thrilled at the idea of creating art using live butterflies, so she approached them and began a collaboration. The following year she produced *Nature?*, which was first exhibited at Ars Electronica in Linz in 2000. For this she inserted a very thin needle coupled to a heat generator into the cocoons of the butterflies. When the butterflies emerged, one wing had a modified pattern, while the other was normal. These asymmetrical wing patterns had never before occurred in nature. As the genes were unchanged, the next generation of butterflies had normal,

7.10: Marta de Menezes, *Nature?*, 1999.

symmetrical wing patterns. This was a work of art with a limited lifespan. De Menezes displayed her butterflies in a specially constructed room.

"My methodologies in creating artworks and pursuing ideas have been deeply shaped by my interactions with scientists," she says. "With them I learned to plan my actions, discuss the ideas, and be as clear as possible in explaining them to others so that my projects can grow. With scientists I learned to keep a lab book to make sure I can retrace my steps. I learned to work on my own project and simultaneously share objectives, to work collaboratively."

De Menezes sees *Nature?* as straddling the frontier between art and science, though ultimately it is art. "I always refer to my work as art, not science-influenced art. It is art in every part of it and very clearly not science. I'm a science-influenced artist that makes art with a strong scientific content," she says. Although her work relates to scientific considerations, it is not science per se. Rather than advancing science, it is influenced by science.

## Biology-influenced art in New York: Suzanne Anker

As Suzanne Anker, doyenne of biology-influenced art in New York, puts it, "Art uses every other discipline as its data bank." Dynamic and bursting with energy and enthusiasm, Anker is chair of the fine arts department at the School of Visual Arts. Her *Aha!* moment came in 1988 when she was working on bronze sculptures of tree forms, egg forms, and flower forms which she linked with man-made objects such as funnels and soccer balls. In one piece, she attached a kaleidoscope to a vase. When she looked through the kaleidoscope, the circular form of the vase produced an image akin to images of cells seen through a microscope. Looking through biology books for interesting images of cells, she realized that the image she had created was similar to chromosomes, the DNA structures found in a cell.

She decided she wanted to have her own chromosomes mapped. At the time it was unclear what this meant and her health insurer became concerned. They asked why she was doing it, to which she replied, "Chromosomes can't tell me who I am, but they can tell me what I am." Unimpressed, they dropped her health insurance. It took her a year to construct a three-dimensional model from the two-dimensional images of her chromosomes. Anker realized that she had finally found the connection between art and science that she had dreamed of as a teenager.

Born in 1946 in Sheepshead Bay, Brooklyn, Anker studied chemistry at Brooklyn College. The subject had always intrigued her—"how parts fit together, the alchemy involved in mixing stuff." The mathematics, however, was too much and she changed to her other interest, art.

At the time, Brooklyn College was a center of abstract art. Anker worked under one of America's foremost Abstract Expressionists, Ad Reinhardt, and found him a great inspiration. "It was the first time I saw the philosophical underpinnings of art, not just mimesis, rather art that had to do with social and spiritual values." She immersed herself in the 1960s art scene, hanging around Max's Kansas City, the Manhattan watering hole of choice for Robert

Rauschenberg, Roy Lichtenstein, and Donald Judd. The back room was reserved for Andy Warhol and his entourage and the Velvet Underground regularly played there.

Overwhelmed by this exalted company, Anker concluded that it was "too hard to be an artist in New York City," although she continued to harbor the desire. Having graduated, married, and moved to Boulder, Colorado, she did a master of fine arts degree at the University of Colorado. At the time she was drawing, printmaking, experimenting in papermaking, and casting forms in paper pulp. Requests from New York regularly arrived to include her work in exhibitions. Finally, in 1975, she agreed. Her first group show was at the Martha Jackson Gallery. She found herself going back to New York so often that she finally took the plunge, ended her marriage, and stayed.

Anker began to "think about manipulating nature and the way in which science is about discoverable truths about nature and how it can be adjusted to its transformative capabilities." She planned a show on the intersection between art and genetics, at the time not exactly a hot topic or even one well known to artists or scientists. A friend at Fordham University in the Bronx asked Anker to curate a show on the campus. The show, entitled Gene Culture: Molecular Metaphor in Contemporary Art, took place in 1994 and was the first exhibition devoted entirely to the intersection of art and genetics.

Around that time Anker met Dorothy Nelkin, a sociologist at New York University known for her studies on the relationship between science and society. Anker and Nelkin shared an interest in the way that the ethical and social issues of the emerging biology-influenced art affected both scientists and artists, in particular artists working in laboratories trying to represent life forms and molecules not known in nature, or even actually producing these molecules, as would become the case in synthetic biology.

As the twenty-first century dawned there were giant leaps in biology, such as the human genome project. The double helix structure of the DNA molecule was by then iconic. While Anker incorporated it into paintings, sculptures, and installations, Nelkin looked into its significance for society.

Their work led to a book, *The Molecular Gaze: Art in the Genetic Age*, published in 2003 by Cold Harbor Laboratory Press (associated with Cold Harbor Laboratory, where James Watson, the co-discoverer of DNA, was director). The book had an excellent reception but sadly Nelkin died shortly before it came out.

After many years of writing and practicing on the borderline between biology and science, Anker has concluded that "anything to do with the natural world is art and science." For her, "bioart or sci-art is an umbrella term that has no clear-cut geographical area, nor is it media-dominated." Hers is not an anything-goes view, however, akin to an artist declaring that a spaghetti-like structure represents string theory. There must be artistic integrity, as Anker shows in her creations in which organisms are clearly organisms. Examples range from plants grown under LED lights to *Zoosemiotics*, tiny chromosome sculptures laid out in identical pairs, which were exhibited at the Getty Museum in Los Angeles in 2001 (see next page). This is not merely figurative art. A strong sense of fantasy pervades. "That's because artists believe in magical thinking," she says with a smile.

Like many artists involved in biology, Anker is self-taught and knows enough of the concepts to use them effectively in her art. "My interest is not in being a rookie scientist, but a professional artist." She has never worked directly with a scientist. At the School of Visual Arts, her students take biology courses at the Museum of Natural History in which they learn to use high-tech equipment. In this way she hopes they will produce high-grade biology-influenced art.

She has also installed a wet lab, the Nature and Technology Lab, for students to carry out actual experiments to use as the basis for artwork. "Things that are going on in labs are very inspiring—dangerous, but inspiring. In the art world there has been too much of a gravitational shift toward entertainment and popular culture when there are some really fabulous things being done in science. Artists are well poised to enter that dialogue."

7.11: Suzanne Anker, *Zoosemiotics* (*Primates, Frog, Gazelle, Fish*), 1993 (detail). Installation at the J. P. Getty Museum, Los Angeles, California.

Anker is interested in extemporizing from nature, using highly technical approaches such as rapid prototyping and other techniques for producing actual objects. The creative lever is to "present the technology with problems it can't do. That way there's an upgrade in the ways in which one can transform matter." Unity must come from a fusing not only of art and science, but technology, too.

Anker makes frequent visits to science laboratories. She agrees that artists seem to get more out of a collaboration than scientists. But artists "can bring other things to scientific practice. Artists don't work according to scientific methods, they shoot from the hip"—they can be freer in their approach to a problem. They "can come into a lab from the outside and spark a scientist's mind with a chance remark, perhaps coaxing the role of serendipity in scientific research."

In 2007, Anker recalls, the artist Mark Dion, whose work focuses on ecological issues and our perception of nature, was asked to create an installation at the Natural History Museum in London to commemorate the three hundredth anniversary of the great naturalist Linnaeus. Many of the scientists involved were angry that funds were going to a mere artist. So Dion decided to do something that an "artist might think up, but not a scientist." He attached flypaper to the top of his car and drove at high speed down a large street in London. Then he sent the flypaper to a laboratory, where scientists found three new species of insect. After that, says Anker, "the scientists changed their tune."

She turns the conversation to interdisciplinarity, to the importance of overlap among art, science, and technology. A course has just been implemented at the School of Visual Arts which is taught by an artist, a designer, and an architect—a very encouraging development. "Going back into one's ghetto doesn't work any longer."

## Speaking up for nature: Brandon Ballengée

Brandon Ballengée is another New York artist who steers clear of the art ghetto. Ballengée is an artist, activist, and ecological researcher. He uses nature as his palette, recording the ups and downs of ecosystems as civilization impinges on woodlands, streams, and oceans. This is a brand-new field of art, still trying to find its way, often resulting in little more than banal photographs of icebergs.

Ballengée uses actual science in making his artworks. For him art and science are one.

A tall New Yorker who sports a cap and neatly trimmed triangular stubble sideburns, Ballengée speaks with calm assurance. Born in 1974, he grew up in central Ohio with nature as his playground. As a teenager he witnessed the incursion of lumber companies, followed by housing developments. One by one streams disappeared to be replaced by sewage systems, and shopping malls sprang up. "Much of my early work dealt with this sense of loss and was infused with the classical theme of man versus nature. The earth, once again, became my studio."

At the Art Academy of Cincinnati, Ballengée studied both fine arts and biology. He was particularly inspired by Ernst Haeckel, the nineteenth- and early-twentieth-century German artist and naturalist famous for his drawings of animals in his book *Kunstformen der Natur* (*Art Forms in Nature*). "While Haeckel drew, he learned the organism's functions," says Ballengée.

Ballengée is particularly concerned about the decline of the frog, toad, and salamander populations and the increasing numbers of deformities. "Amphibians," he says, "are a 'sentinel' species, the environmental 'canaries in the coal mine.'" The walls of Ballengée's Manhattan studio are covered in drawings of frogs made using cigarette ash and coffee and fixed with rabbit-skin glue and latex paint to give a distinctive texture. Sometimes he cuts up the drawings and reassembles the parts.

Ballengée also uses "clearing and staining," a technique by which he makes a preserved specimen completely transparent while dyeing certain bones and tissues. The delicate and almost magical result, as in the piece *DFA186:Hades*, is achieved using a mix of over ten chemicals and dyes (see Insert). As Ballengée puts it, referring to Joseph Beuys, the subversive twentieth-century German artist, it's a "Beuysian idea. The materials that artists use can have alchemical properties." He exhibits the preserved amphibians in all their glittering colors, or photographs them. The beautiful images are both art and scientific studies, a crossover from Ballengée's collaboration with scientists as working partners.

One of Ballengée's ongoing concerns is with the continuing deterioration of biosystems in the Gulf of Mexico after the BP Deepwater Horizon oil spill in 2010, a disaster that continues to devastate marine and bird life. His interest in frogs "relates to ecological problems not necessarily caused by chemical impurities in the water, as well as by injuries caused by changes in the environment. The wetlands have become ecologically stressed as a result of chemicals, nutrient runoff, lawn runoff, garbage, predators, and parasites." Part of his work involves "citizen scientist" activities, taking people into wetlands to collect species, assist in biodiversity surveys, and clean up. He has taken inner city children to polluted urban rivers where they see amazing amphibians living among the broken glass and plastic bottles, which they begin to look on with disgust. This is the way, says Ballengée, that "eco-artists [can] transform the society and environment in which we live."

Ballengée currently runs the Nature and Technology Lab at the School of Visual Arts. Certain art dealers, he says, don't consider his work genuine art, a not unusual response to artworks that involve science. He has found a few dealers who will sell it. "Frogs sell better in Europe," he says. He has had several exhibitions in Britain. And private collectors and museums sometimes buy actual specimens. The art critic and activist Lucy Pippard describes him as one of the "'eco-artists' whose art may be unrecognizable as art because it achieves a social function."

## The "mad scientist": Joe Davis

Joe Davis is not concerned with social function, activism, or political agendas. "If only the government would support truly innovative, fascinating art, rather than vaguely imaginative, mostly placid already-been-done disappointments currently sucking up the allotments," he says. Well put! He makes this remark after lamenting that he himself receives no governmental support. A hard-drinking, fast-talking, peg-legged artist and motorcycle mechanic from Mississippi, Davis is a one-man publicity machine. In addition to braggadocio regarding past girlfriends and how much alcohol he can consume, he likes to tell the story of how thirty years ago he walked into the MIT Center for Advanced Visual Studies and demanded to speak with the director. When he was refused, he trashed the receptionist's desk, after which the police were called. But the fracas ended amicably. He stayed on as a researcher. "They called the cops and never let me go."

With his bizarre outfits and peg leg, like a Caribbean pirate of old, Davis is a walking work of art. He has been known to use his hollow metal leg to open beer bottles, or to play it like a bugle in his local bar. Amazingly, Davis actually went to college. Having been expelled from various high schools and colleges, he ended up at an experimental college in Oregon and did a degree in creative arts, then lectured in architecture at MIT for ten years. He also has an honorary PhD from the Bauhaus University in Weimar, Germany. Davis's father, a chemist, often talked to him about the latest scientific developments. But Davis was also fascinated by art and was torn between the two.

For the past twenty years he has been an unpaid researcher at MIT's department of biology and, as of three years ago, a researcher at Harvard Medical School, also unpaid. He likes to talk about his precarious living situation. He has had numerous evictions for nonpayment of rent and often has to sleep in his office or his car.

For Davis, genes are his medium. From scientists at MIT and Harvard, he learned how to synthesize DNA and insert it into the genome, the repository for biological information in an organism. The information in the genome is encoded in the DNA, which contains instructions for maintaining and building the organism.

Davis's big idea was that "DNA can encode any information, not just genetic sequences." He dubs it an infogene. In 2000 he created a work of symbolic art which he called *Microvenus*. For this he used the bacteria E. coli, which exists in everyone's gut and can survive extreme temperatures and radiation, even those of outer space. To make *Microvenus*, Davis created a visual icon—the letter Y with a vertical line drawn through the middle, representing the female genitalia. This is also, coincidentally, an ancient Germanic rune for life. He then digitalized it and, with the help of scientists, created a string of nucleotides—biological molecules, the building blocks of DNA—out of this string of zeroes and ones. He inserted the new molecule, imprinted with the *Microvenus* symbol, into E. coli, which multiplied it into billions of cells each carrying the same imprint.

Davis's objective was to amplify information on the plaque sent up in the *Pioneer 10* spacecraft in 1972 which showed images of a nude male and female. Only the male's genitalia were shown, not the female's. Davis looks forward to NASA transmitting the digitization of *Microvenus* to correct the oversight. Meanwhile, galleries refused to show *Microvenus,* citing the dangers of biologically engineered bacteria. Might they be superviruses? *Microvenus* was eventually displayed in 2011 at Ars Electronica, inside a pressurized containment vessel.

Davis also equipped ballet dancers with a transmitting device inserted into the crotch of their tutus, as a cross-disciplinary project involving artists, architects, and electrical and mechanical engineers to build a contrivance that recorded the sounds of their vaginal contractions and beamed them into space. The sounds were transmitted for several minutes from MIT's Millstone Hill radar station until the United States Air Force shut the transmission down. Like many biology-influenced artists, Davis has a sense of humor.

Another of his creations is *Bacterial Radio*, based on the crystal radio. The simplest type of radio receiver, the crystal radio has no batteries or connections to external energy sources. The basic circuit uses a crystal detector that converts radio signals into electrical signals that can be listened to through headphones.

To grow his *Bacterial Radio,* Davis put E. coli and other organisms into a petri dish in which he had laid out a circuit. As the E. coli grew, they produced chemical reactions that generated an electrical current that could be heard through earphones. Basically, the *Bacterial Radio* is alive. Davis points out that nature is more efficient than industry, so it makes sense to make use of biological processes.

At the 2012 Ars Electronica Davis won the top prize, the Golden

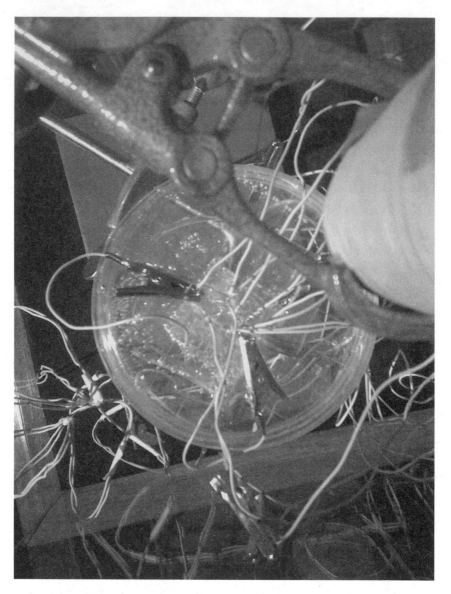

7.12: Joe Davis, *Bacterial Radio*, 2011.

Nica, in the hybrid art category, for *Bacterial Radio*—for the concept, that is, as the radio didn't work. Biosafety regulations prohibited the use of recombinant materials outside the laboratory. So, for the first time in his research, Davis tried to grow a circuit by metallizing non-recombinant bacteria, to no avail.

Davis is regularly invited to address science departments and symposia, though he finds that "often I speak a language that nobody understands." He feels passionately that artists should not pander to market tastes. "Artists should create things that hold an idea, no matter what it is. That's what art is."

## The firefly brain: Jun Takita

Jun Takita is another outsider, dreaming unimaginable dreams. A highly contemplative Japanese artist, he is inspired by Zen gardens which depict water using only sand. "The concept of a garden is in your head. The Zen monk watches over it."

Born in Tokyo in 1966, as a child Takita was enthralled by natural landscapes and deeply affected by urban development leading to the death of plants, animals, and insects. "Most of all, the little fireflies that I loved were disappearing." He studied arts at Nihon University, then moved to Paris and did his master's degree at the École Nationale d'Art, while also reading books on science, "because without science there is no way to visualize the world." In Paris he studied under the unconventional Polish artist and architect Piotr Kowalski, who looked for ways to express natural phenomena through the senses. His projects included explosions, electronic devices, plant growth, and gravity. After he died in 2004, Takita took over his large garden.

With the help of physicists and geophysicists Takita has continued Kowalski's work exploring gravity and how we keep our balance. He muses about whether there could be a connection between gravity and light, as both penetrate everything. He is particularly interested in bioluminescence, the power of living organisms to produce light, and what that tells us about our relationship to nature. Very few species glow in the dark. Among them are fireflies, glow worms and Dinoflagellata, a kind of plankton which has characteristics belonging to both plants and animals.

Takita believes that "artistic action" is necessary in the modern world where "nostalgia for paradise [and] the impact of a technocratic future" clash, a world where even fireflies may become extinct.

7.13: Jun Takita, *Light, only Light*, 2004.

Through his art he tries to create a landscape that is "a direct extension of the body, of my body, the landscape of a dreamed reality."

To further this aim, he produced a work which he called *Light, only
Light*. First he had an MRI scan made of his brain, then used a computer to convert the two-dimensional scan into a three-dimensional
sculpture. Then he turned to scientists at Leeds University's Centre for Plant Sciences. He was looking for bioluminescent moss;
they developed a new, genetically engineered strain of it for him, an
excellent example of the way in which art can affect science. Takita
then covered the sculpture of his brain with the green, bioluminescent moss. He exhibited the result at the exhibition Sk-interfaces

in 2008, at the FACT art center in Liverpool. The brain is a glowing garden, which has to be looked after like an ordinary garden. Takita says it was a singular experience for him to see a living replica of his own body, going "against the conventions which hold that there is a barrier separating the inside and the outside of the body." He was especially enthralled by the brain's crumpled surface, "in which light creates a strange relief." Whereas a garden is usually regarded as a place that catches light, Takita's gardens create light. His *Light, only Light* is a beautiful, intriguing, and rather disturbing work.

Takita's musings on the nature of gardens have led him to conclude that there is a vast difference between the traditional garden, which each viewer sees differently, and a garden of transgenic organisms "acting as plants and nonplants," emitting light, transgressing against the laws of nature. When he sees the light he has himself created, he thinks, "this is the expression of man's impossible desire to possess light."

Sadly, to exhibit his work in Liverpool, Takita had to use ordinary nontransgenic moss with no glow. The British authorities were afraid of unleashing a pandemic if they exhibited transgenic organisms in a public place. As well as the moss-covered brain, Takita also showed photographs of the original in all its glowing glory. It was exhibited at the top of a lighthouse at the Article biennale of electronic art in 2008 in Stavanger, Norway, and at Sk-interfaces in Luxembourg between 2009 and 2010.

Takita's thought-provoking pieces are exhibited worldwide but cannot be sold due to the biogenic material. Photographs, however, sell quite well. In Japan, he says, they are considered art and media, rather than art and biology, which shows the importance of media art and the wider context in which these artworks are seen.

Takita has a gut reaction that tells him when the work is completed, he says. Aesthetics is "beauty in proportion," which is what he seeks. Another aesthetic moment for Takita is that of discovery, when it occurs to him "how to do the work." "Process rather than result" is the basis of his method.

Some people find biology-influenced art shocking and question whether it is deliberately intended to shock—whether it is "shock art." Takita's moss-covered brain may be shocking but this, he says, is very far from his intention.

## Half-woman, half-horse: Marion Laval-Jeantet and Benôit Mangin

Quite a few biology-influenced artists, such as Stelarc and ORLAN, experiment on their own bodies. Even closer to the edge is the collective Art Orienté Objet—Marion Laval-Jeantet and Benoît Mangin— who inject themselves with horse blood. "We are interested in blurring the boundaries between species and between art and science," they say. They call their experiment-performance *May the Horse Live in Me*.

The first performance took place at the Kapelica Gallery in Ljubljana, Slovenia, in September 2011. Laval-Jeantet had previously spent many months with scientists receiving preparatory injections of horse immunoglobulin in order to ward off allergic reactions. Mangin, her husband, joined her in the process. This period of "process" involved a communication ritual with the horse, in which Laval-Jeantet walked around with it wearing prosthetic hooves.

On the day itself, Laval-Jeantet tells me, as she lay down to receive the first injections of horse blood, "I had the feeling of being extra-human, I was not in my usual body. I was hyper-powerful, hyper-sensitive, hyper-nervous, and very diffident, the emotionalism of an herbivore. I could not sleep. I probably felt a bit like a horse." By the end of the performance, a high proportion of her blood content was from the horse.

Afterward, Laval-Jeantet had samples of her hybrid blood extracted and freeze-dried. Two series of artworks have been produced in the form of engraved aluminum cases containing blood samples. None are for sale. What adds to their value is that the experiment cannot happen twice, the curator of the show, art theorist Jens Hauser, tells me. He writes, "[Their performance] explores

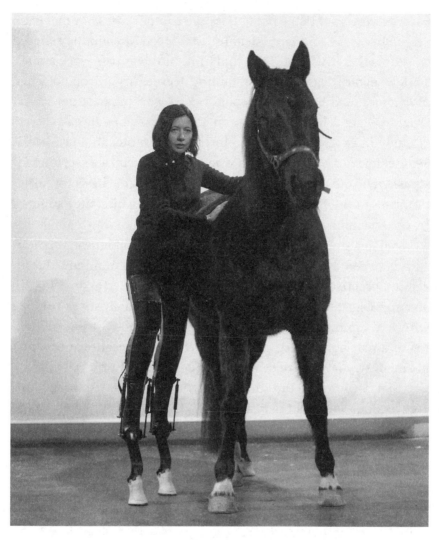

7.14: Marion Laval-Jeantet and Benôit Mangin, *May the Horse Live in Me*, 2011.

the new hybrid organisms, and in the utopian, artistic, poetic way anticipates the possible forms of coexistence between animals and humans." In 2011, Ars Electronica awarded them a Golden Nica for hybrid art—an appropriate category.

A very attractive couple, Laval-Jeantet and Mangin were both avidly interested in art and science—especially Laval-Jeantet, whose parents were scientists—but both opted to study art. They met in an art history class, and Laval-Jeantet convinced Mangin they should work together. "After all," she told him, "no scientists work alone." They went on a degree spree, obtaining degrees in anthropology, art, psychology, and biochemistry, and began working together as Art Orienté Objet. Their theme was to use their own bodies as objects of study in experiments on the forefront of biotechnology. Laval-Jeantet had done this as a child when she picked up a wasp, found it squishy, and out of curiosity bit it. Both participated in secret initiation rites of the Baka pygmies in Gabon, from which they suffered severe liver ailments for four years afterward.

Laval-Jeantet and Mangin think a great deal about the relationship between art and science. "Why restrict one's mind?" Laval-Jeantet practices what she preaches, having worked as an artist, an anthropologist, and a psychologist. They see scientists as trapped within disciplines, while an artist can offer a different way of looking at a particular piece of research. The pair recall working in a biochemistry lab where scientists tried to explain their work in a way that bordered on the dumbed-down. Each scientist thought his own explanation was the best. Finally they asked the scientists to draw pictures to show what they did. All drew drastically different images and were unable to explain their role in the research project in simple terms even to their own colleagues. It seemed they didn't really understand the conceptual basis of what they were doing. The artists' request for a visual representation had forced the scientists to look again at their work.

Nevertheless the two disagree with scientists who claim that collaborating with artists widens their thinking. "We know that's bullshit," says Mangin.

The pair see a role for the artist in our changing world. "Space for freedom is shrinking in our society and the artist has the responsibility to take this into account. The artist must know about science and be able to transmit scientific information to inform the populace. The artist is a go-between." They can't yet envision the long-term effects of their experiment. They "like the idea that it's never-ending" and could affect surgical methods which involve animal matter. Their horse blood transfusion work is a reminder that standard surgery often involves animal parts—pigs' hearts, for example.

Thus biology-influenced art pushes the boundaries of our everyday world and makes us look more closely at what is around us—while also fascinating us, shocking us, and occasionally making us laugh.

# 8

# Hearing as Seeing

U nlike visual art, sound art surrounds us and has the capacity to create an atmosphere—a world. Sound art ranges widely. A very early example is Yoko Ono's *Cough Piece*, "scored" in 1961 and recorded in 1963, consisting of thirty minutes of coughing varying in rhythm, intensity, and volume, complete with background sounds. At the other end of the spectrum is today's computer-augmented music.

Most people agree that sound art probably began in earnest with John Cage, many of whose scores consist of random sounds from random sources, such as his performance at the now legendary *9 Evenings: Theater and Engineering* in New York in 1966, not to mention his groundbreaking *4'33"*—4 minutes, 33 seconds of total silence, carefully divided into three movements of unequal length, at the end of each of which the pianist closes the piano lid and opens it again.

At Supersonix: Celebrate the Art and Science of Sound, an event in London in 2012, the sound artists' "conference papers" are their performances. "Sound art has the shortest tradition," says Angus Carlyle, a London-based composer, speaking into a microphone linked to a computer. There is a lengthy silence while translation software translates his words into French. Solemnly Carlyle listens to the French translation through earphones, then retranslates it word for word back into English. That's it. Aura Satz, another London-based sound artist, takes the stage. "I will be deliberately slow," she says, and indeed she is. She reads each sentence of her paper extremely slowly. Once again, silence is a key element.

In the question and answer session, the panelists insist on the connection between art practice and art research—that art *is* research. The chair, Salomé Voegelin, a London-based sound artist and writer, emphasizes the importance of the "independence of the sound image from the visual image," of engaging separately with sound as distinct from visual art.

The speakers banter around the term "aesthetics" with no clear definition. This is surprising, because a key issue in sound art is how to rethink the visual aesthetic toward evolving an aesthetics of sound, how to move from seeing to hearing. Sound performances sometimes involve a sensitive table on which performers place objects and move them around to make sounds, often accompanied by computer-generated images. Sometimes performers play musical instruments along with electronic devices. Sometimes sound can even be sculpted by otherworldly objects. The sound artist Katie Paterson translated the score of Beethoven's "Moonlight Sonata" into Morse code, then transmitted it onto the surface of the moon. When it was reflected back to earth, parts had been lost on the moon's rough surface, absorbed in its shadows and craters. She called this the "moon-altered" version and retranslated it into a new score with the gaps represented by intervals. In her performance the new score is played by a self-playing grand piano.

## Sculpting sound: Bernhard Leitner

"What does it mean if you really move sound in space?" asks Bernhard Leitner, a pioneer of sound art.

Leitner is one of those people who looks much the same as he must have done in the 1960s. His shoulder-length white hair, tucked behind his ears, enhances the look. For me, meeting him is a high point of Supersonix.

Born in Austria in 1938, Leitner was, to begin with, interested in classical music. In the late 1950s and early 1960s he studied architecture in Vienna, where he was swept up in the New Music by composers who were turning the music world upside down: Karlheinz

Stockhausen, Luigi Nono, Mauricio Kagel. To him, this New Music seemed to open up space. From Vienna, Leitner traveled to Paris, then attended the cutting-edge International Summer Courses for New Music in Darmstadt, Germany, in which tonal writing was virtually banned. France and Germany were the European poles of the New Music.

What appealed to Leitner was the way Stockhausen manipulated sound. His *Piano Pieces* (1952) epitomized the "reigning aesthetic of pulverization: sounds ricochet from the top to the bottom of the piano, as if the instrument were a pinball machine," as the music critic Alex Ross puts it. Then Stockhausen moved into electronic music, which further expanded the range of sound available to composers.

Leitner was intrigued by the notation for the New Music, the way the complex chord rotations were expressed graphically. Seeing this with the eye of an architect, he was drawn to their spatial characteristics.

Leitner pays homage to Cage, whom he sees as breaking down the concept of music by widening it, much as Duchamp did with art. The French-born composer Edgard Varèse, in his opinion, was even more important than Cage, in that he "re-envisioned sound itself," constructing installations for experiencing how sound moves through space. Varèse is celebrated as the father of electronic music. In addition, the Greek musician and music theorist Iannis Xenakis appealed to Leitner's architectural side, in that he attempted to sculpt sound. Xenakis made important contributions to the New Music by injecting into it concepts from mathematics, such as theories of randomness.

Finally, there was modern dance, which Leitner had always loved.

In 1968, Leitner was mulling over these three strands—architecture, which dealt with space, music, which dealt with time, and dance, which concerned movement in space and time—when he took a flight to the US. He had stopped thinking of this problem. "This subconscious detachment or liberation," he recalls, was the key. The passionate, intense desire to solve a problem keeps it alive in the unconscious. Then it came to him—"a spark-like confluence

of thought, leading to an idea whose content was a kind of manifesto: Yes, why not use sound as building material?"

"I could not have developed my work in Vienna because it was too
close, everyone on top of everyone else," Leitner recalls. "I needed
a city that is really pulsating"—a city where he could be anonymous
but that was also inspiring. New York was the answer. He settled into
the Lower East Side, "where most things happened." Among those
he met was Billy Klüver, and there was plenty of intellectual stimulation from hanging around with writers and artists connected with
the magazine *Artforum*. He soon found work as an urban designer in
the city planning office, and in 1972 New York University appointed
him head of the study program in urban design. This gave him the
financial security to afford spacious studios where he could pursue
his "theoretical investigations" toward the idea which he had come
upon on the plane. He needed a language to deal with "sound as
architectural material." In 1968, "no one talked about sound."

Art and architecture are spatial, while music is temporal. The
different disciplines had to be put together into a metaphorical
four-dimensional language which would not be static, as it is in
architecture. Such a language existed already for dance, and there
were also the graphic representations of the New Music.

The new language Leitner came up with was a representation of
sound with sound lines that crisscrossed, bunched up where sound
was more intense, and changed direction, whose characteristics
could be altered by using different materials, taking into account
his belief that sound is material and can be sculpted. In this way he
could plan sound spaces which could support different motions,
such as linear or spiraling or in the shape of an arch or a pendulum.

Between 1968 and 1970 Leitner produced a series of intri-
cate architectural-style sketches—meticulous drawings on graph
paper—for placing speakers in relation to an observer. Now he was
finally ready to start on his "empirical studies."

In these arrangements of speakers he found a "corporeality of listening. Ears are fine, but we have sensory instruments all over us.
On our skin there are 'acoustic cells.'" He realized that we hear dif-

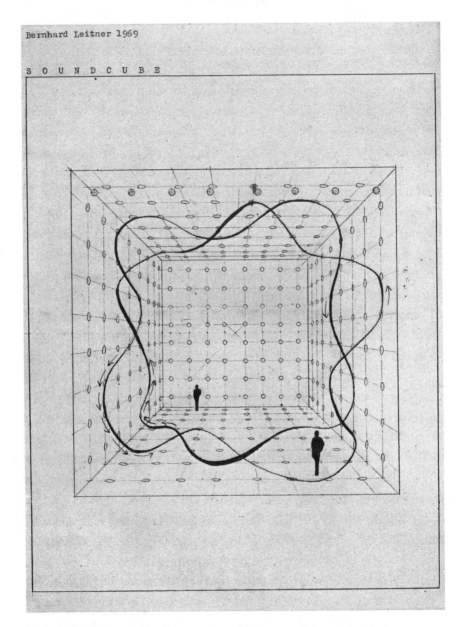

Bernhard Leitner 1969

S O U N D C U B E

8.1: Bernhard Leitner, *SOUND CUBE. Two sound spaces, in countermotion, intertwined*, 1969.

ferently depending on our constitutions—the amount of weight we carry affects the sound waves reaching our skeleton. "I hear with my knee better than with my calf." The soles of our feet are especially acoustically sensitive. In fact, we are made up of sound spaces, yet know little about these interior spaces.

In the early days of his empirical work, in 1971, Leitner had twenty speakers controlled by a voltage-changing device called a potentiometer, which altered the volume on each speaker. Two years later, he was using over forty speakers controlled by an electronic device driven by punched tape. The output came not from musical compositions but from sound recordings taken from the environment. His installations were huge, up to 130 by 130 feet. They filled his studio. Nowadays he uses computers to control the speakers.

In recent years Leitner has been working with haptic sound: sound you can touch. Using parabolic mirrors he can "bundle up" sound and bounce it off walls. Visitors can hear as well as sense the sounds moving up and down the walls, and even reach out and try to touch them. The visual part of the artwork is the speakers and the actual art is the sound. It's an extraordinary, immersive experience.

In 1975, Leitner designed the sound chair, a reclining chair with six speakers distributed across the back. The speakers are hooked up to make the sound move back and forth and up and down the body. "In the sound chair, sound doesn't leave the body. Sound spaces can go through the body." Leitner recalls that at an exhibition at MoMA's PS1 gallery in New York in 1979, Cage lay back in the chair and expressed amazement at feeling the sound moving over and within his body.

In the 1980s, doctors at the Bonn University Clinic tested the sound chair. They found that after sitting in it for twenty minutes, many preoperative patients were more relaxed, describing the experience as "a kind of holistic thinking." Perhaps they meant it was a kind of meditation. Some doctors noticed that patients who had received the sound chair therapy had a faster recovery rate than those who had not. "It was a medical research project that developed out of an artistic research project," Leitner says. It is indeed a very interesting example of science benefiting from art, or, as Leitner puts it, "medical research following an empirical aesthetic approach."

Leitner interprets aesthetics as a minimalist concept characterized by clean lines with clear-cut patterns, as in his meticulous drawings which set down sound lines and speaker positions. As for the intuitive side of aesthetics, it is the experience gained from experiment. For Leitner art is research.

He sees a difference between his own research and that of scientists, for whom the parameters are more clearly defined. In his case, he says, the unconscious plays a larger role. Constructing diagrams for speaker configurations requires a great deal of rational input— the measurements of the room, the electrical characteristics of the speakers. But irrationality chips in to the finished artwork as he feels out the best configuration of sound lines to produce a harmonious response in the body and the senses.

He does not consider that irrationality plays a large part in scientific discovery—an opinion with which I disagree. But we agree on the intriguing distinction in German between "to invent"— *erfinden*—and "to discover"—*finden*. People invent automobiles but they discover scientific theories, in that they are plucked from the cosmos.

In 1981 Leitner left New York for Berlin. He is now professor of fine art at the University of Applied Arts in Vienna. For forty years, he says, galleries took no interest in his work. He did, however, exhibit in museum shows and at international festivals like Documenta 7 in Kassel, the Venice Biennale, and Ars Electronica in Linz. Sometimes museums bought pieces. Fortunately, Leitner had a steady income as a university professor.

"The art market," he says, "goes through galleries which establish a certain kind of aesthetic evaluation." For many years, sound art was not shown by galleries because of the widespread bias against art associated with technology. Recently galleries have begun to take an interest in his work, though, he says, "White Cube and sound art just don't work. Collectors are not interested in something you can't put on the wall." One problem that galleries or buyers anticipate with art that involves science is durability. If it breaks down, can you call the artist in to fix it? Leitner points out that there have been enormous advances in technology, from two-channel tapes to CDs to chips in which the only moving parts are electrons. This means that durability is no longer a problem. People buy sound art to live with it. Leitner recently sold a large piece to a friend who put it in his courtyard. It's not only eye-catching but a conversation piece.

## Beyond music: Sam Auinger

Rather than creating sounds, Sam Auinger harvests them.

In 1991, Auinger was in Trajan's Forum in Rome with the American composer and sound artist Bruce Odland. The two often work together, signing their works O+A. As always, they were alive to their environment, listening to the sounds. Trajan's Forum was a hideous cacophony of cars, buses, horns, and sirens from passing traffic. The shape of the forum actually amplified these sounds, resulting in cognitive dissonance between what the two saw and what they heard. On an upper floor of the market they found piles of amphorae, large ceramic vases which were used as shipping containers in ancient Rome. Applying their adage—"We, O+A, listen to everything. Everything."—they lowered a microphone into an amphora and listened to the results through headphones, expecting to hear not very much. In fact it was a "hallelujah" moment. They had struck gold. What they found would affect their research for years to come.

Instead of the barrage of street noise, the sound they heard had been mysteriously processed so that it was "jaw-droppingly profound, deeply mystifying, and very real." Actually they had rediscovered a basic effect in physics. The amphora acted like a harmonic resonator, a vessel with an opening into a closed space which selects frequencies—sound vibrations from the sounds bombarding it—according to the size of the opening and of the vessel.

Auinger and Odland selected several amphorae with compatible harmonies, put microphones into each, amplified the sounds, and bounced the sound beam off the walls and ceiling. The soundscape in the forum was completely transformed. Not only that, but the mood of passersby seemed to brighten: they smiled and their steps became lighter. Auinger and Odland called their work of sound art *Traffic Mantra*.

I met Sam Auinger at Ars Electronica in 2011, where he was the featured artist. Born in Linz in 1956, as a child one of his favorite toys was an iron bar which he could strike to make different sounds. His father wanted him to go into business and in college he studied economics and mathematics, but his real interest was music. He learned to play several instruments, including the saxophone, guitar, and drums, and also sang and experimented with early synthesizers. In the 1970s, he was a member of a band and went on tour to San Francisco. "We were a big hit," he recalls nostalgically.

Back in Austria, Auinger studied computer music at the Mozarteum University in Salzburg, where he was taken on as a special student since he had no formal music background. To bring him up to speed, his tutors suggested he listen to music beginning with the Middle Ages and carrying on right up to the present. This crash course helped him to realize "how music changes have a lot to do with developments in science and society." Auinger was particularly taken with Renaissance music and the Baroque "besides the most famous Mr. Bach, Corelli," and others of that era.

When I ask him which modern-day musicians inspire him, he replies, "Morton Feldman." Feldman came of age in the heady days of the 1950s' New York art and music scene. A close friend of Cage, his compositions emphasized the worlds of sound rather than of music per se. His friendship with artists such as Robert Rauschenberg and Mark Rothko inspired him to write music which focuses the ear on resonant sounds, just as these artists emphasize texture and pigment, the physicality of a painting. "Morton Feldman could organize time in a way I never heard it before," Auinger says.

Besides Feldman, Auinger passionately admired Cage and "was completely in love with Edgard Varèse and Erik Satie," as well as the

Greek avant-garde composer Iannis Xenakis. Auinger also admires Miles Davis and John Coltrane, whose music he describes as "everything perfect, not one note too much."

While Auinger was at the Mozarteum, an instructor took the students to church and asked them where the perfect place was to put a clarinet—an instrument that creates a sound close to the human voice. The answer was obvious, although no one thought of it: the pulpit, where the priest speaks. Auinger realized that the "church becomes a sound box. This really got me thinking about sound events and architecture."

Meanwhile, he was attending events at Ars Electronica, where he finally heard Xenakis—"amazing"—and the wild American downtown guitarist Glenn Branca. By altering the speaker setup Branca produced an "unbelievably distorted barrage of sound." "Never heard sounds like that before," recalls Auinger. After a while Auinger began to pick melodies by the Beatles out of the distorted sound. It occurred to him how your "brain and the precondition of everything you have ever heard in your life is important in this whole context because by listening you can be connected with all you've ever heard."

Auinger had found the path he wanted to pursue: "to go deeper into sound than music."

In America, amphorae were in short supply, so Auinger and Bruce Odland found other objects to act as harmonic resonators. They settled for long lengths of metal tubing with a microphone strategically placed inside. This eliminates hard tones while magnifying others, and transformed discordant street noises into pleasing sounds, inducing an almost meditative state in listeners. It was like a giant didgeridoo played by the city itself.

Then in 2004, they set up five cube-shaped speakers in the World Financial Center Plaza, fed with sounds from tuning tubes of three different lengths. People could actually sit on the speakers, relax, and listen in real time to the sounds made by the Hudson River, at low and high tide, superimposed with all the sound of the harbor with its daily rituals of ferries and helicopters. O+A called this wide-open plaza a sonic commons. "Sound alters mood," and these sorts of installations "unite people to listen." In fact, people approached Auinger and thanked him. They had never before been aware of the tides in the Hudson.

"Is it possible to hear Bonn's history and architecture?" Auinger asks. In Bonn, with support from the city council, he set up an installation beside the railway station, in a rather dingy area full of harsh noises, threateningly empty and quiet at night. He called it *Grundklang Bonn*, which he translated as *Contemporary Soundscape of Bonn*. Once again, he used sound tubes to refine the ambient sound, such as the heavy traffic and other infrastructure sounds from railways, highways, and ships on the river Rhine. These refined sounds are played by a speaker in the cube at the foot of the main staircase of the plaza (center right of photograph). A second speaker is in the cube at the top of the left-hand set of stairs, just above. From time to time it played sounds from a water harp actually set in the Rhine, to remind passersby that the city is shaped by this river. "Spaces are reanimated through the energy of sound," he finds. It was not only the sound that was improved;

8.2: Sam Auinger, *Grundklang Bonn* (*Contemporary Soundscape of Bonn*), 2012.

the area also appeared softer to the eye. Like most sound artists, Auinger emphasizes the connection of eye and ear, whereas too often the eye has primacy.

Thanks to his work, there is now sound tourism in Bonn. People come to listen to his installations. Auinger advocates that instead of city architects building sound insulation walls beside busy highways, they should make sound receptacles to shape the soundscape.

As for aesthetics, Auinger mentions his "personal perceptions and struggle with form. What is the scope, what is the room for sound, what is the canvas on which I paint my painting? Where is the space in my canvas [for sculpting sound]?" The intangible essence of aesthetics is the "time and quality of research" he puts into a work. In planning a project, he "knows where to begin, but not where to end," a quality he shares with other creative researchers, especially scientists. The scientific component of Auinger's work is clear in his use of speakers and harmonic resonators and generally in his applications of acoustics. For his work he consults with architects, whom he looks upon as scientists, and, more recently, with neuroscientists, with whom he discusses how sound is perceived.

At the end of the interview, Auinger looks around the rather bland room we are sitting in—a classroom with a chalkboard, table, and chairs, and nothing on the walls, giving off a flat sound. "Completely uninspiring," he says. When someone gives a lecture, he continues, they should be able to arrange the room as regards light, temperature, and sound. "It's a question of aesthetics. How far should you go to have the conditions in which you can actually transform something? I can't transfer information in this room." For a face-to-face talk, however, our room suffices, concludes Auinger, ever alert to the visual and sonic possibilities of spaces.

## "We can close our eyes, but we cannot close our ears": David Toop

Besides collaborating with Brian Eno, Max Eastley, and Scanner, among others, David Toop has been described as one of the "three most influential British music writers of the last three or four decades" by the pseudonymous *Guardian* music critic Magotty Lamb, who goes on to write that the "prophetic undertow of land-mark volumes such as [Toop's books] *Ocean of Sound* and *Exotica* has become harder to resist with every passing year." Dressed entirely in black—black T-shirt, black jacket, and black trousers—Toop looks like the archetypal middle-aged artist, but he is anything but typical. He exudes calmness and depth and is at ease talking about his journey through music and literature, which many have found inspirational.

Born in 1949, he describes his art education, in which he quit two art schools, as "catastrophic." He attributes this restlessness to his "desire to cross boundaries—quite promiscuous, much more of that contemporary attitude that to be an artist you could work with any material in any environment." Music turned out to be "the right form—its routines and framework."

In 1971, Toop was interested in ecology and bionics—how bio-logical systems could be used in technological design, such as how a robot could be designed along the same lines as a cockroach with its many legs. Visiting Dartmoor with an artist friend, he hung a microphone on a fence and recorded himself playing the flute. Then he played back the tape, expecting to hear his flute. "But actually it was the flute operating in the sphere of the environ-ment. Actually, recording nature was the discovery." Toop's subse-quent music research flowed from that realization: that anything can produce sound and that that sound can transcend music. For this way of interpreting what happened that day on the moors, he was inspired by the quasi-scientific thinking of the iconoclastic British artist John Latham.

Latham was from the auto-destructive school of art, in company with Gustav Metzger and Jean Tinguely. He was famous for burning books and sometimes eating them. The concept behind his art was the "least event" that defines the "zero moment," from which every-thing else follows. One of his early works was produced by using a single burst from a spray can to pepper a white surface with black dots. Art reached its least event in 1951, he asserted, when Robert Rauschenberg displayed a plain white canvas on which the only form was the viewers' shadows. The following year Cage's *4'33"* premiered, and a few years later the anti-art, anti-commercial Fluxus movement sprang up with its happenings, in which the audience participated, making the work change with each performance. Fluxus can be described as Dada with a purpose, and in its heyday its ranks included Yoko Ono, Joseph Beuys, and Nam June Paik.

Toop's zero moment occurred on the moors. It was the origin of his timeline, defining where he stood in three-dimensional space. Realizing that "any device can make sound," he felt "decentered" from traditional music in both space and time.

"In 1971, there was no such thing as sound art, which was a good thing because it's a contentious term that makes no sense," Toop says, adding that he once "spoiled the day" at a panel discussion when he said this.

To open up his zero moment, his least event, Toop looked to Cage and to jazz and ethnic music from Africa, Indonesia, China, Japan, New Guinea, and South America. He was intrigued by the instruments of these cultures, especially the alien sound of the sacred flute from New Guinea. The flute is a long piece of bamboo open at both ends and played by blowing across a hole drilled near one end. The sounds it makes play havoc with temporal scales, harmony, melody, and pitch relations as well as sound, "which is what music is all about." There was a musical vocabulary for all these but, in Toop's view, "no coherent language for texture, the main feature of twentieth- and twenty-first century music." Texture in music is intuitive, the overall quality of a piece, a combination of all the above properties.

The sounds the New Guinea flute makes are "based on straight physics," he says. To understand the physics, he looked into the great nineteenth-century German physicist Hermann von Helmholtz's *On the Sensations of Tone*, where he found out about the Helmholtz resonator, also used by Sam Auinger. It makes an ethereal sound akin to that of the New Guinea flute. But texture still escaped him.

I mention to Toop the problem in quantum physics of trying to describe the ambiguous objects of the subatomic world, which are both wave and particle at the same time, using language from our daily world where objects are localized and tangible. Toop agrees with this analogy and adds that we need to get away from object-based thinking and localization. John Latham, he says, believed that it was all an illusion, that events just spring from a "zero moment."

Toop recalls that in the 1970s people interested in experimenting with music started to move away from the security of straight music and art. The ensuing movement was toward music as episodic and structureless, as a series of events. He mentions Wagner's huge cyclical operas, which mean more as pagan rituals than as staid opera. Toop wanted "to work as simply as possible in a complex world and produce instruments that can take you beyond the world of senses." "Now," he continues, he "works with the computer and bamboo." The high tech and the low tech, "a polarity of means, quite healthy."

"Working with a computer, I can create sound works not possible forty years ago," he says. But even in the age of computers, Toop is delighted that, to some extent, he can still work as he did years ago—as, for example, in his performances in which he uses a laptop together with dry leaves rustling on an amplifying surface.

"Technology is extremely important because it is the expression, or articulation, of the system," he says. Technology is the reason why the flute produces perfectly tempered scales, while also enabling the player to go beyond this, to work between the notes. "There are many wonderful examples of music that goes beyond that, such as Japanese shakuhachi music played on a shakuhachi flute made of bamboo, both ends open, with finger holes on its length and blown at the end. In contrast to most wind instruments, competent musicians can play at any pitch they choose." The New Guinea sacred flute, on the other hand, produces its variety of pitches by blowing across a hole. Both instruments operate by straightforward physics, which inspired Toop to join other musicians devising new instruments better suited to "sound making."

In 1975, along with Max Eastley, who works in kinetic sculpture, he produced an album, *New and Rediscovered Musical Instruments*. Eastley produced kinetic devices driven by wind or water or some other force of nature to produce variations of the sounds we hear around us. He called these instruments centriphones and hydraphones. The advent of laptops with the software to produce any sort of sound rendered these unnecessary. The laptop itself became a musical instrument of infinite variety.

"Electronic music," Toop says, "goes back to the sixties, to Stockhausen, for example. Now it's out of the box. There's no more programming. The computer pretty much acts like an instrument." In fact, he continues, "in live performances, performance programs allow you to improvise as you move along."

Toop uses a computer, electronics, musical instruments, a sound table, often images, and sometimes voice. The performances elicit multilayered responses—aural and imagistic as well as bodily responses to the rhythms. His pieces can be mournful dirges, evoking eerie images in the mind. *Flat Time* is a good introduction to his work. In it, Toop explains his musical techniques while playing.

8.3: David Toop, in performance using flute, laptop, and amplified percussion, 2012.

In *Flat Time*, he explores the connection between John Latham's work and music, a subject which Latham himself considered the best way to express his cosmology, his view of space and time. This, Toop says, can be explained in terms of a two-dimensional representation. The horizontal axis represents the structure of events—of "least events"—passing through every possible universe, while the vertical axis represents clock time. Occurrences in music are dots in this space that, when projected onto the horizontal axis, have a "zero moment," and on the vertical axis a time read off a clock.

Musicians live in the event-structured world, says Toop. Improvisation, he feels, is an excellent illustration of Latham's cosmology because of its lack of formal structure in space and time and its many potentialities. "Each moment has the potentiality to direct it anew," he says. There is a string of "least events," commencing with the opening notes. *Flat Time* is performed by an ensemble of violin, saxophone, and flute, accompanied by Toop on the computer as well as the flute and manipulating material on a sound table—crinkling cellophane, stroking a drum head with a piece of wood, and moving leaves back and forth—not to mention a zither-like stringed instrument amplified electrically and a phonograph record played DJ style.

The listener moves from here and now into the worlds of fantasy and mystery, as well as experiencing the music of very different cultures whose sounds possess an intriguing musicality. The musicians play off one another, sparking a series of "zero moments" in the time and space of the studio.

What he is doing, says Toop, is "putting sounds under a microscope and looking at the grains of the sound." Today, mixing sounds from different cultures as well as the immediate surroundings has become commonplace among electronic musicians. As the music critic Simon Reynolds wrote, "We are all David Toop now."

"Toop's writing always meant for me the freedom—as a writer—to actually leave aside Deleuze, or any theorist *du jour* as canonized frameworks of legitimization, and to explore instead unexpected, incorrect, and incidental references," writes Daniela Cascella, who researches sound art.

In the 1980s, Toop turned to writing and produced several key books, including *Rap Attack,* the first serious study of hip-hop, and in 1984, *Ocean of Sound,* a meditative, nonchronological, stream-of-consciousness study of the emergence of avant-garde music, from Debussy's discovery of Javanese music to the present. These are the books, as the music critic Maggoty Lamb wrote, "you'd be most likely to find on Bjork's bedside table." Indeed, the Icelandic singer has recently turned to science and technology for themes.

*Rap Attack* led to invitations for columns and features, and Toop plunged into journalism which for a while pretty much took over from music. Then, just after the publication of *Ocean of Sound,* his wife committed suicide. Shattered, he drastically altered his life, fed up with journalism and what to him seemed a "relentless cycle of compromises and triviality." He focused on sound and from then on wrote only the occasional book.

When I ask Toop about aesthetics, he replies, "Intuitional." He explains, "Intuition is when you're not thinking about what you're doing. At a certain level you don't know what you're doing. But at a deeper level you do." In other words, intuition comes with experience, which is essential to improvisation. Performing, Toop adds, "I am always thinking about aesthetics without knowing I am thinking about aesthetics."

"I have a split personality between critic, in which I am analytical, and musician, in which I am illogical," he says.

He gives a good example of the intuitive side of music: one does not play a note at a time but clumps of notes, neither does one recall a melody note by note. Music is the perception of organized wholes, or Gestalts. "The analytical faculty," he says, "can be detrimental to music."

Toop feels that it is difficult to open out the concept of the creative musician from Bach, Mozart, and Beethoven to include musicians today. The "methods of working have changed dramatically and have become so diverse." His own musical inspiration began in the seventies with the urge to move toward simplicity and away from the complexity around him. His creative development, he believes, was nurtured by advances in technology and by his writing, which forced him to become more analytical about his music.

We discuss how Toop views connections between art and science, on which he has an interesting take. Duchamp, he says, was a key figure, a "fantastic quasi-scientist." Surrealism "deeply fascinates" Toop, especially as he has always been intrigued by the unconscious. Among the artists he has studied are André Breton—with his automatic writing, supposedly springing from the subconscious—and André Masson, who championed automatic painting, where the hand paints as if possessed.

As with these artists, says Toop, the irrational, the unplanned, is what lies behind his music. I ask him how he composes and what sort of notation he uses. He replies:

> I don't use any notation in the conventional sense. My experience as a musician is founded in improvisation, where the term "composer" has the meaning of "bringing together." In other words, I bring together events through the medium of the computer and its software applications, or I use means such as texts, images, and directed improvisation to affect the conditions of performance. In both cases, improvisation is the key, so if a work involves other people, then the act of composition is collaborative.

To him, improvisation is a loosening up of the mind toward experimentation. The closest Toop comes to the rational is his notion of "directed improvisation," that "least event" or "zero moment" which directs the irrational unfolding of the music.

But, Toop cautions, "today demands proof, despite being surrounded by irrational behavior; there is a pull toward irrational behavior even though we live in a very rational age." He wonders "what a human being can produce spontaneously." He sees "a tendency to apply logic to everything, including art," reducing human behavior to the laws of physics and chemistry, which is indeed what today's neurocognitive scientists attempt.

Toop sees his own work as combining logic and illogic, science, anthropology, and the theory he uses in his books to explore the fantastic, the mysterious, and the unconscious.

## Nature's own music: Jo Thomas

In 2011, Jo Thomas spent six months at Diamond Light Source at the Harwell Science and Innovation Campus in Oxfordshire. Diamond Light Source is the UK's national synchrotron facility, a vast circular building like a mini-CERN housing a synchrotron that accelerates electrons to near the speed of light, generating brilliant beams of light which are used for academic and industrial research. Thomas, a sound artist, noticed something else: the tiny rhythmic sounds made by the machine as it monitored the electron beam. To capture them, she put a microphone in the synchrotron, then combined that recording with "ambient sounds such as the synchrotron machinery, people moving—in other words acoustic space, pitches

8.4: Jo Thomas, live performance of *Crystal Sounds of a Synchrotron Storage Ring, Diamond Light Source*, October 2011.

and dissonances, internal rhythms beating." The result was a thirty-eight-minute sound piece called *Crystal Sounds of a Synchrotron.*

The mesmerizing sound can lift the listener out of the here and now. The five tracks move from a hum to staccato sounds as the electrons are injected into the beam, creating extraordinary images in the mind. In the sometimes harsh, dry synchrotron sounds, Thomas says, she heard "so much of classical musical forms."

*Crystal Sounds of a Synchrotron Storage Ring* won the 2012 Golden Nica for digital music and sound art at Ars Electronica, which is where I met her. Thomas is a bubbly woman with an infectious laugh, contagious enthusiasm, and a great swirl of blond hair. As a child, she tells me, she studied the violin and the viola and learned to compose. She was interested in the potential of recording, too. One Christmas when she was a teenager, she was given a Yamaha synthesizer. She took it to her room and didn't emerge for days. "I was obsessed with the ways I could change sound with it." From that point on she was "not really interested in playing tunes but in how sounds change," as well as the technology of how sound works. To encourage her talent for sound art, her parents arranged a composition tutor for her. She went on to study at Bangor University in Wales which had "incredible sound studios. I was just in there, forever. I never came out."

I ask Thomas what composers or schools of music inspire her. A long hesitation follows while she anguishes over how to reply. Her answer is intriguing. She can't put her finger on any musician in particular, although she mentions Bernard Parmegiani, a highly experimental French composer known for his electronic music, along with Karlheinz Stockhausen, who got her interested in "how sound moved."

Both were born in the late 1920s. No more recent composer comes to mind. Then she confesses that what really interests her is "any music by anyone whatsoever, played on whatever, as long as the performance exhibits skill."

When I ask how she composes her work, she says that she "sees sound." She can "paint and draw music, a kind of synesthesia."

To create *Crystal Sounds of a Synchrotron Storage Ring*, Thomas worked with two scientists, Guenther Rehm and Michael Abbott, who acted as guides, showing her the audio output used to monitor the beam. She found that her take on sonic information was very different from theirs. One set of sounds which she found intriguing, the scientists considered "rubbish—just problems in the beam line," malfunctions in the machinery. Nevertheless, she decided to include everything in order to provide as complete as possible a sonic space.

Eventually each came to respect the other's viewpoint. "We always thought incredibly big. Nothing was ever turned down. This tells you something about science and art. There are no boundaries." She continues with great enthusiasm, "This is a universe and this is what can happen. It was so nice to think big. I always think big. It's nice to be accepted for thinking big." At Diamond Light Source, Thomas was in her element.

When she asked one of the scientists if he had learned anything new from their collaboration, he replied, "No," but added that it was a "nice union of ideas." The scientists involved, she says, felt "very creative when they worked with me."

She has an intriguing take on aesthetics which applies to process as well as finished product. At first she talks around the subject, saying that she associates aesthetics with being "interested in an exchange of ideas." Then skill enters the picture—how skillfully a piece is made. Then comes the artist's feeling for the work. "A work has a belief in itself and must be able to exist in its own right." Finally, the product must involve a "transfer of knowledge, that is, of beauty and skill." Ultimately, the aesthetic is to do with intuition.

## Sound and image: Paul Prudence

Paul Prudence looks every inch the avant-garde artist. He sits hunched over one or two computers, in jeans and hoodie topped off with a wool cap, driving sounds and images. I see him perform his magic in the vault of a deconsecrated church at Goldsmiths College in London. The audience is in heavy coats, breathing steam. As we watch, he conjures up on a screen universes of whirling, swirling patterns, evoking four-dimensional geometry and diagrams of four-dimensional space-time out of relativity theory, which spin and morph into quasars, black holes, and the Big Bang, all accompanied by booming sound.

Although totally modern, Prudence has a huge respect for his musical predecessors. He started out at the University of Manchester studying textiles, specializing in printed fabrics for couture fashion. "Always a bit of a surprise for people," he says with a wry smile. He soon realized that was not for him and went to Goldsmiths to do an MA in fine arts.

In fact, as he points out, there is a historical link between fabric design and computers. In the nineteenth century, the Jacquard loom used punched cards to control the weaving process, anticipating the role of punched cards on early IBM computers.

Prudence pays homage to his artistic predecessors, starting with the Swiss Adolf Wölfli and the Scottish Scottie Wilson. Both were early outsider artists who produced highly complex, fantastical designs using pen and ink. A bit later came Oskar Fischinger and

8.5: Paul Prudence performing *Cyclotone*, 2013.

John Whitney, pioneers in computer animation, Victor Vasarely, whose work had a strong element of Op Art, with emphasis on geometry and changing forms, and John Cage, a particular inspiration in his efforts to get rid of authorship. Cage's works continually evolve, each performance differing from the last.

Prudence soon discovered the Web and found that he was rather handy with computers. "I learned animation packages and started programming, using code and algorithms"—no easy task when you have a minimal technical background. Early on he sensed the "tension between how much randomness within a system can create something that is aesthetically acceptable." The element of randomness, determined by algorithms, allows him to move away from authorship. Once the performance starts, it is the system, not the author, that produces art.

Prudence often uses two computers linked together, one for sound, one for image. He works with one at a time, moving back and forth. "This creates an element of tension," he explains. "I can control the amount of randomness, I drive it, but it also has a mind of its own, in a way that creates a performance. I like to be surprised during the performance," he adds.

He is worried that too many digital artists are concerned exclusively with the technical aspects of their work, ignoring "old conventional ideas of story, of narrative—a beginning, a middle, and an end. Music needs to have a sense of drama." Audiences sometimes think that the video portion of his performance has been prepared beforehand, rather than being a real-time performance, generated before their eyes. "Quite belittling," he complains. This differs from the audience's shock at Cage's *4'33"*. Artists have "a bit of a conundrum," which is whether "at the start of the show you should announce this is not a video, or just let the audience experience the show." "Was it important to know that for many of Cage's open-ended works he had consulted the *I Ching*?" Cage wouldn't have cared, Prudence concludes, and Prudence doesn't either. The work can stand on its own and there's no harm in giving the audience a bit of a challenge, leaving some questions unanswered.

To program sounds, Prudence makes field recordings in different environments, including industrial areas and kitchens. Then he mixes them, bringing in music from diverse cultures. In this way, Prudence says, repeating the famous words of the music critic Simon Reynolds and with a nod to his eminent co-artist, "We are all David Toop." "This access to sound, and to software tools, gives rise to nested layers of infinite possibilities. [In this way] I try to create sounds you can see. The product of binding them together is greater than the sum of their parts."

Although "I don't interact with scientists all that much," Prudence continues, his work interests them because it is science-influenced. A fascinating example is the images that his algorithms create. These are generated dot by dot, with each dot used as input for another image or another aspect of the same image, using video feedback loops which mimic what mathematicians call recursive functions that can create fractal diagrams. Thus Prudence discovered that he can produce visual representations of complex mathematical functions using controlled signals. Using a spectrograph, he converts the sounds he produces into visual patterns of intriguing beauty. He has compared these patterns to emergent phenomena, referring to the creation of an ordered image from disordered origins. He has even lectured on this work to mathematicians. This is an example of an algorithm written for artistic purposes, yet capable of generating mathematically interesting results.

When I ask his opinion of the proliferation of disciplines cropping up—video art, digital art, computer art, data visualization, media art, and so on—he replies, "Media art, that's the catchall term these days. It's everything, it's a scary term and doesn't help." He adds, "I'm not a category person. Even the art/sci thing confuses me a little bit; I always thought that artists were a bit like scientists and vice versa. My community is very much based on the synthesis of art and science."

"Aesthetics is complex," he goes on. "It's not just about new media experience, but about much deeper modes of understanding and the history of cultures."

He shifts the discussion to "aesthetic sensibility," about which he is intentionally vague. "It's some sort of formal way of putting together sound and visual material that seduces you on a number of levels," he says. "It doesn't come out of nothing, it's based on experience."

## Hyper music, hyperinstruments: Tod Machover

Tod Machover radiates enthusiasm. His very presence seems to raise the temperature of his laboratory. With a broad smile and a halo of woolly dark hair, he seems possessed of endless energy. A dynamic speaker, his expression can change instantaneously from a smile to a frown when deep in thought, and just as suddenly back again.

"If I had all day every day, I'd just play Bach," says Machover. In fact, he spends his day composing avant-garde music and opera at the MIT Media Lab, sometimes in collaboration with Joe Paradiso, director of the Responsive Environments Group, and members of other groups. For this he uses advanced interface devices which employ algorithms to enable sensors to respond to gestures and generate the sounds of musical instruments.

Machover also invents musical instruments which he calls hyperinstruments. These use electronics to provide extra power and finesse for professional performers, enhancing the electronic scope of kcyboards, percussion instruments, and strings. He has even invented an electronic conductor's baton.

Machover began thinking about hyperinstruments after hearing the Beatles' *Sgt. Pepper's Lonely Hearts Club Band*. "It ideally balanced complexity and directness," he recalls. The problem was that music like that could only be created in a recording studio. He dreamed of inventing instruments that a musician could use to play such complex music anywhere, combining performance with the limitless creativity of a recording studio.

The first instrument he developed was the hypercello. In this, sensors on the player's wrist, along with a pressure sensor on the

8.6: Tod Machover, Playing the Hypercello, circa 1993.

bow which responds to his grip, are linked to a radio transmitter, which records the bow's movement over the strings. The finger-board is also wired with electronic strips. All this information is fed via an antenna on the cello's bridge to a Mac computer.

The hypercello had its debut at the music festival at Tanglewood, Massachusetts, on August 16, 1991, played by the virtuoso cellist Yo-Yo Ma. The piece Ma performed was composed by Machover, whom the critic Edward Rothstein describes in the *New York Times* as "a virtuoso at computer manipulation of sound." He writes of the complicated nature of the piece and the sounds emerging from Ma's instrument as a work "of extroverted eclecticism and sincere self-examination raised to an almost feverish electronic pitch."

Twenty years on, Machover has largely overhauled the original hypercello. He describes the resulting instrument as able to "morph the cello sound, fracture or congeal complex structures, and shape a magnificent lighting installation, all through subtle changes to cello gesture, touch, and timbre." Indeed.

Another of Machover's electronic musical instruments is the

hyperpiano. The piano itself is a Yamaha Disklavier, a baby grand equipped with solenoids and optical sensors. Besides storing data, it can receive input from a Musical Instrument Digital Interface (MIDI) as well as from compact discs. It can also be hooked up to a computer. Machover and his group put the "hyper" into the piano by adding a Mac minicomputer to do software processing and a parallel visual system that translates the music and the performer's movements into imagery.

At first glance it looks rather eerie. Besides the keys being played by the pianist, other keys go up and down for no visible reason. The hyperpiano also generates pre-programmed effects, making "a repertoire of colors, lines, and shapes that can morph in and out of textural focus," as Machover describes it. This "image choreography" responds to what is actually being performed live.

The musical notation for the hyperpiano includes an indication of where the computer enters the performance. When it does, it does not interfere with the pianist. The pianist can also alter the hyperinstrument's software by using an extra keyboard beside his left hand. "If you're immune to raw awesomeness, it might be time to listen to the tone clusters based on Gaussian distributions," Cooper Troxell, reporter-composer, writes in the Kickstarter blog, referring to Machover's compositions.

Machover's hyperpiano is a powerful example of the interplay between music and engineering. Not everyone finds Machover's compositions melodious but, as with Stravinsky, Schoenberg, and others, his day will come. As Machover puts it, "Visual art is easier to grasp than music. Consider, for example, Schoenberg. People complained it had no melody and could hurt your ear."

Troxell writes of Machover's opera *Death and the Powers*, "Tod Machover is one of the strangest composers alive today. While many are content to continue the grand classical tradition of string quartets and symphonies, and others have retreated into the world of purely digital instruments and programs, Mr. Machover has pioneered a unique intersection of electronic and acoustic music across a complex landscape of shifting idioms and textures. His

8.7: A scene from Tod Machover's opera *Death and the Powers*, 2011.

latest landmark achievement was the opera *Death and the Powers*, which tackles the role of humanity post-Singularity head-on. After the protagonist uploads his consciousness onto a computer, robots take to the stage performing alongside their human counterparts like it's no big deal. (Note: it is.)"

Besides electronics and music, the crossover between music, psychiatry, and medicine excites Machover. He envisages "music and medicine as a sort of prescription. Pick the right piece and it homes in on the sweet spot." Music, he suggests, can be used for early detection of Alzheimer's and to help Alzheimer's patients. "Music can tell you something about memory and can be used for pulling back memories." Machover's lab also provides ways for physically challenged people to make music, by using gestures or by blowing into tubes. An important concern of the MIT Media Lab is to help people overcome physical problems.

As technology and music come together, Machover sees a blurring of boundaries between different genres with, for example, elements of rap showing up in string quartets. All in all, there will be a greater participation in music, which could result in anything from chaos to mediocrity, as in the era when Bach came on the scene, looked around, and said, "I love this, I love that. I can knit them all together." "There are so many ideas out there now," Machover concludes. "A new language is needed to bring them all together."

## Making the computer improvise: Robert Rowe

From early on, Robert Rowe recalls, "Nothing interested me more than music." As a child, Rowe sang Handel's *Messiah* in a choir. He studied classical music and also played in rock bands. While a student at the University of Wisconsin, he came across a Moog synthesizer, which kick-started his interest in electronic music. In graduate school at the University of Iowa, Rowe began to work with computers—indirectly, however. In those early days of computers, he was rarely able to touch one, and instead had to submit his programs on punched cards and return the next day to get the results.

In 1979 he moved to the Institute of Sonology in Utrecht, Netherlands, and was finally able to have hands-on experience. To his good fortune, he was even able to work with one of Stockhausen's assistants. There Rowe learned to program in devilishly difficult assembly language, which works at the control level of the computer—its architecture, its guts. It's "an excellent way to learn programming because you really learn what a machine is," he recalls.

From Utrecht, Rowe moved via the Royal Conservatory in The Hague and the ASKO ensemble in Amsterdam to the Institute for Research and Acoustic/Music Coordination (IRCAM), beside the Pompidou Center in Paris. There he studied with Pierre Boulez. Equally comfortable with music and computers and with a strong bent toward experimentation, Boulez was the ideal mentor. Rowe used his skill with assembly language to experiment with IRCAM's powerful real-time synthesizer, the 4X machine, and began composing music for musical instruments and computer.

The problem was, Rowe recalls, that the "computer didn't know what it was doing. It was not taking into account any information from the player. If the player left the stage the computer would go on as programmed." Computer and player were not interacting. This was difficult, he knew, because performances took place "in a very demanding environment. Working onstage there was zero tolerance." He wanted to "make the computer musically aware—the computer should make decisions that depended on what the player was doing."

Thus began a research program that has taken up his life. Says Rowe, "my long-time interest is to have computers involved in real-time performances." After a ten-year stint in Europe, Rowe enrolled as a graduate student at the MIT Media Lab in 1987. There he was able to pursue his thoughts on interactive systems capable of linking computer and performer into a single unit.

At MIT he worked with Tod Machover, who had also spent time at IRCAM, as well as with Marvin Minsky, one of the pioneers of artificial intelligence (AI). Rowe's was the first PhD awarded by the MIT Media Lab in experimental music. In particular, he focused on music and cognition. He published his first book, *Interactive Music Systems: Machine Listening and Composing*, in 1992.

In person Rowe looks like a typical midwesterner, as indeed he is. The bookshelves in his office at New York University are filled with computer science books and he comes over like a scientist. But then I notice the piano in a corner and the music texts. In fact Rowe is now professor of experimental music at NYU, a highly interdisciplinary academic whose passion is to understand how our brains are structured to experience music.

Rowe walks over to a bookshelf, takes out a copy of *Interactive Music Systems*, and shows it to me. My response is, "This looks like a computer science text." "Indeed it does," he replies. "It's hard-core computer science." After all, he continues, considering the zero tolerance conditions of working onstage, any scheme that links computer and performer requires mathematical precision.

Traditionally there are two elements in music-making—the player and the instrument. Rowe adds a third, the computer which responds to the sounds and/or gestures (via sensors) of the performer, using an algorithm which is—amazingly—capable of improvising in specified keys so as to complement the performer's own "live" music. Rowe describes this intriguing interactive process thus:

> The computer is (at least in my music) making decisions itself as to what and when to play, but there are large-scale changes in behavior (state changes) that are triggered at certain points in the piece. I have done that automatically as well, where the computer is looking for certain cues that tell it when to change state, but often just advance through those manually, usually one every 30–90 seconds. Other composers play a more active role in changing parameters, or may do even less than me and just be watching the software to be sure it is operating correctly.

This way of working requires a change in musical scoring to indicate where the computer enters, with an algorithm for particular notes. For example, with the Yamaha Disklavier, the pianist depresses keys while other keys play in response to the computer. Sometimes the performer will take his left hand off the keyboard to manipulate a laptop to alter the algorithm, which at first sight looks rather surreal. Rowe's music is a remarkable give and take between performer and computer, each responding to the other in complementary ways. It is a giant step toward integrating human and computer, removing completely any trace of an interface.

"Are there fundamental rhythmic structures in the brain?" Rowe wonders. Rowe's interest in music and cognition and computer models of thinking has led him to explore whether we are born with structures hardwired into our brains which possess rules to process input information provided by our environment. This could be why we develop a preference for music from our own culture while finding music from other cultures less accessible.

Today so many people around the world are immersed in Western music that it is difficult to find these basic structures, these human universals. This question attracts the attention of scholars across a wide spectrum, including specialists in musicology, computational modeling, perception, cognition, neuroscience, and mathematics. Rowe insists that the tools are there in other disciplines. To avail himself of them, he works with Godfried Toussaint, a mathematician whose speciality is mathematical and computational aspects of rhythm, and Gottfried Schlaug, a neuroscientist. Both are also musicians, Toussaint a percussionist and Schlaug an organist. The three have written a series of papers putting together a mathematical model of rhythmic similarities among different cultures, searching for something in common among them. The neuroscientist's input includes analysis of the brain using $f$MRI. A key issue is how to program a computer to seek out basic structures in the brain that respond to music of a particular culture, says Rowe.

This seems a bit reductionist, bearing in mind previous attempts to simulate thinking using a computer. "Correct," he says. "But you have to start somewhere. These are classic AI problems," AI being a field of study in which he detects a resurgence of interest.

I bring up A. Michael Noll's computer program which produced Mondrian-like paintings that 59 percent of the public preferred to the real Mondrian. Rowe notes that this ought to be looked into, that there is something in Noll's computer-based art that must trigger an aesthetic response in the brain neurology of the observer.

A similar example is the work of David Cope, an American researcher looking into what AI can tell us about musical intelligence. Cope spent many years trying to model Bach's music. "It was a valiant attempt," Rowe says, in that "many people who had not heard Bach before couldn't tell the difference" between computer-generated music produced from Cope's model and real Bach. What AI does, Rowe continues, "is to reduce the distance between where you are and some sort of goal, e.g., aesthetics in a music program."

In his work Rowe collaborates with scientists with whom he feels a rapport as one of the new breed of artist, being very much a scientist/musician himself. Clearly scientist and artist alike benefit from this relationship. Rowe goes even further, insisting that there is "more of a payoff for the scientific side of the collaboration."

Rowe sees a great future for work such as his because "more and more people are thinking computationally about what they do." He passionately believes, however, that "truly creative thinking will always remain beyond the power of any machine." In his faculty profile he states his position thus:

> Music is and will always remain a fundamentally human endeavor, but technology has frequently had the power to significantly change and shape how we experience music, and I believe the process of trying to help a computer become musical enables us to better understand the true essence of the art form.

## Symphonic form in a digital world: Tristan Perich

Tristan Perich is a maestro of electronic music, though in the case of his compositions, sonic art can also have a strong classical input. Slight of build with long sideburns, he is somewhat intense but not manic, easy to converse with. He grew up surrounded by art created by his father, Anton Perich, a filmmaker, photographer, video artist, and pioneer of computer art, who was part of the art scene that revolved around Studio 54 and Max's Kansas City in New York.

Tristan focused on the piano, but from an early age he wanted to become familiar with a wide range of instruments including the cello, violin, guitar, and flute. He "wanted to see what they are and understand the mechanics of vibrating strings." He was also fascinated by science, inspired by Brian Greene's book on string theory, *The Elegant Universe*, and Douglas R. Hofstadter's classic on computer science and its relationship to music and mathematics, *Gödel, Escher, Bach*. Torn between computer science, mathematics, and physics, he went to Columbia University and studied them all. He even attended one of Greene's courses, where he was so carried away with enthusiasm that he suggested a collaboration with Greene on mathematics and music. There was no response.

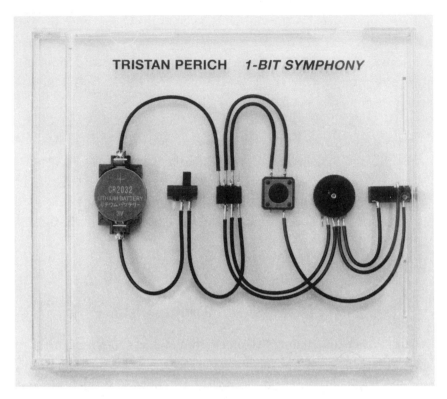

8.8: Tristan Perich, *1-Bit Symphony*, 2009.

Perich then enrolled in the Interactive Telecommunications Program at New York University, which focuses on imaginative ways to use communication technologies, particularly in artistic work.

"It is possible to create a rich sound composition with 1-bit audio, a 1-bit symphony," says Perich. In his electronic music he uses 1-bit circuitry, "bit" being a contraction of "binary digit," the minimum unit of information in electronics, which can take the values of zero and one, off and on. Circuits made up of 1-bit electronics go back at least to the early digital Casio watches. *1-Bit Music*, produced in 2004, consists of a single electronic circuit set in a CD case with a built-in headphone jack. Perich programmed the microchip, assembled the circuit, and marketed the whole package as a limited art edition. The rhythmic electronic riffs are pleasing, even soothing.

Five years later he produced the more ambitious *1-Bit Symphony*, a symphony in five movements, an electronic composition on a single microchip housed in a transparent CD case. You open the case, plug your headphones into the built-in headphone jack, and the music, which is written in code on the chip, begins to play. Strangely enough, it sounds rather grandly symphonic. "With electronic sound I can create any sound imaginable," he says.

Why one bit of information, the smallest amount possible? As a student, Perich had studied the history of mathematics and science, and was particularly intrigued by early attempts to reduce mathematics to the fewest statements couched in numbers, the atoms of mathematics. In quantum physics he saw a similar minimalism in the atoms of nature. In music he was inspired by the minimalism of Philip Glass. "Just two notes oscillating back and forth, that's the musical material." He also admires the work of Steve Reich. He speaks of his "phasing process. Things go in and out of phase; takes ten minutes, that's all you've been listening to."

He expands: "You can see the aesthetic from afar, but then you can also see how it's built up. I find that inspiring, how music applies to the same part of the brain as science. All similar stuff, music, art, mathematics—dependent on basic building blocks." All this is embedded in Perich's 1-bit music, built up of information atoms. "Simple stuff is the most pristinely beautiful stuff," he says dreamily. This is also why Bach's music so inspires him—it's clean, pure, and sublime.

For a while Perich turned away from electronics and began composing in the classical style. "Electronic music seemed to have nothing tangible to it, no characteristics of, say, the piano," he recalls. Then he began to think about putting actual instruments together with electronics.

By electronic instruments he had in mind speakers. After much experimentation, he worked out a unique way to do this. He connected the speakers to a circuit board with a single chip which sent electrical signals, making the membranes vibrate. In other words, he found a way to convert electricity into sound. He also composes the instrumental music. His understanding of the basic physics and mathematics involved in vibrating surfaces and also of computer science enabled him to do all this himself.

He now composes for speakers together with pianos, clarinets, flutes, and percussion. He programs the electronics driving the speakers to work in tandem with the music, so that what he encodes in the circuit boards correlates to what the musicians play. The results are more like music than random sounds, in that they can be melodic and the circuit input is not just rhythmic accompaniment.

Perich tells me that his compositions spring from improvisation, the mind at play—usually at the piano, which is his main instrument. He contrasts this with other composers who use algorithms, which introduce complications. "There's a difference between process being part of the inspiration or the tool set that you have, and process being a determinant." He prefers the former.

Perich thinks out each stage of his work. Although the code he writes for the chips is logic, he writes it "with music in mind." The scientific input, however, ends with the programming. The music itself is intuitive.

# The Art of Visualizing Data

Scientists have always looked for patterns in data, reflecting patterns in nature. In 1913, the American astrophysicist Henry Norris Russell found himself in a quandary. The Royal Astronomical Society in London had invited him to discuss the properties of three hundred stars in the night sky in just fifteen minutes. He quickly concluded that what he needed was a visual representation. He settled on a graph, plotting each star's temperature along one axis and its brightness along the other.

To his surprise, many of the stars congregated along a diagonal band. He recognized immediately that this representation actually chronicled the life and death of stars, in that each star grew hotter and brighter, then dimmer, as it grew older. For historical reasons this famous representation became known as the Hertzsprung–Russell diagram. It plays a critical role in our understanding of the universe and of why we are here.

Data are collections of numbers taken from surveys or collected in the laboratory or even collected oneself, such as how many times you went to the movies last year. Such information can be more than dry numbers. It can hold deep secrets about the world or nature or one's personal habits. The question is, how to draw this information out of the raw data. The visual display of data—to make it more accessible and, even more, to draw out its deeper significance, as Russell did—is an issue in many fields, business and psychology as well as science. In the past, methods for describing data were usually straightforward, with no particular need for an eye-catching aesthetic or representation.

One notable exception is Charles Joseph Minard's representation of data charting the fate of Napoleon's army as it advanced into Russia in 1812 and then retreated. Drawn in 1861, Minard's representation combines text with drawings, showing dates, battles, the number of men lost, the temperature, the direction of the army's movement, and the distances covered. It is effectively six-dimensional. The pioneer of data visualization, Edward R. Tufte, wrote of it, "It may be the best statistical graphic ever drawn."

Tufte was writing in 1983. Up until then, no one had taken the time to survey the field of data representation, put together ways to accomplish this, or critiqued them. The topic was considered so humdrum that Tufte had to second-mortgage his home in order to self-publish his book *The Visual Display of Quantitative Information*, based on a course he gave at Princeton University. The book subsequently became a bestseller.

Today information is everywhere. Google, Amazon, and Facebook use your website and book preferences and your age and interests to build your profile and offer you products. Subjective responses to events such as the Olympics carried in texts and tweets can be put together to communicate the general mood. All this is possible using increasingly sophisticated software to analyze and represent data.

What is sought in data mining is not a cause-and-effect relationship between data exchanges, such as my emails with a friend, but correlations—patterns. For example, should I or my friend decide to communicate with a terrorist website, red lights go off at the National Security Agency, and the patterns of both of our electronic communications will be investigated. The NSA cannot tap into each of the trillions of phone calls and SMS messages sent every day. Patterns are what they are looking for; phone taps follow.

In the twenty-first century, data has taken on another dimension—as a medium for art. In mining data in a way that is aesthetic or beautiful, perhaps a deeper meaning can emerge.

## Nanoart: Mike Phillips

Mike Phillips works in nanoart, as part of a research project with artistic aims, exploring how technology can help us become aware of the world beyond our senses. He is concerned not with the very large but with the very small. Like many cutting-edge artists, he deals with the fact that what we see is not necessarily what is "out there."

To him the world is a "landscape of data" and his task is to visualize that data. One of his projects is entitled *A Mote it is to trouble the mind's eye,* the words spoken by Horatio in *Hamlet* after he and his comrades have seen the ghost of Hamlet's father. Although they have seen the ghost, not everyone will believe them. Was the ghost real, or did they all imagine it?

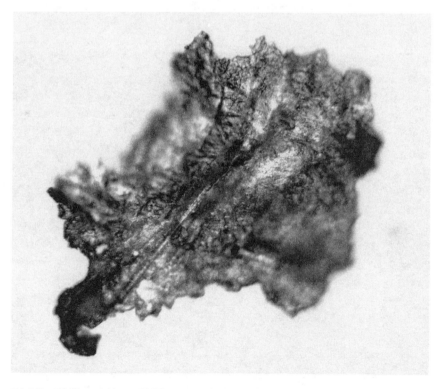

9.1: Mike Phillips, *A Mote*, 2009.

Phillips begins with a mote, a speck of grit, which he actually found in the corner of his eye and removed with tweezers. It's almost impossible to see, but then he magnifies it many times over to reveal that it is ragged and twisted, with a distorted surface, like a piece of shrapnel. Then he uses an enormously powerful atomic force microscope (AFM) to magnify it still further. The AFM reveals the landscape of its atomic structure, a world beyond sight, knowable only by data (numbers), processed with software to produce images. Phillips chose a blurred image to suggest the invisible world of atoms which actually make up the mote: the "ghost" of the mote is the "landscape of data," be it from an optical microscope (data measured through our senses—sense data) or the atomic realm (accessible from data measured in numbers).

The term "mote" runs deep into history. It comes from Middle English, with Indo-European roots and Christian Masonic over-

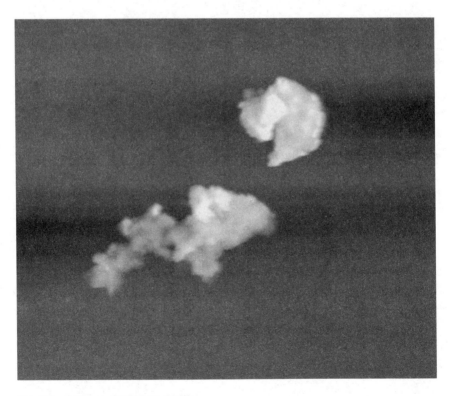

9.2: Mike Phillips, *AFM Mote*, 2009.

tones, and represents the tiniest element possible, which nevertheless gives existence to things. That it has this power and yet is
incredibly small "makes it a powerful talisman of nanotechnology."

With his close-cropped salt-and-pepper hair and black T-shirt,
Mike Phillips fits the deliberately dressed-down image of the artist. He studied at the University of Exeter, the University of Massachusetts, and the Slade School of Art at University College London
(UCL), where he did postgraduate work. When I ask him why he
went into art, he replies that he was good at physics, bad at mathematics, and excellent at pattern-matching. "Then the computer
came along and offered me the vehicle to create real situations."

There were other reasons, too. Widely read in the fledgling area
of technological art, he had run across Robert Rauschenberg, Jean
Tinguely, Billy Klüver, 9 Evenings: Theater and Engineering, and
E.A.T.—"heroes of mine." He was excited by the collaborations, the
synergy of it all. He feels strongly that "digital art today makes the
mistake of ignoring" these early pioneers. It's typical of developing
fields in science and technology to cut themselves off from their origins, akin to trying to understand the UK or the USA today while
ignoring their roots.

While at Exeter, Phillips took an interdisciplinary course
called 4D, named not after the fourth dimension but the number of the classroom where it was taught. Part of the course concerned computers, and just the right machine happened to be
available: the ZX81, which was very popular at the time. Besides
being inexpensive, it was relatively easy to assemble and program, perfect for the beginner. Phillips's lifelong love affair with
computers had begun. "I sometimes dreamed in BASIC" computer code, he says.

By the time he got to the Slade, it was phasing out its work in
computer arts. He spent as much time as he could in the physics
and computer science departments at UCL. He even played around
with interactive programs and light shows for nightclubs. He also
met Roy Ascott and became involved in the pre–World Wide Web
network of artists through UCL's then infamous email program,
EUCLID. After taking a higher diploma in fine art in mixed media at

the Slade, Phillips was offered a position at the University of Plymouth as professor of interdisciplinary arts, a title he had to negotiate
for fiercely to ensure the interdisciplinarity he wanted.

In 1992, Phillips set up a course at Plymouth called media arts,
based primarily in the computing department. He began spending more and more time there and recalls that "in the computing
department I talked art more than in the art college." Currently,
after several years of negotiation, his work is more fully integrated
into the art department.

Presently Phillips's main interest is in nanoart, dealing with the
unimaginably small, at scales a mere ten times the size of an atom.
At the nano scale scientists have been developing new materials, a
field which has become known as nanotechnology. One spectacular
result was the production of graphene, an exceedingly strong material made up of ultrathin graphite layers, engineered on the nano
level. Nanotechnology may also play an important role in detect-
ing cancer and as a vehicle for drug delivery systems, in addition to
opening new vistas in electronics.

Nanoart began with scientists who exhibited some of the strange
structures revealed by the atomic force microscope (AFM), an
instrument with a rounded silicon tip approximately one nano-
meter across (0.000000001 meters). The AFM can scan an apparently smooth surface at the nano level, producing a topographical
map much like a mountain range as seen from outer space. These
maps are as beautiful as a Mandelbrot set—the collection of points
whose boundary is the familiar fractal pattern—or the images of
stars from the Hubble Space Telescope. They are so strange that
to the layperson they seem like nature's art. Soon artists began
to manipulate these images and a new form of art emerged, connected with technology.

An early example of nanoart was *Nanobama*, created in 2008 by
John Hart, a professor of mechanical engineering at the University
of Michigan. Using nanotechnology, he inscribed an image of President Obama on 150 million carbon microtubules—cylindrical structures made of carbon atoms, tens of thousands of times thinner than

a human hair yet with the strength of steel. He produced the image of Obama's face by scanning the microtubules with an electron scanning microscope. Hart intended this as a political statement of Obama's high regard for science, as opposed to that of his opponent for president, John McCain. It was a straightforward use of nanotechnology by a scientist.

In 2000, Phillips participated in the search for terrestrial intelligence as part of his work as director of i-DAT, a branch of the Centre for Media, Art and Design Research at Plymouth University. The Wellcome Trust, the Gulbenkian Foundation, and the Arts Council supported this research. Whimsical though it sounds, it was actually a serious exercise in pattern recognition. The question was whether, from scientific data taken from space telescopes, patterns could be recognized indicating life on our planet. Phillips's group wrote algorithms for perusing—"flying over," as he likes to put it—a database of over 3,000 photographs. This was, of course, before Google Earth.

There was also a citizens' science angle, in that all the photographs were on the Web and could be perused, and input was encouraged. The verdict was, "Not conclusive." It would be difficult for aliens, scanning our universe from afar, to conclude that there was definitely intelligent life on earth. Similarly, it would be equally difficult for us to identify intelligent life on other planets using the space telescopes at our disposal.

Phillips believes that the scientists he has worked with now see data differently as a result of his input. "There are not enough collaborations," he laments. Moreover, such relationships are fraught with difficulty, and he has noticed that "a lot of collaborations have not borne fruit." It is difficult for artists and scientists to meet. "Speed dating," where scientists and artists pair off for a few minutes to decide if they will get on, usually doesn't work. Either the partnership "doesn't root properly" or is short-term due to time constraints, usually the scientist's.

When Phillips took up his position at Plymouth, he found a glass wall separating him from the more traditional artists, who chafed at the idea of introducing computers into the curriculum. "I always liked to operate outside of art per se; but operating in that way seemed threatening to traditional art." Phillips has always tried to steer clear of the art world. When I ask if he sells his work, he replies that he isn't dependent on that because he is a salaried academic and occasionally receives commissions and grants.

"I've always resisted galleries because I thought they were part of the problem," he says. "But at a certain age, you go for the easy life, embracing the gallery, betraying your belief system." Recently Phillips has designed installations specifically for galleries. His mote is displayed in the John Curtin Gallery at Curtin University, Perth. Visitors have responded positively. "They engaged with it."

When Phillips's involvement with computers first began as a young man, he probably never imagined he would spend his whole life working with them or what advances in technology there would be. "Whenever I fly over data fields, I relive that first dream in BASIC," he says with a smile.

## Experimenting with evolution: William Latham

"When scientists have data, sometimes they are constrained by it," says William Latham. "But not artists." Latham's spectacular 3D mutating computer graphics have been the basis of games, have been featured in movies, and have been exhibited worldwide. Since 2007, he has been a professor at Goldsmiths College in London, teaching a master's course in creating and programming computer games. He is certainly not constrained in any way.

At first glance, Latham's office at Goldsmiths is more scientist's den than artist's studio. Books on computing codes and reprints of articles written for scientific journals are strewn around, including a copy of *Evolutionary Art and Computers*, the book he coauthored with the mathematician and computer graphics expert Stephen Todd. Brimming with computer code and computer-generated scientific drawings, it "brings together the worlds of modern art, geometry, and computer graphics, employing the underlying themes of natural evolution and artificial life," according to its Amazon description—an extraordinarily ambitious brief.

Latham cuts a dramatic figure. With piercing eyes and sharp features accentuated by long, tapering sideburns, he sports a carefully coordinated geometrically-patterned shirt and tie. His father was a chemist at Harwell Nuclear Research Centre, his mother a composer and musician, so he had a "mixture of art and science from an early age."

At Oxford he studied fine art and painted his way through the twentieth century. He took a particular interest in Russian Constructivism and appreciated the scientific element in it, particularly the focus on chance, which he also saw in the work of the British artist Kenneth Martin. There he also discovered computers and realized their potential to create art, but was frustrated at first by the difficulties of programming and the "apparent mismatch between how artists think and the implicit constraints of software."

He moved to the Royal College of Art and frequently visited the nearby Natural History Museum where his interest was piqued by the theory of evolution. Could you apply this theory to abstract shapes? he wondered. What other forms might have developed? He worked out a system using rule-based art, where the artist selects objects, then lays down procedures for assembling them. "With just a simple set of rules, it was like a Pandora's box. The wealth of possibilities was endless."

He thought of two forms working their way up the evolutionary tree and imagined what might result if they came together. At first he used forms evolving out of geometrical shapes—cubes, tetrahedrons, icosahedrons, and funnel shapes. He drew up a set of rules for reshaping them and bringing lines together, making them bulge, twist, scoop, grow tendrils, stack, merge, or clone. He included deformations decided on beforehand, such as adding a bulging nose-like appendage. He even "married" certain forms to produce more complex forms that still bore a resemblance to their "parents." All these forms were abstract; they did not occur in nature. In this evolutionary world, he was God. He decided which were the fittest on the basis of aesthetics—which he would allow to survive—and discarded the rest.

Latham graduated from the Royal College of Art in 1987. On a whim, he phoned IBM and asked for some sort of sponsorship. To his amazement, they offered him a post as a research fellow in the computer graphics unit at their UK research center in Winchester, to do whatever he wanted. At the time IBM financed research centers akin to the Bell Telephone Laboratories in Murray Hill, New Jersey.

There Latham met Stephen Todd, a computer graphics expert. The two set about expanding Latham's evolutionary drawings. Their goal was to derive a new form of art from the evolutionary process.

Latham's evolutionary drawings were "brainstorming, plagiarizing the natural world." Todd applied science to this process, setting the evolutionary process into code for a computer. The next step was to discuss whether the resulting shapes were what Latham had had in mind.

They developed a powerful algorithm, Mutator, which contained the element of randomness, enhancing Latham's rules for his original evolutionary drawings. With computer graphics, they were no longer confined to the original geometrical shapes but could generate a cornucopia of fantastic forms seemingly from an alien world, in a spectacular array of colors. They also built in a means to ratchet randomness up and down, and included dynamic elements such as going to sleep and waking up, letting the mutations turn themselves on and off. "The artist became like a gardener," as Latham puts it.

The resulting forms unfolded in beautiful and strange ways, fantastic organisms from "possible worlds." "Some artists," Latham and Todd wrote, "feel that [Mutator] provides a genuinely new way of working, and it has certainly led to the creation of forms that would not have been created by other methods." Nor, they might have added, could they have been imagined either.

The economic depression of the early 1990s forced IBM to shut

9.3: William Latham, *Mutation X Raytraced. Bump mapped/Blue*, 1993.

down their research centers. By then, Latham had developed an interest in computer games and had to make a choice between academia and the world of business. His art wasn't selling, the core of the problem being, as we've seen before, that "the art world was never very comfortable with computer art." Finding the art world rather dull, Latham opted for business. For fourteen years he successfully developed computer games, and for ten of those years was CEO of his own company. To his amusement, he found that his evolutionary films showing otherworldly organisms unfolding, created for artistic–scientific audiences, were popular at "drug-fueled raves."

Nevertheless he still wanted to make a difference in the art world.

In 2007, he became professor of computing at Goldsmiths, while retaining his games business. He also resumed his collaboration with Stephen Todd. They have streamlined their original systems, including Mutator, and are currently working on scientific visualization in protein folding with the bioinformatics department at Imperial College. Bioinformatics is the study of methods for storing, retrieving, and analyzing biological data, particularly in sequences of proteins in DNA. By interacting with each other, the chains of amino acids that form proteins fold to form the organs in our bodies. Unfolded proteins can be toxic, while defective protein foldings can produce degenerative diseases.

After a year, Latham and Todd succeeded in linking Mutator with codes used by the Imperial College scientists. They called the resulting code FoldGrow. Using scientific data, FoldGrow can create real protein structures as well as foldings not found in nature. Once again, Latham played the role of aesthetic filter, using the guidelines of "symmetry, elegance, and balance." He describes this as "tech-inspired art." The foldings FoldGrow produced were amazing, beautiful, and mesmerizing. "They looked organic but were completely artificial—they might, however, be on an alien planet," says Latham, who also on occasion writes science fiction.

As the artist of the team, Latham had more freedom to use scientific data to produce new protein foldings. Scientists, he says, have to beware of "peer disdain." "Scientists don't want to be seen as doing something absolutely mad. But if you call it art, then it's not mad!" While scientists have to take on board ethics and morals when they use data, for the artist anything goes.

"What I'm trying to do is to change the definition of art," Latham says. "Art is running out of track. It's time to go back to a Renaissance collaboration of artificial intelligence, geometry, and math [with art]. Otherwise, artists are just doing the same thing over and over again. Another white painting, perhaps slightly whiter, on a slightly heavier canvas. Meanwhile, the middle classes are just discovering art." When you bring in computers, "then art becomes a research project." This is why he returned to academia and continues to work there, in the computer science department. "The most stimulating debates," he adds, "are with scientists. It's better than working with other artists."

When I ask whether he considers himself an artist, a scientist, or some combination thereof, he replies, "75 percent artist, 25 percent scientist."

## Visualizing data through sculpture: Erik Guzman

For three months, between October and December 2010, Erik Guzman's *Weather Beacon* lit up the Winter Garden at the World Financial Center in New York City (see Insert). Shaped like an elegant assembly of large cogs, it occupied an eight-sided transparent box 16 feet long, 8 feet high, and 12 feet across. Its averred purpose was to beam out the weather forecast in a code of lights, using highly polished plates driven by weather report data input from the Internet. The plates, differently shaped but each distinctly human-looking, rotated as well as emitting flashing lights ranging from white—cold—to red—hot. When the light pulsed, it meant rain was on the way. Says Guzman, "It took three years' real hard labor to plan and assemble!"

Erik Guzman inherited his curiosity from his grandfather, Fletcher Hanks, "a weird maverick inventor" who built a hydraulic digger to dredge up soft-shelled clams from the coast of Maryland, organized and participated in marathons and triathlons, and, as a pilot in World War II, made countless hair-raising trips "over the hump" in Burma. His advice to his grandson when he reached eighteen was to learn how to weld, scuba dive, and fly a plane. "Got the whole thing done," Guzman says proudly.

Tall, lanky, and relaxed, Guzman studied sculpture at the School of Visual Arts, then went on a tour of Japan which "completely changed everything." He was struck by the delicate scale of the architecture, the joinery, the anime films, and by Buddhism and Shinto. Back at the School of Visual Arts he took a master's degree in sculpture and learned all he could about computers. He gradually developed an interest in digital art and, more specifically, in combining kinetic sculpture with data visualization.

Guzman's kinetic sculptures are made of intricately machined, interconnected parts, a bit like the inside of a clock, and can be very large. As an undergraduate, he had the experience of building by hand, before digital methods became available. These days students start off immediately using digitally driven tools. "Very few artists understand the nuts and bolts" of what goes into producing a piece of sculpture, he contends. They have no appreciation of industry or the hands-on credo.

Guzman is inspired by architecture, especially in the way it deals with "humans reacting to spaces." He admires the British sculptor Anish Kapoor, who has an architectural background. In computer art, he mentions the American digital artist and painter Jeremy Blake, whose animations were based on big fields of color. Of Jean Tinguely he says, "Anyone who sets the MoMA on fire is okay with me. Kinetics is a really scary world." As for Alexander Calder, famous for his mobiles, "he really understood weight, gravity, and movement; there are parts in [his work] that are hard to explain."

Guzman is fascinated by what he calls "digital as material, [the fact] that zero and one can carry the same weight as a piece of wood" in setting a kinetic sculpture into motion. This, he adds, is the link between macro and micro. He constantly thinks of unification in his works—of zero and one catalyzing the flow of electrons that turn metal wheels, as in *Weather Beacon*.

*Weather Beacon* was Guzman's first public piece and he was delighted to be able to display it at the World Financial Center, where there are always people passing through. People liked *Weather Beacon* and expressed sadness when it was taken down in a special three-day event. One woman demanded, "What are you doing with my weather beacon?" He sees *Weather Beacon* as a data visualizer, "big enough that people can see it." "Beyond apps, computers, iPads, and all that technology, there is a real need to have something tactile; visual tactility triggers tactile things in the brain." He is keen to "educate the public as to what art is."

When I ask Guzman whether he would call himself a data visualization artist, he replies with a chuckle, "It's like the space race. Everyone is trying to coin the right terminology which will enter the canon of art per se." Guzman is critical of categorization. He sees it as frivolous, adding sarcastically, "I wish I was sharp enough to figure out one of those lines. I think I'd be much further ahead."

As for funding, Guzman points out that the federal government is trimming expenditures and has virtually cut off funding for the arts. Large high-tech companies such as Norton, Microsoft, and Google are beginning to step in and offer increasing amounts of funds. This could be to the "advantage of a culture" in which science, art, and technology are considered as one and the same, he says. Fortunately, some galleries in New York's Chelsea have expressed interest in projects that combine art with science and technology. This is encouraging, Guzman says, because the private sector often looks to "Chelsea for signals."

He sees a problem in that "most galleries go for flat paintings," for which there is definitely a market. An artist approaches a gallery with an idea based on a combination of art and technology. As he describes it, "The gallery says, 'Yeah, make this really cool thing, then come back to us and we'll put you in a show.' Then when you come back they say, 'No, this is a little too much because we need insurance for the space and so, no, we can't make this happen.'" The major galleries have a yearly overhead of about one million dollars—to cover maintenance, payroll, rent, and insurance, among other expenses—so why deal with something that may not be sellable? Galleries won't buy the idea if they can't sell the product.

## Turning data into sculptures: Jonas Loh

Jonas Loh has found a way to bring data to life by expressing it in the form of extraordinary sculptures. Loh describes himself as a designer and inventor who searches for "new aesthetic forms while also researching how to communicate them." For his undergraduate thesis in 2009, he and Steffen Fiedler, a fellow student, recorded the digital identities of eight people, using data from Amazon and Twitter and including their interests and listening habits. Using data visualization software they sifted out the similarities, aiming to produce a snapshot generated by the connections in this complex of digital identities.

They produced the final shape, or Gestalt—the essence or shape of an entity's complete form—by rapid prototyping, a way of making a three-dimensional computer-generated form. The result is an extraordinary extruded sculpture that looks like a fossil or a seashell or some sort of alien life form, in line with Loh's interest in synthetic biology, a theme in the Design Interactions Programme at the Royal College of Art. Loh and Fiedler entitled their thesis project *The "Gestalt" of Digital Identity*.

Loh, who is German, sports a shaved head and immaculately trimmed beard and speaks with absolute conciseness. He studied at the University of Applied Science in Potsdam, where he focused on data visualization, information visualization, and generative design. He then joined the RCA's Design Interactions Programme, where he became interested in science through conversations he had with scientists at Imperial College. He began exploring complex data sets and ways to visualize them and communicate their meaning, as well as studying their ethical implications.

His key inspiration was not a person or an art movement but a software tool, open-source software developed by Ben Fry at the MIT Media Lab. This software is available free of charge and can be modified for one's own use. It "made programming very easy for artists and engineers."

What you need to begin with, says Loh, is an "interesting data set that communicates something," such as, for example, the digital identities of the eight people he used to create *The "Gestalt" of Digital Identity*. First he applies data visualization software, such as a computer graphics algorithm called Marching Cubes which looks for eight neighboring points and forms a cube, then goes on to generate more complicated polygons. With luck, interesting patterns emerge. If not, Loh writes new, more complex software to represent the data. This is the point, he continues, where data visualization gets really interesting, and something brand new emerges. "What is produced is art."

During the 2012 London Olympics, Loh and his colleagues captured the day-to-day excitement through Twitter. They collected and processed around 12.5 million Twitter messages, analyzing them for content and emotional tone and producing a real-time picture of the daily ups and downs. They called the project *Emoto*.

Once the games had finished, all the data was transformed into seventeen interactive sculptures, one representing each day. What emerged was a "new sort of angle on the Olympics in real time [in which] the data becomes explainable in a new way." The data—numbers registering emotions—was given substance, becoming something you could touch and sense. The finished sculpture looked like an abstract mountain range, a landscape of peaks and troughs on a table 10 feet long, 2.5 feet wide, and 2 feet high. The dimensions of each plate, representing the data for one day, were 17 inches long, 6 inches wide, and 1 inch maximum in height.

Can visual techniques such as these help scientists analyze data?

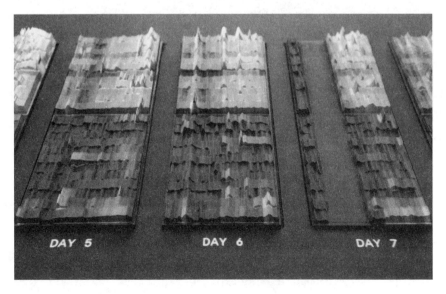

9.4: Jonas Loh, *Emoto*, 2012.

What he does is not really different, says Loh, from what scientists do. But what about the analysis of data that requires certain laws of physics to be applied, as in the Hertzsprung–Russell diagram of the life of stars? Could Russell have accomplished this today using data visualization software? "Absolutely!" says Loh.

Loh's view of aesthetics involves the communication of facts, which is, after all, what data visualization is all about. "Communicating data in an interesting way gets people involved to inquire what's going on." Aesthetics is a visual thing, tied in with information content. The higher the information content, the greater the aesthetic content. A concrete example is Harry Beck's map of the London Underground—a visual representation communicating a great deal of information in a minimal graphical style. The Tube map looks beautiful and has become iconic.

Loh believes that "artists can only aid scientists" to find better methods of visualizing data. Often "scientists find it hard to actually think about their research in a different way," he says, referring to his experiences with scientists at Imperial College. "Artists use technologies in an absolutely different context."

"The border between science and art will melt away, enabling research to be dealt with in a more artistic way," he goes on, adding that fields such as data visualization are in the forefront. "Blurring the line will lead to interesting new developments and to interesting new times."

## Data on a huge scale: Benedikt Gross

"The amount of data around these days is so big that it can't be handled manually," says Benedikt Gross. Take global warming. We hear about it daily, yet it's difficult to visualize. Gross set out to do just that. He realized that NASA space missions had scanned the entire earth and produced hundreds of photographs with excellent data on elevations. But what about the seabed? Then he learned that the Scripps Institution of Oceanography in San Diego, California, had done a thorough survey.

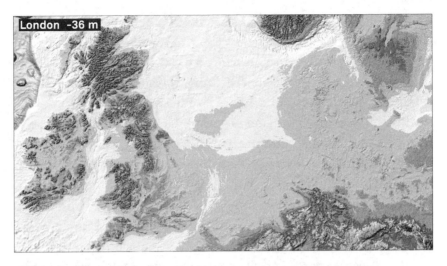

London  -36 m

9.5: Benedikt Gross, a frame from *Speculative Sea Level Explorer*, 2013.

Running the data from these photographs with his own data visualization software, Gross produced a dramatic image—*Speculative Sea Level Explorer*—which can be adjusted so that the viewer can observe seas rising and the resulting effects on geography: coastlines change, countries shrink and sometimes even disappear, other regions increase in size, and economies collapse.

Around 2007, Gross noticed that "there were no books on computer design, on how to program an image." So he and three colleagues sat down to write *Generative Design: Visualize, Program, and Create with Processing*, which was published in 2009. Before this, he had thought about computational design and what else it could be applied to. He started off with the idea that "it would be cool to program things to create visual representations." "A logical next step," he thought, "would be to merge computational design with accessibility to large amounts of data as an information visualization"—in other words, to make large amounts of data more meaningful by finding a way to represent them in visual form. "Programming to create a visual output is a new development," he says.

Computational design—applying computational approaches to design problems—began in the 1960s, with Frieder Nake and Max

Bense in Germany and Bell Labs in the USA. But it was not until the twenty-first century, when new programming tools and processing appeared, that it really got moving. This was when Gross became interested in it. "Visualization of data always interested me from a scientific point of view, long before it emerged as an art form."

Gross started out studying geography and computer science at Stuttgart but decided that information and media were more exciting. He went to the University of Applied Sciences in Schwäbisch Gmünd, Germany, where he completed his thesis, which was actually a proposal for his book, in 2007. The book itself, the first of its kind, was published two years later to a glowing reception. Gross's specialty is using computer art to create visualizations of data. He has won several awards and since 2011 has been in the Design Interactions program at the Royal College of Art, studying for a PhD. He and Jonas Loh seem cut from the same mold, from their physical characteristics to their terse way of speaking, interspersed with wry humor.

While a graduate student, Gross continued to consult. One such job was at MIT's SENSEable City Lab, on the border between urban planning and architecture. The brief of the researchers is to study the life of the city in real time, with the aim of radically transforming it. In New York, particularly Manhattan, the vast majority of vehicles are taxis. The question is how to use them to move people around efficiently without creating traffic jams. There is a huge amount of data. Each taxi has a GPS system, and additional data includes tips and whether passengers pay with a credit card. At first sight, it seemed to the researchers to be data overload—and this set of data was for only one year. To deal with it, they hired Gross to work on a project funded by New York City and General Electric. "Visual design started out producing logos, etc.," he says. "Now it has gone way beyond that."

On arriving at MIT, Gross was confronted with a huge amount of abstract and uncorrelated data. He began by looking for a dispatch algorithm for routing taxis, seeking the best pickup and dropoff points. To do this, he worked with computer scientists, mathematicians, and urban planners. A big challenge was how to encode all the data. The researchers divided the city into rectangular grids of a fixed size. The upshot was a beautiful display of traffic running efficiently. One way they accomplished this was to model an ad hoc taxi service, running only along efficient traffic-jam-free routes, with multiple dropoff points—taxis of the future will be shared. Some of the computer calculations that went into the visualization took up to twelve days.

"As a designer I don't care about certain things that concern scientists," says Gross. Among these is whether a certain visualization of a set of data is unique. This can be of importance in science because it shows that nature really acts in a particular way and no other, but need not be for the designer. There can be lots of different sorts of Coca-Cola bottles, but only one Hertzsprung–Russell diagram. In that sense he believes that designers have more freedom than scientists.

I ask him whether the taxi project was a genuine collaboration, with each person benefiting from it. "Yes," he replies, without batting an eyelid.

"Programming is going to be a new cultural technique," says Gross. As far as programming is concerned, "we are in a time like photography was in the 1950s, expensive and expert-oriented. Now everyone takes photos with cameras and telephones. Photography is just a tool." By this he means that we are entering a period in which people will very rarely program with code, but with apps. There is "no longer a programmer and an artist," he continues. "They are one and the same." Gross's "coding style," he says, differs from an engineer's only in that it creates an aesthetic product that conveys information in a visually attractive way.

So what is the difference between computational art and design? "Art raises the question, and design solves it," is his answer. He is careful to add that both are creative. "I think I'm somewhere in between."

"The diversity is huge at the RCA," he adds. "We play around with technology and create scenarios in nanotechnology and synthetic biology, almost sci-fi-like. We make prototypes which provoke discussions about whether we want them, whether they are beneficial to the planet. Instead of Design Interactions, it should be called 'Speculative Design.'" Like others in this field, Gross is actively involved in designing all our futures.

## Computers can have soul: Scott Draves

"I believe that computation can reproduce the whole creative process, and that ultimately computers can have soul," says Scott Draves. If anyone can communicate that soul, he can. His almost sci-fi presence on my screen (I interview him on Skype)—shaved head with large features and the easygoing yet authoritative way he explains his work as a computer artist—gives the impression that he is definitely at one with the computers running his software.

Draves's spectacular images are everywhere. He is the magus of software art. He is the inventor of Fractal Flames, the first ever open-source graphics software, creating ever-changing images which conjure up brilliantly colored feathers, galaxies, coral reefs, and all sorts of symmetrical and asymmetrical natural forms. Versions of Flame appear all over the world in advertisements and on book covers, including the cover of Stephen Hawking's 2010 book *The Grand Design*. "A graphics package is like a language, and Flame is a visual language, to use the term like Kandinsky did," says Draves, referring to Kandinsky's idea of a "language of form and color." Draves is also the creator of Electric Sheep, run worldwide as a screen saver.

When Draves was ten, in 1978, his parents bought him an Apple II and he "dove right in." He was a natural at programming, quickly moving on from games to constructing graphics. When I ask Draves what he did before computer art, he replies, "There was no before. I've been a math computer programmer guy pretty much my entire life." And an artist too, in the new sense of the term—not wielding a paintbrush or a sculptor's chisel, but a graphics package.

At high school Draves continued doing computer art, "at home and alone," he recalls. He didn't think it was anything that anyone would be interested in. Studying computer science at Brown University, he found himself spending more and more time with the large computer graphics group there and realized he was not alone. When he showed them his art, they said it was not only cool but "pretty exotic." Nevertheless Draves "still didn't think of himself as an artist."

That happened in graduate school, at Carnegie Mellon University, in Pittsburgh, where he was working on a PhD. Ever since his early days working on an Apple II, Draves had been interested not in using the computer to calculate but as a way to produce the unexpected, to make complex images using algorithms. At Brown he became interested in fractals and wrote a program that produced images which could be used as input for further computations. He called this the Flame algorithm, "creating art out of math."

The calculations, however, turned out to exceed the power of the computers at Brown and also at Carnegie, forcing him to simplify the complex equations he was using. The resulting images, however, were not so good. Then in 1992 he spent the summer at the Nippon Telegraph and Telephone Corporation in Japan, where he had access to a supercomputer. He was finally able to solve his equations in all their glory, "to reveal the beauty contained in them." What emerged were ravishing images, some geometric, some organic, eerily evocative of natural forms.

Back at Carnegie, he showed the images to his PhD supervisor, who advised him not to publish them but to enter them in the competition at Ars Electronica in Linz. It was 1993. He won an honorary mention in the Prix Ars Electronica for the image *Flame #149*. Draves's immediate reaction was, "Holy shit, this is art." He was ecstatic, he had discovered his calling. "This was really what I wanted to do."

Flame was probably the first open-source digital art. Draves took this route, instead of selling it, because he believed that scientific results should be freely shared. This accorded with his support for the GNU Project, begun in 1983 by Richard Stallman at MIT, the

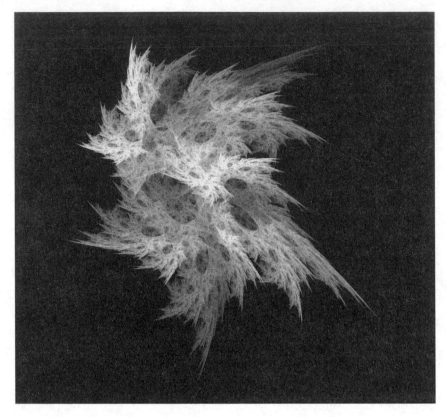

9.6: Scott Draves, *Flame #149*, 1993.

premise of which was that the best way to free computer users was to allow them access to all software, enabling them to share, copy, and modify it.

This seems enough to satisfy the requirements for a PhD, but in fact Draves's PhD topic was "much more complicated," as he puts it. It was related to metaprogramming—the theory of computer languages, of giving meaning to logical statements in a program, and of how people communicate with computers. He finally completed his thesis in 1997.

Two years later he came up with Electric Sheep (see Insert). When a computer is asleep, Electric Sheep comes alive and the computers communicate with each other, producing elaborate abstract forms that continually change when animated. Each user can design their own screen saver or sheep. The animations appear on everyone's screens and users can vote for their favorites. Following the principle of the survival of the fittest, popular sheep live longer and can reproduce and mutate via a genetic algorithm. Draves sees this as a system combining humans and machines, a cyborg mind made up of hundreds of thousands of computers all working together and all based on mathematics. Computer scientists call it artificial life—mathematics generating biological phenomena that emulate Darwinian evolution.

Draves names as one of his key inspirations Alan Turing, the great British pioneer in computer science and the mastermind behind Bletchley Park, where he designed and built the first digital computer in order to break German codes during World War II, contributing significantly to the Allied victory. Then there was science fiction, which fueled Draves's dreams as a boy, together with concepts like artificial intelligence and artificial life, which led him to wonder, "What is life in the abstract sense? Can it exist in nonbiological and chemical substrates?" Big questions.

Draves's artistic inspiration is Karl Sims, an American computer graphics artist who played an important role in using genetic algorithms which mimic Darwinian evolution. Computer artists may use lines and geometry, as in architecture, while algorithmic artists use graphics packages to create images that constantly change according to the algorithm. What Sims did was to make creatures out of assemblages of blocks, move them around, make them fly, and subject them to the laws of evolution. Certain shapes did better than others. These were saved and constituted starting points for further evolution, passing on the good traits to future generations.

Around 1995, Draves recalls, he was "struggling with the definition of information and meaning." He had come across many definitions by people like Erwin Schrödinger, Claude Shannon, "the father of information theory," and Gregory Chaitin, an important contributor to computer science, especially algorithms. Draves felt uncomfortable with their definitions because "they kept trying to quantify [information]." Then he came across the anthropologist, semiologist, and cyberneticist Gregory Bateson, who famously said that information is "a difference which makes a difference." Draves recalls, "It was a sign to me that information (and hence beauty) could not be quantified, or reduced to an equation. But I was also reassured that I could trust myself to recognize it. And so I was able to continue to pursue creativity."

"I want to be surprised by the computer. I want to give up control. I want to exceed my imagination," says Draves of the images his algorithms produce. It's the public, not him, who makes aesthetic judgements on creations such as Flame and Electric Sheep, and the evolutionary machinery in his software makes variations on that basis. Some of the choices may not be to his liking, but that makes it democratic.

I ask him whether there are times when he knows what he wants and writes the software for it. "You're right," he says, "it's not black and white." Naturally "for sellable items I select what art collectors pay for." He will then go through the sometimes thousands of images produced and "pick out the best ones, in my opinion, of course."

"I have a love relationship with technology and computers, and my artwork tries to express that," says Draves. This wonderfully expresses the almost symbiotic relationship between Draves and the computer. I ask whether he believes that computers can be creative and even produce art. "Yes," he continues, "although collaborative. Some comes from the computer but most comes from myself and my colleagues. [The computer plays] a minor part, but I feel its presence."

Recently the celebrated futurist Ray Kurzweil, who is also a computer scientist, inventor, and Google executive, predicted that by 2045 there will be robots whose intelligence will surpass that of humans. This is part of what he calls the singularity, when progress will outstrip human ability to comprehend it. Working together with humans whose intelligence and constitution have been enhanced by nanotechnology, machines will dominate life on earth.

Like many computer scientists, Draves thinks Kurzweil overoptimistic, especially as regards the date. But he does think that "there's something going on. Computers and mathematics can capture the essence of life." This view may seem materialistic but, after all, he says, the brain is ultimately made up of atoms which obey the laws of quantum physics which, in turn, can be programmed into a computer. So, "in principle computers can think." He agrees with Kurzweil on "the arc of history." Computers are becoming faster and faster, he points out, taking over more and more tasks, like translating and finding information, using powerful search engines like Google. He sees nothing to stop this trend. "This is really getting spooky. I see more of a merging of man and machine. If you approach AI with fear," he adds, "this will slow down advance and could precipitate just those conflicts we are afraid of."

Draves "hates science fiction in which there are wars between man and machine. Somebody has to take the first step forward to promote friendship. Giving up some kind of control is okay." Losing control completely would certainly be a disaster, and probably generate a *Terminator* scenario.

When I ask Draves whether he considers himself an artist or a scientist, he replies, "Both." His business card says "Software Artist." I mention that people at MIT and NYU describe themselves as researchers. "Everybody wants to be cool," he replies.

He likes the term "scientist," he says, but insists he isn't one because he doesn't do experiments, and a lot of computer science is really engineering. Yet it is here that you find a blurring between art and engineering.

Draves has won many awards. His work has been shown at Moma .org, the Museum of Modern Art's website, and has appeared in *Wired* and *Discover* magazines. It is also an official skin for Google Chrome—a personalized visual appearance for a Google Chrome page generated by purpose-built software that changes the image at intervals. *Dreams in High Fidelity*, which he considers his masterpiece, produces an infinite variety of patterns in the lobby of Google's headquarters and has been bought for corporate and residential collections nationwide. He also recently produced a work he calls *243*, commissioned by the Gates Center for Computer Science at Carnegie Mellon University. But he has yet to crack the establishment art world.

"The art world doesn't care about technology and vice versa," says Draves. The problem with gallery curators, he goes on, is that they "don't understand the material and can be fooled," which might result in them trying to sell inferior works. There is also the layperson's fear of anything to do with science or technology.

Furthermore, there is the problem of reproducibility. "Anything that's digital is copyable and so in conflict with the concept of sole copy or unique artifact." Materials also wear down, requiring repair and in some cases modernization, which again runs up against the concept of original artifact. "Barriers to acceptance are still numerous."

"The electronic world has created its own world, a ghetto, really," he says. He is optimistic that barriers will come down, though within his world opinion varies from "It's over, we've won," to "Never, they suck," to "I like being separate." One way out may be via today's twentysomethings, who are entering the art world equipped with technological savvy—though "we'll have to wait another twenty years for them to become forty and take over."

Draves discusses all this with his wife and business manager, Isabel Walcott Draves, who is also an Internet strategy consultant. In 2009, in an attempt to break down barriers, she founded Leaders in Software and Art, to put people working in the field into contact with each other. These include software and new media artists, curators, collectors, coders, and collaborators. She organizes monthly salons and occasionally daylong conferences featuring the best artists and speakers from the salon.

Says Draves, referring to Charles Saatchi, the powerful and influential London art dealer and gallery owner, "We have to get past irony and get Saatchi on board!"

## Breathing life into data: Aaron Koblin

"I create art with data," says Aaron Koblin. "We live in an exciting time and we must take a step back, look at that data, and try to understand it."

One of Koblin's first projects was *Flight Patterns* (see Insert). He describes it as being "visually exciting while also related to our lives, giving us new perspectives on how we are living." To put it together, he used a huge amount of data assembled by the Federal Aviation Administration, gathered by monitoring aircraft across the US for twenty-four hours on August 12, 2008, including altitudes, makes, and models of more than 205,000 different aircraft. The result is a fascinating, ever-moving animation showing the flight paths of all these aircraft over the United States. As we watch, first the East Coast lights up as the red-eye flights come in from Europe and the early morning flights leave, then the West Coast comes alive, in a complex, elegant, and beautiful spiderweb of activity, like the flights of a myriad of fireflies.

Viewers can home in on cities, observing the density of flights above New York's three airports, for example, or isolate the different types of airplanes or the different altitudes. Certain areas of the US remain stubbornly black, as if they are no-fly zones—a fruitful area for conspiracy theorists. Apart from the sheer beauty of the animation, it's also a very effective way to communicate a huge amount of information, drawing out the patterns in a vast amount of data.

"I'm 50 percent artist, 50 percent nerd," says Koblin—a self-effacing self-portrait of the creative director of the data arts team at Google.

Born in 1982, Koblin has won many awards for work that uses data visualization to explore how we interact with systems of our own creation. His work is in the permanent collections of the Museum of Modern Art in New York, the Victoria and Albert Museum and Tate Modern in London, and the Pompidou Center in Paris, and was exhibited at the Japan Arts Festival for several years running. Tall, boyish, with floppy dark hair, he radiates confidence and expertise combined with a laid-back Californian manner.

Koblin was interested in computers from an early age. At the University of California, Santa Cruz, he studied computer science and took a minor in film studies. He then did graduate work at the

Department of Design and Media Arts at UCLA. One of the professors there was the artist Casey Reas.

Reas was the creator of Procession, an open-source programming language which could be used to create patterns. Koblin was intrigued by "those systems and [wondered] how [he] might generalize them." He also crossed paths with Mark Hansen, a professor of statistics, who had become disillusioned with studying data just as numbers. He preferred "doing artwork about live data using algorithms for sorting and managing data," connecting this to the way these numbers relate to our lives. "I was excited about visuals coming out of procedural programming and also about the context of working with data to understand our humanity," Koblin recalls.

*Flight Patterns* was enormously successful, so much so that Koblin was approached by the SENSEable City Lab at MIT and asked to participate in a research project to create a visual representation of the way New York communicates with itself and the rest of the world. Using data provided by AT&T on SMS messages broadcast out of New York, Koblin and his colleagues at MIT produced a spectacular multicolored animation which they call *New York Talk Exchange*, with spikes of yellow leaping up at certain times of day when the volume of SMS messages suddenly increases. The work was shown at the Design and the Elastic Mind exhibition at the Museum of Modern Art in 2008.

Koblin did a similar project on SMS messages sent from Amsterdam. At midnight on New Year's Eve, there is an explosion of yellow as millions of "Happy New Year" messages are sent. These are dazzling examples of the power of data visualization to create a story out of what is at first sight an overwhelming mass of data.

Koblin is also famous for his crowdsourcing projects, made up of input from tens of thousands of online volunteers. The first of these was *The Sheep Market*. He put out a request online for drawings of sheep, facing to the left, to be drawn with software provided by Google, in order to investigate the way each one was constructed, and collected through a Web service created by Amazon called the

Mechanical Turk (after a medieval chess game supposedly played by a machine but in reality operated by a human being).

Koblin's inspiration was projects like the Search for Extraterrestrial Intelligence (SETI), in which thousands of people contribute idle time on their computers to help sort out the huge number of signals from outer space striking the earth. Instead of idle computers, Koblin used idle brains, people who were happy to contribute online for a fee of two cents per drawing. They were allowed a maximum of five. Ten thousand sheep arrived in forty days, after which Koblin closed the project down. Still more arrived. Clearly people were driven by motivations other than money. Koblin viewed and approved the sheep, rejecting 662 which "didn't meet sheeplike criteria." He then assembled the drawings in a grid. Animations of how each sheep was drawn can be seen online.

Data visualization is very different from scientific visualization. Type the words "black hole" into Google and what comes up are classic images of a black hole as a round opening in space, surrounded by swirling lines representing gas particles snared by the black hole's enormous gravitational attraction. These are drawn by artists on the basis of mathematical models based on data of what a black hole is supposed to be—scientific visualizations, not art. Scientists sacrifice artistry in order to explain their views simply to colleagues and the public, Koblin says, whereas "I try to reembody data into a context that makes it relevant and emotionally satisfying in some way."

Koblin starts off by using a set of data with already existing software. "If I find something interesting then I ask further questions. For me the pursuit is really about distilling, about simplifying, about making a very clear argument. What I'm interested in is more explanatory. Sometimes it's more about asking questions than about pursuing them."

I tell him about the Hertzsprung–Russell diagram and ask whether he could make a similar discovery by using his data visualization tools to examine scientific data. Koblin replies that his group at Google Sky, which harvests data from several space telescopes including the Hubble Telescope, NASA satellites, and the Sloan Digital Sky Survey, could potentially make new discoveries.

He draws an interesting distinction between the two main ways of representing data: infographics and data visualization. Infographics is a means for conveying statistics and ideas, as in the map of the London Underground. It is a means of storytelling, he continues, similar to what Edward Tufte, the pioneer of data visualization in the early days of computers, advocates when he writes about clear, articulate representations which are also beautiful. Koblin prefers data visualization, which is "time-based and so opens up different sorts of narratives, whereas what Tufte proposes is static." Data visualization methods "allow interfaces enabling you to ask new questions, to toggle on and off parameters, reading the data differently." Thus, in *Flight Patterns*, you can home in on air traffic in specific cities at specific times of day or night. He hopes that the software he and others in his team write will be able to "pull out a pattern like the Russell thing."

As far as beauty, he says, it's a gut feeling on seeing something that is "cleanly designed." It's about "telling a story in its form, it resonates properly with what I'm interested in creating. It's a reaction that you must have." Then there is "the emotional test: is it exciting?" Koblin takes a set of data and applies some standard software to it. If something interesting emerges, he writes new software and hopes for something to appear that stimulates a gut feeling, thereby passing his emotional test for what is aesthetic.

At Google, Koblin interacts with engineers, as he did when he worked at MIT. He believes that at present artists profit more from collaborations than scientists. How the artist benefits "is tangible [while] what scientists get out of it is less tangible, but perhaps the pursuit is as valuable." In his opinion, scientists and engineers need to learn to "think in terms of shifting perspectives and enthusiasm." The problem is that "in the worlds of science and engineering some magic disappears because they are so wrapped up in what they are doing." Then "there is the communication problem," which goes beyond the issue of artists merely working for scientists as illustrators. His own work with engineers involves "pushing technology to the limits," so automatically the engineer gains some benefit.

The next issue is to find a way to "help esoteric scientific pursuits to be communicated properly," how to make his work more accessible to others in data visualization as well as to the public. He also needs to satisfy administrators "cutting checks for those working in such niche, highly esoteric realms," who want to know where the money is going.

Koblin has exhibited in galleries and art museums around the world. "I have been in and out of favor [and] am not particularly concerned" with the establishment art world. In his opinion, "currently galleries have a love of relics in which everything is commodifiable; it is a difficult thing for art to exist in a world of capitalism." Koblin rarely sells his work. "It's more for collectors."

"I've been studying art for about twelve years, but I'm not sure I know what that means because the term 'art' is in flux," he says. There are various names for fields that have arisen skirting art, science, and technology, among them, computer animation film, digital music and sound art, interactive art, hybrid art, and digital communities. "A lot of people in the 'maker culture' are not concerned with debating over where their work fits," Koblin comments. "Any distinction between art and science is becoming increasingly irrelevant." "New terms continuously arise, like 'media' and 'media art.' Perhaps we should just call the whole thing 'media art.'"

It's worth quoting at length his thoughtful view of art as research, which is the crux of his work:

> I do see art as experimentation. It's fundamentally the process of combining things that have never been combined before. I'm fascinated by the process of translating ideas from one reality into another and ways that systems and structures can influence and guide new creation. To some extent art is an excuse. It's a word that we use for yet-to-be-defined progress— necessary experiments in thought space which help us define

and test limits. These are experiments that allow us to question, reflect, discover, and appreciate.

Koblin vividly describes his mode of research at Google—testimony that lateral thinking can sometimes be effective:

> At Google, I think we have a healthy mix of team work and solo time. We quickly and comfortably share as a team, creating a rich pool of ideas that we can discuss and experiment with on our own. I've seen great ideas come about in casual conversation, and greater ideas be proposed by individuals and then be taken to even greater heights while being "implemented" and experimented on by the team.

Together with a colleague, Chris Milk, Koblin has managed a rare achievement in data visualization or, for that matter, any sort of art that borders on technology: an exhibit at Tate Modern. *This Exquisite Forest* has a room to itself. It is a spinoff of the Surrealist game Exquisite Corpse, in which players add to a drawing or a sentence

9.7: Aaron Koblin and Chris Milk, *This Exquisite Forest*, 2012.

without knowing what comes before. It was a hit in Parisian cafés and exported as the game Consequences.

*This Exquisite Forest* is played online with software provided by Google. As the artists describe it, "users create short animations that build off one another [resulting in] a collection of branching narratives resembling trees." Viewers at Tate Modern can see the evolving image on a huge screen; zeroing in on each "leaf" with a cursor reveals the animations. Like *The Sheep Market, This Exquisite Forest* is a crowdsourcing event.

"This is very much the age of big data," Koblin concludes.

## Visualizing the world of science: W. Bradford Paley

W. Bradford Paley turns data into visualizations that can actually be read to provide an enormous amount of information, from charts for clients on Wall Street to complex diagrams that lay out different aspects of scientific research. They are also beautiful. His most famous "map," *Relationships among Scientific Paradigms*, is a schema relating the number of times scientific papers are cited in different fields of research (paradigms) and the key words and concepts they use. The nodes are linked not with straight but with curving lines, with tiny words swirling off at the edge like the wisps of a peacock's feather. For all its seriousness of purpose, it's a beautiful image that almost gives the impression of flying (see Insert).

Paley designed *Relationships among Scientific Paradigms* together with Kevin Boyack and Dick Klavans, using data provided by the Institute for Scientific Information, an academic citation indexing service. The image first appeared in *Nature* magazine in 2006. Paley describes it as follows:

> The image was constructed by sorting roughly 800,000 scientific papers (shown as white dots) into 776 different scientific paradigms (red circular nodes) based on how often the papers were cited together by authors of other papers. Links (curved lines) were made between the paradigms that shared com-

mon members, then treated as rubber bands, holding similar paradigms nearer one another when a physical simulation had every paradigm repel every other: thus the layout derives directly from the data. Larger paradigms have more papers. Labels list common words unique to each paradigm.

*Relationships among Scientific Paradigms* was so popular that it was featured in *SEED, Discover,* and dozens of other magazines around the world. Paley was so touched by the size of the response and the number of requests for copies that he gave it away for the cost of shipping and handling. Those who have seen it have been struck by the delicacy and beauty of the differently colored nodes and curved lines of varying width that connect the nodes. Paley says he chose curved lines to connect the links because straight lines would have obscured one another, hiding the density of links where it might matter most. The printed edition, 42 by 43 inches, is eminently readable.

But, iconic though the image is, Paley himself was far from satisfied with it. "Today such a representation would never leave my lab," he says. "My driving goal is to find beauty in the *world*, not in my translation algorithms." Part of what he disliked was the wisps of "hair," made up of key words from papers in a certain paradigm, and how "seductive they are—they really grab your eye," whereas Paley meant them to have only a "playful meaning." This effect was "at odds with the informative goal. The wisps of hair are actually distracting." He concluded that the "meaning or sense of the information transmitted by the representation was a mess [in the first version]." The tiny white dots representing scientific papers needed a darker background.

A perfectionist, Paley set to work to improve the image. He transformed the nodes representing paradigms into transparent colored circles, making them into single visual elements, and cut out the individual white dots he'd used to represent each paper. The lists of words still looked like hairs, but he pushed them into the background by printing them in gray, emphasizing the colored nodes;

this gave the data more prominence. He also used a more modern typeface. (See Insert for the new versions.)

Paley's colleague Dick Klavans, an expert in scientometrics—the art of measuring the impact of scientific publications in one field of study on another—suggested a variation on the data to discover what sort of science major nations focused on. Paley responded by setting up his representation in such a way that he could peel data off the main figure to produce representations showing the number of scientific publications on a particular topic in a particular country, shown in the bottom row of images. Thus the USA and the UK focus research on medical sciences, while France and Germany are weighted more toward chemistry and physics.

"Photography was very formative for my future," says Paley. When he was fifteen and already an accomplished photographer, he was printing commercial photos of grapes when he lost track of whether he had printed a particular negative. He noticed that the photographic paper was not properly aligned and moved it into place. When he developed it, he found that he had double-printed. He had moved the paper by the width of half a grape. The result was extraordinarily "visually engaging."

Excited by his discovery, he went to his local library to inquire whether any other photographer had ever done this. Instead of books on photography, the librarian sat him down in front of books about Cubism and art history. "Studying the Cubists led me to wonder about the structure of the mind," about how Picasso and Braque had created Cubism, and then to look into cognitive science.

When he reached the age to apply to a university, "Dad wouldn't pay for an art history and philosophy degree, so I studied economics." As a minor subject he did computer science and found he was pretty adept, enough that he ended up working at Citibank and then Lehman Brothers as a coder. But his real love was his evening job, where he developed tools for animators. There were few in the early 1980s. To his surprise, he found designing the tools more interesting than making animations.

Following up on this interest, in 1982 he founded Digital Image

Design Incorporated, creating user interfaces for data visualization—in other words, ways for the user to access representations of data. Paley was especially interested in data issues in the financial sector because of the cognitive complexity of the task and lack of an inherent spatial structure in the data, and subsequently established the first significant financial data visualization team at Lehman Brothers; night and day jobs had merged.

"Personally and emotionally, my driving goal is to find beauty in the world," including in the data we move around in, Paley says. Here his experience as a photographer comes to the fore, in that to obtain the proper image you have to "select it and draw it out of the world." The rules of selection are not arbitrary. To Paley, the most important thing is "the meaning from information process." Aesthetics comes second. By "meaning," he means how well a visual representation can be understood, as opposed to one that verges on the abstract.

He also believes that visual representations must not be over-simplified or conventionalized. As an example, he takes the customary visualization of a black hole—a well in space represented by a deformed grid defining the depression inside which a star is collapsing. "This is dangerous," he says. "The layperson could take this as real and think there exist actual grid lines defining the fabric of space, or that the warp is only two-dimensional, like a piece of graph paper."

When he came to create his improved version of *Relationships among Scientific Paradigms*, Paley tells me, the key was, quoting Bateson, that "information is the difference that makes the difference." Now he had a representation with meaning, one which drew the eye to the most meaningful parts. Finally the new representation was aesthetic, an object of art unto itself. Paley says with some satisfaction, "I am a designer. I design tools for everyday use. As a secondary goal I want to make them attractive because that improves engagement with the tool, and I like to make pretty things. I like to make people happy." He was still, of course, not completely satisfied. It was still too "hairy"-looking, some letters were upside down, and there were still too many overlaps. He is fiddling with it to this day.

Paley is also involved in designing interfaces for the New York

Stock Exchange. Here there are no images to be evoked. It is "not meant to pull on the retina, but on the mind."

Basic to all of his work in data visualization is, as he puts it, "How do you make a mark on a page that makes somebody think?"

Paley has worked with the great pioneer of data visualization, Edward Tufte. "As a curator," he says, Tufte "brought together hundreds of years of the most inspiring work ever [and] the way he categorized these things is pretty damned good." Nevertheless, in Paley's view, much of the work Tufte collected has become too abstract; the representations worked more as a result of their retinal effects than their effect on the mind—their understandability. Paley's own work tends in the opposite direction.

"If I create something and someone considers it art, then it's art," says Paley.

He quotes James Joyce's *Portrait of the Artist as a Young Man*: "If a man hacking in *fury* at a block of wood . . . make there an image of a cow, is that image a work of art?" Stephen Daedalus's reply is negative. Paley replies conversely that it is, even when the man does not intend to produce a work of art; it becomes art by the act of the viewer or curator, not the creator. Much of the work "blessed" by curators and thus turned into fine art was created to be tools.

His one counterexample is *CodeProfiles*, commissioned explicitly as art for Codedoc, an exhibition curated by Christiane Paul at the Whitney Museum of American Art in New York in 2002. "Codedoc" refers to the code documents that generate a work. For this exhibition, Paul commissioned miniature works from twelve artists, one of whom was Paley. Their task was to write code (program) to connect and move three points in space, a brief that could be interpreted in any way they wanted. What made the exhibition unique was that each artist had to display their code. The software itself was part of the exhibit.

Without the code there would be no work of art, just as without paint there would be no painting. To take this analogy further, the curator emphasized that what makes a painting unique is the artist's brushstrokes. In Codedoc, the artist's brushstrokes are their codes.

Code is written in arcane symbols and each artist has their own style. Like music notation, the symbols carry instructions. Writing code, the New York new media artist John F. Simon Jr. argues, is a form of creative writing. So the works on display at Codedoc could be enjoyed on two levels: watching the animations, and reading the code and appreciating the creative process that went into making it.

Even in the usual case, when the tools are appropriated as pieces of art, Paley sees his work as "artifacts out there in the world." He is delighted when people invite him to exhibit. "I'm the maker of Duchampian pre-readymades, then the curator turns them into art." Paley, however, turns down most invitations to exhibit because he disagrees with the aesthetic agendas of the curators.

Does he consider himself an artist or a data analyst? He replies, "Neither. I'm a toolmaker who tries to understand how analyzers analyze and how artists represent. I'm a designer, a 'cognitive engineer' applying the findings of science and art to perform the engineering task of building tools for analysts. I'm programming the human mind, using human-recognizable *visual* code."

# Comrades-in-Arms: Encouraging, Funding, and Housing Artsci

Artsci is "not only sexy, it's seductive, it goes places where nobody has really gone before," says Gerfried Stocker, artistic director of Ars Electronica. "Artists are like scavengers working from trash to gold, foraging deep into science and technology." The radical Austrian performance and media artist and critic Peter Weibel goes further. According to him, the age of artists working in partnership with scientists has passed. Rather, art is dependent on the input of science. In his words, "Today art is an offspring of science and technology," an extraordinary statement and one worth pondering.

In 2008, reviewing the exhibition Design and the Elastic Mind at the Museum of Modern Art in New York, which looked at the relation between science and design and covered bioengineered organisms, nanotechnology, and the powers of technology, the *New York Times* critic Nicolai Ouroussoff wrote perceptively, "The exhibition's overarching theme, the ability to switch fluidly from the scale of the atom to the scale of entire cities, may sound the death knell for the tired technological divides of the last century. It should be required viewing for anyone who believes that our civilization is heading back toward the dark ages."

At present these divides are still alive and kicking, and artists whose work deeply involves science and technology are affected by them. The movement we've been looking at, which seeks to fuse art, science, and technology, has lofty goals but marked difficulties in finding a market for its products. As we have seen, private galleries tend to shy away from showing works created by these artists. One path is simply to ignore the galleries, as many artists in the artsci world do.

But there are places that see this as the art movement of the twenty-first century and specialize in it—places such as Ars Electronica in Linz, Le Laboratoire in Paris, the Science Gallery in Dublin, the Wellcome Collection, Kinetica, The Arts Catalyst, and GV Art in London, and Weibel's ZKM in Karlsruhe. So who are the magi who work behind the scenes to encourage this new art, find it, and showcase it?

In 1988, aware of the lack of contact between artists and scientists, the New York artist Cynthia Pannucci founded Art & Science Collaborations, Inc. ASCI provided a much-needed forum for discussion in the form of symposia and exhibitions but lacked a sponsor with clout. In England, the Wellcome Foundation plays a similar role—with ample funding.

## The Wellcome Collection: Ken Arnold

The 1990s saw dramatic developments in biotechnology, such as in vitro fertilization, gene therapy, organ transplants, and functional magnetic resonance imaging ($f$MRI). As Ken Arnold, the Wellcome Collection's head of public programs, puts it, "Artists could not afford not to be interested in these ideas of what it meant to be alive. They began to show a genuine interest in something they hadn't thought of before." One of the first artists to spot the new trend was Damien Hirst, in 1992, with his installation *Pharmacy*, made up of shelves of medicine bottles.

With Arnold at the helm, the Wellcome Foundation has become one of the most important showcases for science-influenced art, specifically biology-influenced art, in Britain.

Henry Solomon Wellcome's life is a great American rags-to-riches story. Born in a log cabin in Wisconsin in August 1853, he was a brilliant inventor and businessman who was eventually made a Knight of the British Empire. This came about through his interest in medicine, which was always commercial. He was particularly struck by the poor way in which medicine was packaged. At the age of twenty-seven and by then a British citizen, working with Silas Burroughs, another American expat, Wellcome came up with the idea of using a capsule to dispense drugs in precisely measured quantities. The two formed Burroughs, Wellcome and Company, which soon became a giant in the pharmaceutical industry.

A philanthropist passionate about advancing medical research, Wellcome left a huge amount of money on his death in 1936, which was used to form the Wellcome Trust. This became the largest non-governmental source of funding for scientific research in the UK. A portion was earmarked for a division called the Wellcome Centre for Medical Science, whose brief was to inform the public about advances in medicine. Laurence Smaje was the director. An affable, easygoing man with a carefully coiffed beard, Smaje kept a firm hand on the helm, guiding the division across the administrative hurdles at the Wellcome. With an MD and a PhD in physiology, he moved easily between the practical and research sides of the medical profession. He also had a keen interest in education.

In 1993, the Wellcome Trust sponsored a highly successful exhibition entitled Science for Life. The curator was Ken Arnold, who had arrived a year earlier. Tall, with an air of authority, a sharp dresser with a wry sense of humor, Arnold is the archetypal patrician American midwesterner. He studied natural sciences at Cambridge where he was particularly attracted to the "physical, visual, material aspects of science and how it interacts with society," and also the history of science. With an interest in science but no great desire to become an expert in any one field, it seemed obvious that museums were the place for him. Having decided to become a curator, he did a PhD at Princeton University on the history of English museums before taking a post at the Wellcome. He was in the right place at the right time.

Science for Life included interactive exhibits, computers, microscopes, holograms, and a giant model of a cell as part of a depiction of the story of life. David Attenborough, the eminent naturalist, writer, and broadcaster, described it as "one of the most exciting and innovative exhibitions on science that you will find anywhere in this country." Versions of Science for Life went on tour around Britain. But Smaje felt that something more was needed.

Working with the designers and artists who set up Science for Life, Smaje was struck by the parallels between art and science. He also realized how delicate it was to work with such people. "Creativity, that crucial element—you just have to let people do it. Just tell them what you want to achieve and then let them do it." The answer, he decided, was to engage with the arts to put together a project that linked art and science.

Wellcome's first engagement with the arts entailed using theater to generate discussion. Smaje organized a workshop where geneticists informed dramatists about genetic disorders and genetic engineering. Then the Wellcome announced a competition for story lines, with entries judged by an advisory board of playwrights, geneticists, and educators. In 1995, the board commissioned a play by Nicola Baldwin entitled *The Gift*, about a family's reaction to the news that one of their children has a lethal genetic disorder. Smaje insisted on the involvement of the advisory board to make sure that the "science was right and the drama compelling." Whereas Science for Life targeted adults, Baldwin focused on children, too. Production companies toured schools, setting up post-performance workshops to generate further discussion, and versions were performed at the Edinburgh Festival Fringe, with professional geneticists engaging with audiences.

Theater and education were one way of spreading information and stirring up debate about advances in medicine. "Another way was pictorial art," recalls Smaje. In 1994, the Wellcome Centre for Medical Sciences had moved into a spacious new building at 210 Euston Road, across the street from their former building at 183 Euston Road. Smaje had been pondering how to use the large space on the building's ground floor and brainstormed with Terry Trickett whose company, Trickett Associates, was designing the building's interior. Smaje had an epiphany. "Why not have an exhibition?"

It was at that moment that Smaje was approached by Matthew Holley, a researcher at the University of Bristol specializing in the inner ear, whose work was funded with a grant from the Wellcome Trust. Holley is that rare breed of scientist who is also an artist. As a boy, his mother had encouraged him to do enamel art. At Bristol he took evening classes in enamel art, studying under the enamelist Elizabeth Turrell.

"The sensory organ of the inner ear has a remarkable architecture and it is fun to explain how it fits together and senses the smallest vibrations with such astonishing sensitivity," he explains. He put together lectures for the public and for art schools. Artists frequently asked why he was so adamant about the beauty of the inner ear, given that we can't see it. He replied, "Because it works, [and] all things that work well in the natural world have an inherent beauty in the balance between form and function." The question was how to show this.

All the buzzwords were there: form, beauty, balance. Holley's idea was to combine real images, such as electron micrographs of the inner ear, with ceramics, paintings, enamels, and other objects, into an exhibition of art and science. The art objects would be inspired by Holley's scientific images, sixteen of which were exhibited. One of the pieces he himself contributed was a ceramic design based on the cerebellum (the part of the brain at the back of the head that controls movement). Another was by an artist who created a design using enamel on copper, depicting a section through the cochlea (the section of the inner ear that looks like a snail shell, where sound is received). Ken Arnold curated the exhibition and Trickett designed a background of large scientific images on convex surfaces, with the artworks displayed at the front.

Look Hear: The Science and Art of the Ear opened in April 1995 and was the first exhibition of art and science at the Wellcome. Viewing the juxtaposition of scientific data, models of the ear, and videos alongside the artworks they had inspired, visitors were led on a journey along the road taken by sound into the hidden depths of the inner ear.

The show was enormously successful. One critic described it as "a genuinely innovative exhibition, and a fertile marriage between science and art." Another commentator pronounced it "an extremely intelligent and eclectic show." It went on to tour the country for two years. Rather than being a joint venture in which artists and scientists explore the invisible world, the stated aim of Look Hear was to inform the public about science. As Holley says, they had shown that "scientists can play an important part in going out into art colleges and talking to artists, enthusing them to deal with it [science] in their art."

Flushed with success, Arnold, Smaje, and Trickett dreamed up a funding program for collaborations between artists and scientists: the Sci-Art program. "We began to wonder what would happen if we invited scientists and artists to work together, given a grant from us. Sci-Art was born," recalls Smaje. The first call for submissions went out in 1996 with a deadline of February 1997. Two hundred and fifty high-quality submissions were received and twelve received funding.

Two of the chosen artists, Heather Ackroyd and Dan Harvey, used grass as a photographic medium onto which they projected images. The problem was that the images quickly faded. As a result of the call for proposals, they got in touch with plant scientists at Aberystwyth University, in Wales, who were investigating methods to preserve the greenness of grass. Up until then these scientists had ground up grass in order to study how its chemical composition changed over time. They had lost the ability to look at grass growing, to see its colors and textures, as artists do.

The artists and scientists met in Wales and decided to work together. Looking at Ackroyd and Harvey's work, the scientists became aware of the range of tones in the images on the grass. They realized that there had to be a better way of analyzing how grass lost its greenness than by grinding it up, and that they should pay closer attention to the light around them in order to make a noninvasive analysis of grass. They called their collaboration Fixing the Ephemeral. The result was images on grass that lasted longer, on the one hand, and new imaging techniques for studying the cells in grass without grinding it up. "We wouldn't be doing what we're doing now if it weren't for [Ackroyd and Harvey]," says Helen Ougham, one of the scientists on the project.

The Sci-Art initiative succeeded so well that after two more rounds, Smaje decided that the Wellcome needed partners. He approached the Arts Council, the National Endowment for Science, Technology and the Arts, the British Council, the Calouste Gulbenkian Foundation and the Scottish Arts Council. Finally, after some three years of bureaucracy, the Wellcome Trust took over. Trickett, who was on hand during all the critical developments, summed up the situation and confirmed the critical role played by Smaje:

> [Sci-Art] was an idea whose time had come. But to launch it, it required first of all the genesis of an idea, a person to champion that idea, and an organization willing to back it with no sure knowledge of what might materialize. When I think about this combination of circumstances retrospectively, I am bound to reach the conclusion that, at the time, Wellcome was the only organization in the UK that would have been willing to take the plunge.

The building at 210 Euston Road closed in 2004, and the division moved back to 183 Euston Road. Arnold played a central role in transforming the massive ground-floor space there into a museum (the Wellcome Collection), a café, and bookstore, and rose from head of exhibits to head of public programs. The launch of the new venue was on June 21, 2007. Just before the doors opened, Clare Matterson, then head of policy at the Wellcome Trust, stood with Arnold. Both were rather apprehensive. She turned to him and asked, "Will they come?"

Indeed they did. Arnold expected maybe 100,000 visitors a year, but the number far exceeded that. In 2012 there were over 490,000, and summer 2012 saw the two-millionth visitor cross the threshold.

Today Arnold plays a pivotal role in promoting artsci, focusing on art associated with medicine. He looks for "artists grappling with big ideas, some of which hatched in a lab." He is interested in bringing the arts and sciences together in the public arena. He wants to give scientists the opportunity "to look at things from a different perspective" and bring them together with artists eager to work with them, to create "a mutually shared excitement." Over the years he has arrived at what he describes as a three-step approach to art and science in the public arena:

1. Science needing to find a more sympathetic public understanding by associating it with art; in this way the public will see it as less threatening.

2. Artists taking seriously the almost philosophical questions that science is throwing open in the worlds of biology and medicine.

3. A genuine, open-minded sharing of the intrigue of ideas, a new topic for artists who realize that art is more than being just about art.

The "types of artists who are interested in us are the types of artists interested in investigation," he says. They are researchers. Similarly, "scientists interested in working with artists are also highly investigative." For scientists it sometimes goes beyond interest. "The process of science, owing to [the need to get] grants, has to be fairly safe, which leads me to think that scientists willing to work with artists are interested in risk-taking."

In the academic world, the predominant view has long been that those in the arts work at home, while scientists work in their offices and are expected to be present on site. Scientists are also expected to produce publications which have to do with science or other serious subjects—a brief that does not include art. Some scientists have reported to Arnold that collaborating with an artist has "ruined [their] career" by signaling that they were not serious scientists. As we have seen, Bell Labs permitted engineers to collaborate with artists in their own time, not the company's, though they encouraged joint efforts once the projects seemed feasible and generated good publicity. In the case of David Weinberg, who collaborated with Josiah McElheny on *An End to Modernity*, which produced spectacular glass evocations of the universe, his colleagues commented that as his name had not appeared on the finished work, he had wasted his time. Scientists of Weinberg's stature, of course, are free to involve themselves with artists, especially on projects directly related to science.

Arnold says that while very few scientists are interested in collaborating with artists, there are two categories that are more likely to: young scientists eager to venture into new territory and established scientists with permanent appointments. Both groups are free to be "unorthodox in career structure. So why not grapple with something that might be slightly bonkers?"

Collaborations, he comments, have dramatic ups and downs. "Collaboration is bloody difficult." This applies, too, to exhibitions requiring collaboration among people from many different sectors, from art, science, and anthropology to the press and designers. Often when the staff at Wellcome are sitting around and discussing a show in retrospect, someone will say wryly, "Never, ever, ever again collaborate." But of course they do.

Arnold mentions the "myth" that artists work alone and are interested in their own glory while scientists work as a group. In fact, scientists can be as egotistical as artists. It has always been the case that scientists publish in order to gain personal recognition, not merely to disseminate knowledge. There are many battles in science over who discovered something first, especially when there is the chance of winning a Nobel Prize. It is not surprising that clashes in a collaboration generally surface when only the artist's name appears on the finished work, even though scientific input was essential during the process. If there were no science, there would be no work of art.

One of the roots of this attitude, Arnold says, is that artists consider themselves unique. They feel that scientific ideas are in the air—"scientific ideas are everyone's ideas"—and that Einstein, for example, just happened to be there while Picasso, on the other hand, created his paintings from his own genius. "Artistic ideas are not accidental." Such a simplistic view, of course, ignores Picasso's predecessors, as close in time to him as Cézanne. There was much in the air, and Picasso was well aware of it. But this myth continues nonetheless.

In Arnold's many experiences with collaborations, he has found that "halfway through, [both the artist and the scientist start to feel] that they had the main ideas and misunderstandings arise." The artist begins to feel that the scientist offered a "gift of knowledge."

Collaboration works best, he says, "if both parties are clear from the beginning" as to their contributions, aims, and what the finished product should be. The "power of collaboration," as Arnold has come to understand it, "is that people come together for a short period of time—it's the misunderstandings that stir the interest—and then they go back to their other worlds." He is, however, concerned about "people who serially do this and can't figure out which world they are in."

"The Wellcome sits oddly in all of this," he adds. "In the permanent site there are works of art done in collaboration with scientists in by-and-large temporary exhibitions. But we don't showcase this material and some people are disappointed. These exhibits don't include Sci-Art. They have science, they have art, but no Sci-Art—very infrequently, art that has been inspired by science."

Today we are bombarded with information from across the Web, through email, texts, and social networks such as Facebook and Twitter. Says Arnold, "In two clicks we can go from anything to anything, from a physics department to an art class. Everything is available everywhere. What is the role of the museum in the digital world? Anyone with a mobile phone can look at hundreds of things and get not much out of it. The luxury of museums is that they offer you the possibility of only a few choice things." Museums are oases in the vast and often barren landscape of information.

## The future is now: Ars Electronica and Gerfried Stocker

Dressed from head to toe in black, long black hair swept back into a ponytail, Gerfried Stocker is an immensely personable and knowledgeable advocate of media art—"the art of my generation," he calls it. Trained in technology, he has worked on the borders of interactive art, robotics, and telecommunications all his life. He has been the artistic director of Ars Electronica in Linz since 1995, and since 1996 codirector with Christine Schöpf of the annual Ars Electronica festivals.

Stocker prefers the term "media art" as "a catchall for digital art, etc." After all, as he points out, the term "fine art" is just as general. Media art "encompasses everything not yet settled, not yet found." He has no use for categorization, preferring the old approach of "blurring distinctions, a stunning idea. The purpose of art is like science—to investigate, to find out what is behind borders." He is committed to widening collaborations between artists, scientists, and technologists and helped initiate the artist-in-residency program Collide@CERN, a partnership between Ars Electronica and CERN.

He feels strongly that the "role of art is not just to communicate." Nevertheless, he is in favor of the trend among artists toward using science to produce images that go beyond scientific visualizations, showing that they are capable of a deeper understanding of the subject matter of science. An example is the brain as depicted in $f$MRI. Artists can take these images and turn them into something beautiful, perhaps enabling scientists to see something deeper in them.

Stocker's approach to establishment art galleries and the art market is to avoid friction. You wouldn't expect music or literature in the Louvre. Similarly, establishment galleries sell a specific form of art. While fine art is shown in galleries, artsci and media art feature at places like Ars Electronica, Documenta, and ZKM, which offer prizes and scholarships. "Artists can make a living through festivals," says Stocker.

More and more artists are producing media art and sometimes their work is even bought by galleries and sold to collectors. But this can cause the established art scene to lose the ability to define art, resulting in anxiety, especially when gallery owners see the results of "art and science collaborations reaching people," because it's so new and so appealing. Another reason for friction with the art world lies "in the nature of science-media art, which employs interaction technologies and visual technologies, [making it] much more practical, much more available, much closer to the people, giving them the notion that they can understand it; and so it becomes a much more descriptive sort of art."

One problem with exhibitions of media art, says Stocker, is that viewers see devices that look like computers. This is just surface; it doesn't offer insights into the work itself. More people know how to use art museums than media art exhibitions. "We put our hands behind our backs because it looks cool and walk around." But merely looking at a painting, at the surface, doesn't provide an intimate understanding of it. Rather, the painting ought to inform us about "how one should think differently." This requires a sophistication, gained from familiarity with art.

Nevertheless, things have changed since the mid-nineties, when "no one from art history was interested in the archives at Ars Electronica." Artists that visit nowadays consider themselves "in between [and are] eager to call themselves researchers, and many scientists are eager to do art." In the future there will be no more pure art and pure science, says Stocker. He speculates on a "new aesthetics of the future" emerging, particularly from data visualization. He is all for blurring of borders. He is encouraged to see that interdisciplinarity "has improved drastically over the last twenty years" and mentions that teamwork is ever more necessary in scientific research, often using networks of members in different locations.

Today's children are growing up in a technological world, rather than learning about computers as adults. They are "hybrids" who will operate easily in a world where there is "no more pure art and pure science."

## ZKM: Peter Weibel

Stocker's predecessor as artistic director of Ars Electronica was Peter Weibel, the prolific media and performance artist. In 1988 he founded the Zentrum für Kunst und Medientechnologie (ZKM) in

Karlsruhe, Germany, as a parallel but larger institution with more facilities. It was still in the planning stage when it was hit by a wave of hostility driven by skepticism toward technology that seemed to challenge the conventional way that art is categorized.

ZKM bulldozed over these issues, becoming "a role model for art, technology, and science," says Weibel. Its principal interest is in media art, covering visual media, sound art, interaction art, electronic art, and whatever else is new in the twenty-first century. It houses a media library, a museum with archives for collections of twentieth- and twenty-first-century art, laboratories, and exhibition spaces, and offers stipends for study. In short, "the ZKM is a performative museum, [a] digital Bauhaus," like the famous hothouse institution established in 1919 in Weimar, Germany, by the architect Walter Gropius, with the aim of unifying art and technology and being an incubator for young talent. In 2012, ZKM won the Grand Prix of the International Council of Museums in the sound installation category, competing against institutions such as the MIT Media Lab.

While the MIT Media Lab manufactures products, says Weibel, ZKM manufactures ideas. He paraphrases the architect Frank Lloyd Wright's famous title "Machinery, Materials, and Men," transforming it into "Media, Data, and Man." This, he says, is the essence of ZKM.

## Process and product at Le Laboratoire: David Edwards

From outside, Le Laboratoire looks like a boutique, but as soon as you step inside it's apparent that it is not. Paris's answer to the MIT Media Lab, it is spacious, with three floors of glass-walled rooms, laboratories, and a comfortable lecture space with movable partition walls. It would meet the sound artist Sam Auinger's standards of proper climate and sound qualities, ensuring that speaker and audience have the best possible aesthetic experience.

When I visited in August 2011, seventy high school and university students from around the world were participating in a project on water: how to mine it, transport it, and utilize it. All were astute, well-spoken, and enthusiastic. The impressive "graduation" ceremony showcased design technology. Teams of four to seven students gave four-minute presentations on their projects using four to eight PowerPoint slides. Their suggestions ranged from mesh nets to extract water from fog to purification bags that would dissolve in water and could be used as soap, thus avoiding plastic waste products. Their assignment included seeking investment capital, and in this too they seemed to have made inroads. Thus their projects covered science, technology, business, and media art.

"Since the beginning of time artists have been interested in the frontiers of knowledge, the imagination," says David Edwards. Edwards is an entrepreneur of the intellect. He divides his time between Harvard, where he is a professor, and Paris, where he established Le Laboratoire in 2007. Located behind the Louvre, Le Laboratoire showcases work combining art, science, and technology in both its formative and finished stages.

Trim and energetic, sporting horn-rimmed glasses, designer stubble, and dark wavy hair, wearing loafers but no socks, Edwards fits neatly into the elegant intellectual world of Paris. He thinks carefully, closing his eyes, peppering his speech with words like "riffing" and "skill set." An accomplished scientist, he is also a writer and inventor. With great seriousness, he tells me that "Le Laboratoire is based on an educational program like the Bauhaus."

Edwards is passionate about the "wonderful naivete" he sees in young people, for whom "all is possible," and aims to nurture it at Le Laboratoire, where there is "no distinction between art, science, and technology." He also wants to bring their talent to bear in finding ways to solve society's problems, such as climate change and the widening gap between rich and poor.

He is well aware of the social and educational constraints that force artists and scientists apart, following the old adage that science is the search for truth while art is merely frivolous. Rather, he believes that the creative process is the key to the frontiers of knowledge. Edwards defines this as "a kind of experimentation, where the catalyst for change, for movement, for innovation, is a fusion of those creative processes we conventionally think of as 'art' and as 'science.' This fused process, what I call 'artscience,' is the basis of a new kind of culture"—the very culture he has created Le Laboratoire to house.

"As time has gone on," he continues, "getting to these frontiers [of knowledge] has become more and more difficult; providing that access, just seeing what is possible, is what interests me."

Edwards has observed that there is often "too much institutionalization of creativity, which leads to experts and dogma," something he does his best to avoid at Le Laboratoire. "One of the challenges we both have right now is being able to articulate what is going on," he says. "We shouldn't be too careful about it because it's a revolution." I entirely agree. Writing about the new art movement, one has to be forceful in confronting its critics and naysayers.

Edwards passionately believes that the "process of artscience needs to be better integrated into our cultural institutions. [The] *process* of artscience creation, even more than the works that result, is critically relevant to culture today." To achieve this, he invites the public into Le Laboratoire to observe the process of creativity, as it appears in works in progress in a functioning laboratory. This is diametrically opposite to the objects people see in museums, presented "as outcome, as product, dug up, carved down, highly edited, that follow a mysterious process of creative thought and engagement." At Le Laboratoire, the process of thought and engagement is clearly manifest.

Edwards is also an inventor, responsible for products like Le Whif and Le Whaf, both of which are on sale in Le Laboratoire's shop. Le Whif is a method of breathing in food rather than eating it. Using aerosol science, Edwards has created a nasal spray which sprays tiny particles of chocolate—the perfect way to enjoy chocolate without gaining weight. Le Whaf is a tastefully designed flask into which you pour a cocktail. Tilting the flask at forty-five degrees sets off a mechanism that vaporizes the liquid into a billowy cloud, heavier than air, which can then be poured into a glass and sucked through a straw. The user enjoys the taste of the cocktail but imbibes none of the calories or the alcohol. *Wired* magazine wrote of Le Whaf, "Instantly geeks up any cocktail party. Works with soups and sauces, too."

Edwards organizes many public events and private visits and sees a role for Le Laboratoire in instigating interactions between artists and scientists. When scientists visit Le Laboratoire, he says, they behave "like little kids, they become excited." When they talk to artists, "artists sometimes ask questions that scientists are not used to hearing, especially high-level scientists." There is, however, a downside to collaborations, he says. Artists are sometimes just not attuned to working on a complex theme that they know little about apart from what they have learned from a scientist. The resulting artwork can be mediocre. Some of his exhibitions have had unenthusiastic receptions. Nevertheless, he says, generally both scientists and artists come away with something, however indirect, from such interactions.

## Dublin's Science Gallery: Michael John Gorman

"Science Gallery is a platform for scientists, designers, artists, and entrepreneurs—a meeting place," says Michael John Gorman.

In 2007, the venerable Trinity College, Dublin, decided to open a science gallery to showcase and exhibit science, and invited Michael John Gorman to be the founding director. He has made Science Gallery into a powerhouse in the world of art, science, and technology.

Science Gallery is located on the edge of the campus, on a busy street near the inner city, which gives it an additional edge. When Gorman signed up, the building had already been designed. Thinking of the seventeenth-century coffeehouses which buzzed with ideas, he persuaded Trinity College to add a café as a meeting place for discussion. The café also houses a small bookshop with an intriguing collection. Science Gallery itself is a small space, though large enough for in-depth displays grouped close together. "The exhibitions are not really exhibitions but projects," Gorman explains.

With a background in philosophy and physics from Oxford, Gorman "got the art-science bug" while studying for his PhD in history at the European University Institute in Florence and decided to focus on curatorial matters. Slender and boyish, he speaks enthusiastically about the present state of Science Gallery and his plans for the future. "For me the exciting thing is that there is the opportunity to do something different here," he says. What he has in mind is a place totally unlike many science centers, which are meant for children and "infanticize science." Science Gallery, conversely, is linked with a university and can engage with researchers and bring them into the gallery to make contact with artists and designers, thereby "bringing [both artists and scientists] out of their comfort zone." Gorman envisions a situation where the "public can come into contact with them and be provoked and stimulated by ideas about emerging research that are not dumbed down for kids."

One of Science Gallery's financial supporters is the Wellcome Trust. "Wellcome likes a bit of competition," Gorman says with a twinkle. The concept of Science Gallery also struck a chord with engineers working at Google, which encourages its staff to devote 20 percent of their time to non-core work. Google went on to provide one million dollars in seed money for the gallery and is a creative participant.

While museums traditionally focus on the finished artwork, Gorman decided instead to home in on new ideas, filtering them to separate out those ripe for exhibition. He puts out periodic calls for new ideas and also created an additional source, the Leonardo Group. This is made up of some fifty artists, scientists, designers, and entrepreneurs who feed ideas into the gallery as well as coming up with big themes. The members change at least every two years, and so far this arrangement has resulted in eighteen exhibitions that have attracted some 800,000 visitors. This method, says Gorman, helps the gallery focus on creative projects while filtering out the "huge amount of bad stuff out there, providing also an environment for creativity."

Gorman had barely launched Science Gallery when an idea occurred to him. "Wouldn't it be interesting to have a network of galleries associated with universities around the world?" This would be a way to circulate exhibitions that started in Dublin, so that each exhibition need not be a one-off. The idea rapidly grew. Gorman now envisages Leonardo Groups in every city, constantly locating fresh concepts, with each Science Gallery acting as a "platform," a jumping-off point for creative ideas. His strategic view of his Science Gallery empire is that of a network, with each gallery learning from the others, and launching exhibitions which would move between them: a dynamic system with feedback loops.

He plans to launch one spinoff Science Gallery at Guy's Campus, part of King's College London, in 2015 with Daniel Glaser as director. By 2020 he envisages eight galleries in total. He is already in conversation with interested parties in Bangalore, Moscow, Singapore, and New York.

A major advantage is that the experience and machinery has already been tried and tested in Dublin. Gorman's farsightedness has resulted in collaborations of Science Gallery with the Wellcome Trust, Design Interactions at the RCA, Le Laboratoire, Microsoft, and Google, among others.

Gorman is aware that few commercial art galleries in London embrace artworks based on science and technology. But that is irrelevant. "We don't sell work, but we do connect interested buyers with artists," he says. "In a weird kind of way," he continues, "some of this artsci stuff is tipping into the mainstream. For many years it was experimental and rarefied. Now all of a sudden people—for example, entrepreneurs—want to know about artsci. It feels like we're not pushing uphill anymore." He points to outlets such as *Leonardo* magazine, which focuses on art, science, and technology, and to The Arts Catalyst, which provides aid in seeking funds, and CERN, as well as major exhibitors such as the Wellcome, Ars Electronica, and Le Laboratoire. "It's popular, it's sexy. Some people who have been out in the wilderness for years are now getting traction. It's a new cultural area that people feel they have to connect with."

As for art, science, and technology moving together, it's "already happened, but word is not completely out yet," says Gorman.

There is a need, he says, for flexible-minded people who can dive into situations, cross disciplines, and feel comfortable swimming in these waters, "like amphibians." He mentions several such cases. One is game design, another animation. A specific early example was the Defense Advanced Research Projects Agency (DARPA), created in 1958 by the US government for the development of new technologies for the military. These weapons systems demanded highly interdisciplinary input, and the spinoffs have had great impact on the civilian sector, most notably in communications, as well as providing funds for research at laboratories such as the MIT Media Lab. In the 1960s, DARPA instigated a program of research into techniques of deception, involving teams of scientists, technologists, artists, and psychologists. In 2013, Gorman held an exhibition at Science Gallery on this, entitled Illusion: Nothing Is as It Seems, a fascinating topic.

## Kinetica, a museum in the grand tradition: Dianne Harris and Tony Langford

"The distance between idea and invention has shrunk massively, and now it seems anything is possible, one just needs to think about it. For artists working in these realms, it is a golden era," says Dianne Harris.

Dianne Harris and Tony Langford founded Kinetica Museum, focusing on kinetic and electronic art, in 2006. Harris had worked in electronic art, robotics, photography, and the film industry, including special effects, while Langford worked for several years on regulatory aspects of electronic communications. Then he began to find himself more interested in communications that reach out and move people, so he did an MA in interactive art and became an artist.

In 2003, Harris opened a small gallery, the Luminaries Gallery, in West London to show works by herself and others, including Langford, which they describe as technology-influenced art that tries "to tap into unseen forces." The title of one of the shows at the

Luminaries Gallery was Frequency. Human beings are receptive to one specific frequency band, but there are other frequencies which dogs and cats can sense. One wonders what their worlds might be like. Langford's work featured a deep-sea fish that can "see" electronic pulses, while Harris's showed a wave form turning in three-dimensional space.

When Luminaries closed in 2004, Harris and Langford decided to team up and find a way to continue to show the same sort of work. What they wanted to do was to set up a gallery where cosmic questions would be discussed through works that "tap into the unseen, into space-time, that push the frontiers of art." It took a year or two to find funding. Finally, in 2006, the Arts, Humanities and Research Council gave them a grant of £250,000. Thus Kinetica Museum was born. From 2006 to 2007 it had a home in Spitalfields. For a while its shows were scattered around different venues in London, and it now has a permanent home at London Fields.

Harris and Langford planned Kinetica Museum to be part of what they see as the massive resurgence of technological art. They see it following in the grand tradition that began in the UK with *Cybernetic Serendipity*, the groundbreaking show that Jasia Reichardt curated at the ICA in 1968. As Harris points out, the members of the great art movements of the twentieth century—the Futurists, the Constructivists, the Bauhaus—considered that technology should affect art as much as it does everyday life, and used their art to express this belief. In homage to all this, the Kinetica Art Fair of 2010 was partly curated by Jasia Reichardt, and was dedicated to the masters of kinetic art, such as Bruce Lacey with his radio-controlled robots which would kiss you if you lingered too long in front of them, and Edward Ihnatowicz, creator of *SAM*, the *Sound Activated Mobile*. Since 2009, the annual Kinetica Art Fair, held at Ambika P3, a massive basement space near Baker Street, London, has been a milestone event in the world of art, science, and technology.

The artists at Kinetica are well versed in technology and science. Tired of telling engineers what she wanted, Harris says, she learned technology and programming. "But some artists just want to be called artists, rather than sci-artists, which categorizes them," she adds. In her view, artists think "in a more metaphysical way" than scientists. What artists and scientists have in common, says Langford, is "a quest for knowledge." The two describe scientists as dogmatic, and less creative than artists, "less metaphysical." Their view on collaborations is straightforward. Real collaborations should be acknowledged on the work, they say. But the name of a paid technician, no matter how great his contribution, "doesn't necessarily belong on the work."

The two are eager to articulate the qualities that set Kinetica apart:

> Firstly the work is experiential and performative, the works have a vitality or life force that can touch and engage with people on many different levels. Artists that exhibit at the Kinetica Art Fair can be generically termed "transdisciplinary performative users of knowledge," first described by the French theorist [Jean-François] Lyotard where artists are "users," as he calls them, who extract knowledge from science (on the nature of the universe) and through their own artistic process and appropriation, re-present and often simplify concepts with challenging results and new meanings.

In other words, they expect that scientists can learn from the way artists represent (or re-present) scientific concepts, in that both artists and scientists are intent on exploring the nature of the universe. From the viewpoint of the spectator, what sets the works at Kinetica apart is that they are kinetic—they move—they are beautifully made, and they are also often radical, subversive, funny, or just plain crazy.

## GV Art: Robert Devcic

In 2000, Robert Devcic saw an exhibition at the Hayward Gallery in London entitled Spectacular Bodies, a survey of anatomical art since the Renaissance, which examined and depicted the whole human body, inside and out. "It was not just aesthetically beautiful, there was a lot of knowledge behind it and in it," he recalls. Five years later, having learned the curatorial trade by selling at art fairs, Devcic set up his first exhibition in London, in a row of three houses he owned. The show featured nine artists, though none of the works was directly to do with art and science. Four thousand people visited over five days and the show was a sellout.

Devcic discovered that he "likes to be involved in artists' lives" and was already assembling the artists whose work is now central to GV Art. He was particularly interested in artists whose work could "generate discussion." Thus art and science became Devcic's world.

In September 2008 he opened the doors of GV Art in central London, near Baker Street, devoted to showcasing art influenced by science, particularly biology. As a gallery owner, Devcic is way ahead of his time. In the current climate, he is taking a huge risk. He dedicates immense energy and enthusiasm to this effort. The gallery also provides a congenial space for discussion.

Devcic's interest in science began when he was a boy in Australia. Initially he was fascinated by fossils and rocks, then moved on to plants and animals, while also taking a variety of courses, including beekeeping. After moving to London in 1986, he took photography courses and worked on magazines, as well as collecting art and regularly going to art galleries. When he found an artist interesting, he recorded the name in a notebook. Devcic is a fastidious collector of names and facts in numerous notebooks.

Devcic has chosen to specialise in biology-based art—for the moment, at least—because "people want to know more about it," in that biology is more immediately relevant to our lives than, for example, physics. Often, he has found, the "public feel out of their depth" at artsci exhibits. "GV Art tries to change this by displaying objects capable of standing on their own as art," without an explanatory text other than the title. "People gravitate to them—there's a magical quality about them." Recently Devcic has begun expanding the gallery's brief. In 2013 he put on an exhibition of sound art, still rare in the UK. While small, GV Art commands a loyal following. It is the only private gallery in the UK that showcases artsci.

## The Arts Catalyst: Nicola Triscott

Artists producing work on the borderline between art, science, and technology are well outside the mainstream, which means it is frequently difficult to find funding and support. In 1993 Nicola Triscott founded The Arts Catalyst to help solve this problem, and also to make the work of such artists better known and to seek out new talent.

Triscott has a clear strategy for the new art movement. Having started out in the performing arts, in the early 1980s she traveled widely in Africa, particularly in Zimbabwe and South Africa. There she regularly found herself in meetings with groups of artists, listening to tales of their struggles "to understand and assimilate the rapid changes occurring in their countries," particularly the changes in science and technology. She returned to the UK in the early 1990s and decided to follow this up. She pondered the developments in biotechnology and the implications of climate change. "Perhaps," she thought, "a way could be found to commission artworks that explored these rapid changes."

She was well aware that art and science was not a burgeoning field. "When I began talking about art and science, people thought I meant art and technology, computers." The Wellcome Trust's initiatives in this area were only just beginning. She was, however, impressed by the magazine *Leonardo*.

Triscott got to work. By the time she launched The Arts Catalyst in 1993, she had rounded up funding from the Gulbenkian Foundation, the Arts Council, the British Association for the Advancement of Science, and the Committee on Public Understanding of Science. There was, she recalls, "lots of goodwill and enthusiasm for the start-up." There were also criticisms. People questioned whether The Arts Catalyst's brief was rather narrow, in supporting art and science. "Narrow?" she replied. "Just about everything in the world relates to science." It was becoming clear that part of The Arts Catalyst's mission would have to be "to open people's eyes, [to show them] that science is not barren, it touches everybody."

Initially she had planned to pursue her work in artsci for a couple of years and then move on. "But I realized that the whole momentum gathered behind it had built up so much. Even today there is not enough transdisciplinary research. We've just scratched the surface." She and her colleagues spend much time "finding artists by in-depth curatorial research," traveling the world in search of artists and scientists whose work they deem worth funding. They get to know them and encourage them to develop their ideas further, in the hopes that a project will emerge.

Triscott is cautious regarding how an artwork should be credited. She prefers to leave this to galleries, who generally do not mention the commissioning organization or the collaborators, leaving only the artist's name on the piece. She is, however, willing to give "credit where credit is due."

She feels that specialization is necessary. Yet the broader outlook, the "extraordinary breakthroughs," can only emerge from groups of specialists supplemented by "nonspecialist interactions." The "really brilliant scientists are broadly based in science as well as being highly cultured," she goes on. She sees one aim of the artsci movement as being to eradicate the highly specialized cultures that riddle science.

"Everywhere is a hotbed of art-science now."

These organizations and others like them play an enormously important role by providing funding and a framework for a vital support system, making it possible for artists and scientists to work together, research, experiment, and create works of all types and colors. At the moment, they form an alternative subculture. As they become better known, they will bring the new avant-garde into the mainstream.

# In the Eye of the Beholder?

What makes a work of art beautiful? What makes an equation aesthetic? The conceptions of beauty in art and science are quite different yet not dissimilar, in that both strive to strike a responsive chord.

What I've been looking at in this book is a new sort of art, where artists and scientists work together. But does that mean we also need a new definition of aesthetics? Is beauty even a relevant criterion for a work of art anymore? And, to paraphrase Richard Taylor, who spotted fractal patterns in Jackson Pollock's paintings, can science "throw some narrow beam of light into those dim corners of the mind where great paintings exert their power?"

Is aesthetics simply a matter of taste? Is beauty in the eye of the beholder, or is it possible to make an objective aesthetic judgment? Recently neuroscientists have claimed to be able to do just this using magnetic resonance imaging (MRI) to scan the brain of a subject looking at a work of art. It seems that a particular part of the prefrontal cortex sparks when a viewer perceives a painting as beautiful.

These are the origins of the new field of neuroesthetics. One of its proponents, Semir Zeki, a brain researcher at University College London, has suggested that the artist has always unknowingly been "in a sense, a neuroscientist," in that artists instinctively know how to stimulate this particular part of the brain.

Art historians and psychologists have also contributed to the new field of neuroesthetics, but thus far there has been no consensus.

It is too early to know whether, as neuroscientists claim, aesthetic judgments can be explained using the laws of physics and chemistry as we know them today. And even if aesthetic judgments are totally physiological, this still doesn't help us understand what beauty is, why certain works give aesthetic pleasure and others don't.

There is a whole vast body of philosophical works defining and redefining aesthetics in both West and East. But while beauty in art is an ongoing discussion, in science the question of beauty is more cut and dried. The great mathematician Henri Poincaré, whose work also crossed into art, wrote, "The scientist does not study nature because it is useful; he studies it because he delights in it, and he delights in it because it is beautiful." For science, like art, is all about the quest for beauty. In biology there is symmetry of form and proportion, as in classical art. Nature favors the symmetrical, which is perhaps why symmetry is often something that we find beautiful.

When Einstein wrote his first great relativity paper in 1905, he used aesthetics as a guideline for his research. He made it clear in the first sentence that his doubts regarding the scientific theories of his day were not to do with the equations of physics but rather with their interpretation, which led to "asymmetries that are not inherent" in nature. Fired by an aesthetic of minimalism, he shaved away all the redundant explanations and inessential concepts. The result was his theory of relativity—a response to aesthetic discontents.

In physics there is an objective definition of the aesthetic. An equation is beautiful if it retains its form even when certain of its component parts are altered, such as switching right and left. If this switch leaves the equation unchanged, it is said to possess mirror symmetry, meaning that if that experiment were performed in a mirror world the result would be exactly the same. In physics symmetry is a form of aesthetics—a form of beauty.

But what of artists and scientists working today? What do they have to contribute to the debate? With a great deal of struggle, they have encountered the sublime, the aesthetic, the beautiful. What they have to say may be of value in understanding aesthetics and beauty in the context of the new art movement.

In 1933, the famous Harvard mathematician George Birkhoff published a work called *Aesthetic Measure*, in which he reduced aesthetics to a mathematical formula, stating that aesthetics is in the inverse ratio of order to complexity. In other words, the less complex an artwork, the more aesthetic it is. Inspired by this, the German philosopher Max Bense—later to utter the prophetic words to Jasia Reichardt, "Look into computers"—explored the possibility of using computers to generate a scientific notion of aesthetics. He also investigated how one might program a computer with algorithms in order to produce artifacts with aesthetic qualities. He called this framework "generative aesthetics" and the resulting work Generative Art, a term which continues to be used today. The early computer artists Frieder Nake and Georg Nees, whose work is geometrical and linear, were influenced by Bense. Their work fits Birkhoff's notion of the aesthetic rather well.

Neri Oxman, at the MIT Media Lab, links algorithms with aesthetics in a way that touches on Bense's generative aesthetics. She has, she says, an "aesthetic fascination with forms in nature, form generation as given by nature." For one of her projects she investigates the load-bearing properties of calcium, the way calcium distributes itself to form strong bones. She "tries to spec out algorithms that describe this conversation between matter and distribution of loads." The guiding force of the algorithm is aesthetic. Ultimately she wants to find new ways to use concrete in buildings, leading to a bold new architecture going beyond what is possible today. She sees this as encompassing engineering and art: engineering in the use of materials and art as it incorporates aesthetics, which emerges in the process of working.

The computer artist Ernest Edmonds takes as his guideline "minimalism in my means as well as in my aesthetics: simplicity," an aesthetic which seems to echo Birkhoff's. "Writing computer programs helped me in thinking about art problems. The less complicated in doing it, the better" is his take.

Similarly, Bernhard Leitner, a pioneer of sound art, favors a minimalist aesthetic characterized by clean lines and clear-cut patterns, as in his meticulous drawings depicting lines of sound and the locations of the speakers which will produce it. The ability to sense what is aesthetic is learned from experience gained through experiment.

Paul Friedlander, physicist and magician of light, also says aesthetics has to do with simplicity of design, as in ancient and primitive art, which he loves. The pseudonymous Jim Miller of the American artists' group EyeCandy ArtWorks wrote of him, "In a time when so many artists resort to bizarre and shocking gimmicks to achieve originality I take solace in the work of Paul Friedlander and others like him. They prove that beauty still has a place in modern art."

Ken Perlin, who won an Oscar for creating Perlin noise, for rendering animations more natural, agrees that "simpler is better" but adds that in order to be aesthetic, a work also needs to be pleasing to the eye. The question is, of course, whose eye? We are back to the riddle of individual taste.

Rolf-Dieter Heuer, director general of CERN, asserts that functionality is the essence of beauty. To him, functionality is aesthetic and "goes along with beauty." He gives as an example the alignment of cables laid out in parallel, like a work of minimalist art. "If it functions well," he says, "it has to be beautiful."

To Julian Voss-Andreae, whose sculptures evoke the ambiguous quantum world, aesthetics is satisfactory design. "To me, form and function are always a unit and both together make a good design, like in math or engineering," he says. "I cannot separate the experience of discovering or understanding such a solution from a beautiful aesthetic experience."

Rick Sayre, a supervising technical director at Pixar, says, "I don't have a precise definition of aesthetics. For us an element of aesthetics is when there is intention behind the work and the intention is to create a certain emotional response in the viewer and that emotional response is going to be motivated by the story and is also going to be influenced by specific desires of the director." Thus, in *The Incredibles* the aim was to create a simple skin texture that, like real human skin, responds to light. The real test, says Sayre, is whether the audience likes what it sees. If not, Pixar changes it until they do. The viewer's response is the filter that tempers the end result.

Jonas Loh, who creates extraordinary sculptures depicting data, such as his eerily organic *The "Gestalt" of Digital Identity*, or *Emoto*, the mountain range depicting emotions communicated in tweets during the 2012 Olympics, sees himself as searching for "new aesthetic forms while also researching how to communicate them." For him, aesthetics involves the communication of facts, which is what data visualization is all about. Aesthetics is a visual thing tied in with information content, he says. The higher the information content, the greater the aesthetic value.

A concrete example is Harry Beck's 1931 map of the London Underground. Its function is to navigate below ground, so Beck didn't worry about topographical features like hills, roads, or tunnels. He used his extensive experience drawing electrical circuits made up of horizontal and vertical lines and lines at forty-five degrees to the horizontal to produce a visual representation containing a great deal of information in a minimal graphical style. The map of the London Underground is a supreme example of the aesthetic of functionality. It has become iconic, serving as a model for tube maps of other cities.

Scott Draves, maestro of the computer-generated image and creator of Flame and Electric Sheep, sees his work as a collaboration between himself and the computer, in which the computer is an equal partner. "I want to give up control [to the computer]," he says. "I want to exceed my imagination." Electric Sheep is driven by the principles of Darwinian evolution. The public makes aesthetic judgments, choosing from among endlessly generated abstract animations which Draves's software changes on the basis of their decisions, even though their choices may not be to his liking. Thus aesthetics functions as a Darwinian device in which only the most interesting or beautiful survives.

Similarly, the computer artist William Latham, who creates spectacular images of virtual organic life forms, uses aesthetics to select from among the myriad forms he creates, following the guidelines of "symmetry, elegance, and balance."

Aaron Koblin, wunderkind who created the mesmerizing graphic of flight traffic across the US, not to mention *The Sheep Market*, takes data and applies standard software to it. If something interesting emerges he writes new software, hoping for something to appear that will stimulate a gut feeling, pass his emotional test for what is aesthetic: Is it exciting? Does it resonate? Is it clearly designed?

To Bradford Paley, the perfectionist who created the iconic *Relationships among Scientific Paradigms*, clear design is all. "Personally and emotionally, my driving goal is to find and reveal beauty in the world," he says. By beauty he means not aesthetics but meaning, how well a visual representation can be understood. Thus he reworked the *Scientific Paradigms* image, even though anyone less perfectionist would have been happy with it. He found the "wisps of hair" in his diagram, made up of key words from published papers, too aesthetically seductive. His decision was to make them less prominent, to favor meaning over beauty.

Peter Weibel, performance and media artist as well as chairman and CEO of ZKM in Karlsruhe, takes a similar stance. "For me aesthetics starts with the Greek word *aisthanomai*, meaning 'I perceive, feel, sense,'" he says. "Aesthetics is a general medium for questioning the world. For me aesthetics has nothing to do with beauty, but with information. Many of my works are not directly linked to reality, but to aesthetic problems."

In Weibel's universe, aesthetics is something our mind constructs from the complex interplay of our senses. In his film, *Virtual Tetrahedron*, an actor appears and inserts a screen with lines on it into a metal frame of indeterminate shape, and a tetrahedron magically appears. The tetrahedron never really exists. It is an illusion. Thus aesthetics has to do with the way the mind makes sense of the world around it, the satisfying moment when the parts of the puzzle click into place.

Physics-influenced artists Evelina Domnitch and Dmitry Gelfand, who use lasers to create magical works combining physics and chemistry, reach back to their heritage in Russian mysticism. When a physicist who was totally familiar with the technical aspects of a cloud chamber saw their installation, *Memory Vapor*, he was moved to tears, because it took him beyond everyday reality into a vision of the universe far removed from the usual cloud chamber events. "Aesthetics is to be emotionally moved," they say.

They don't deliberately try to make a work beautiful. The "aesthetic emerges; it has to be emergent. We hope the unexpected will appear. This is a most important ingredient, even in an artwork." They seek, they say, "an aesthetics of the ephemeral, of the ethereal."

For the musician, author, and composer of experimental music David Toop, "Aesthetics is intuition. Intuition is when you're not thinking about what you're doing. At a certain level, you don't know what you're doing. But at a deeper level, you do." Intuition is experience, essential to improvisation.

The sound artist Jo Thomas, who recorded the sound of the electrons whirling around in the Diamond Light Source accelerator, includes skill—how skillfully a piece is made—as an important component in the aesthetic quality of the finished piece. The artist's feeling for the work is also important. "A work has a belief in itself and must be able to exist in its own right." And the work must involve a "transfer of knowledge, that is, of beauty and skill."

Paul Prudence, who creates spectacular sound and image shows, says, "Aesthetics is complex. It's not just about new media experience, but about much deeper modes of understanding and the history of cultures. It's some sort of formal way of putting together sound and visual material that seduces you on a number of levels, based on experience," he adds, deliberately specifying experience rather than intuition.

Composer Robert Rowe, the New York experimental musician who delves into the brain's structure, says, "Aesthetics feeds into the compositions I make." He hopes that computer models of music will offer a better handle on the elusive concept of aesthetics by producing pleasing melodies from sets of rules which may offer clues as to the rules hardwired into our minds. This sort of research program is similar to programs in cognitive science which investigate how scientists think through experimental data as they work toward discovering scientific theories.

There's also the mystery of how people perceive beauty, as in the case of A. Michael Noll's algorithm which produced a Mondrian-like painting that many viewers preferred to the real Mondrian, or David Cope's computer simulations of Bach's music, which listeners couldn't distinguish from the real thing. Both created aesthetic objects that triggered responses in the neurology of the viewers/ listeners.

"What else do we need to know to write a computer program that could do something better, that could probe more deeply into how Bach created his works?" Rowe asks. Despite still being in its infancy, Rowe is optimistic that through exploring algorithms for producing music AI may be able to reduce the gap between the present, somewhat fuzzy notion of aesthetics in music to a more quantifiable one.

Bruce Wands, media artist, musician, and composer, homes in on the happy mistake—the ability of the creative mind to recognize the melody in, for example, a musical scale wrongly played. Experience primes the mind to be alert for such unexpected experiences which can occur during improvisation, where "I let the music play me." Speaking of his use of the computer for composing music, he is intrigued by the power of "algorithms to produce new images and sounds," he says. For him aesthetics is intuitive in the most freewheeling sense, playing a role in and sometimes emerging from improvisation and the creative possibilities that it can lead to. Toop shares this view.

The sound artist Tristan Perich, composer of the *1-Bit Symphony*, says of the minimalist music of Philip Glass and Steve Reich, "You can see their aesthetic from afar, but then you can also see how it's built up. It's all similar stuff—music, art, mathematics—dependent on basic building blocks—notes in music, the lines in art, numbers in mathematics, and atoms in physics. I find that inspiring, how music appeals to the same part of the brain as science." "Simple stuff is the most pristinely beautiful," he adds. For Perich, the scientific input of his work ends with the programming, while the music, which contains the aesthetic experience, is intuitive. Thus he separates the rational from the intuitive in his work.

Oron Catts, creator of *Pig Wings* and head of SymbioticA, the cutting-edge bioart laboratory, considers beauty "a strategy of seduction" and sees a "crisis in aesthetic strategies with which artists ought to be struggling." Forms grown out of living tissue, like *Pig Wings*, are inherently aesthetic in that nature prefers pleasing forms.

Jun Takita, who turned a model of his own brain into a moss garden, homes in on the moment of discovery when it occurs to him "how to do the work" as his key aesthetic experience. Process rather than result is the basis of his work.

It's not surprising that such a diverse range of artists offers a diverse range of interpretations of what aesthetics is. Key elements are simplicity, functionality, coherence between form and function, good design, and how well the finished product expresses the artist's intention or communicates information in a new and interesting way. Then there are less tangible interpretations. Most seem to agree that aesthetics has to do with intuition; perhaps it is a quality that can only be intuited, not explained. Or it may have to do with experience. Some artists are aware of the aesthetics of the piece as they work; others see it in the finished product; and for some it's a selection procedure, a way of choosing from a myriad of images, or the way in which the mind suddenly registers the beauty in what at first appears to be a random event or even a mistake, when the computer "takes over," to paraphrase Draves. It also comes with the artist's recognition that the piece is finished. The response of the viewer comes into it too.

Artists with a background in mathematics, computer science, or physics express their opinion more directly and concisely. Most data visualization artists give a direct view of what an aesthetic representation is: the higher the information content, the greater the aesthetics. William Latham gives what must surely be an appealing definition of aesthetics: "symmetry, elegance, and balance."

These artists' comments on aesthetics are dispatches from the cutting edge. They offer fresh approaches to a concept that is usually considered elusive, as well as insights into how the creative mind functions. All link aesthetics with intuition while taking care to note that this is the product of experience, not something fuzzy. Relating aesthetics to algorithms further demystifies it.

Different cultures have different notions of what is aesthetic, as do individuals within a culture. But might it be that there are structures in the mind, hardwired from birth, that generate notions of aesthetics from input from the world in which we live? This is what Robert Rowe and his colleagues are researching in music. Perhaps neuroscience, cognitive science, computer science, philosophy, or psychology, as they are today, can offer only a hint of things to come. Eventually all these subjects will fuse to create a hybrid form of knowledge, electronically based, capable of elucidating brain functions, and offering a new concept of aesthetics and its role in creative thinking.

This is not reductionism in the sense of explaining phenomena using the laws of physics as we understand them today, reducing us all to nothing but subatomic particles, dead matter. We are greater than the sum of our parts. But how can the workings of particles at the submicroscopic level produce inspiration and consciousness? Perhaps explanations derived from a new, wider form of knowledge will leave open a role for inspiration in the production of sublime art and music, as well as producing new explanations as to why the universe is as it is and we are as we are.

In 1913, Guillaume Apollinaire wrote of a "Scientific Cubism" that would move in an entirely different direction from the art of the day. It would be the "art of painting new structures out of elements borrowed not from the reality of vision, but from the reality of knowledge," by which he meant knowledge obtained from going beyond appearances. In a similar vein, Duchamp, wrote Apollinaire, used shapes and color "not to capture appearances but to penetrate the very nature of these forms." One word for the knowledge shared by artists and scientists might be "intuition."

Gerfried Stocker, artistic director of Ars Electronica, a showcase for the new art movement, speculates on an aesthetic emerging out of data visualization, a "new aesthetics of the future." The twenty-first century marks the beginning of the age of information. Our lives are already to a large extent determined by information, by the devices, such as cell phones, that we use to access it, and by the networks we frequent, such as the World Wide Web. It will be exciting to see what new concepts of aesthetics emerge to fit the forthcoming information age.

# The Coming of a Third Culture

n the heyday of the avant-garde, in the early years of the twentieth century, when Picasso and his friends were shaking up Parisian society, one of them, the poet and gadfly of the art scene André Salmon, wrote exuberantly, "Everything is possible, everything is realizable, in all and everywhere"—thrilling words that might also be the credo of today's avant-garde.

Art has always had its rebels, who go against the conventions of both art and society and thereby push the boundaries of art forward. The Cubists' lineage dates back to the Salon des Refusés—"exhibition of rejects"—of the 1830s, set up to show works rejected by the establishment salons. In 1863, the Salon des Refusés was the only place that would show Manet's *Le Déjeuner sur l'herbe* and Whistler's *The White Girl*. In the 1870s, artists like Manet, Renoir, Whistler, and Monet were satirically dubbed the Impressionists and had difficulty making a living.

Then came artists like Gauguin, Toulouse-Lautrec, Pissarro, and Seurat, who flew in the face of the establishment canons of what art should be, and Cézanne, who even threw aside perspective. These artists in their turn were rejected by the establishment and in the 1880s banded together to form the Salon des Indépendants, where they could show their work freely. It was a long time before the work of any of these artists was accepted into the mainstream. Now, of course, the Impressionists and their successors are a much-loved part of the canon.

The Cubists were more radical still. Their work emerged from dramatic developments in science and technology and was highly geometricized and minimal. Unsurprisingly, it too was rejected by the establishment. Only one lone dealer, Daniel-Henry Kahnweiler, took a chance and agreed to handle them. Single-handedly he established a market for their work, enabling them to continue to explore the worlds beyond perception in dramatic new ways.

Artsci, the new movement I've been looking at, is more extreme yet. Not only is it science- and technology-influenced, as was Cubism, but its artists use scientific and technological media, whereas the Cubists used only their interpretations of the new ideas. Today's artists often work together with scientists, which the Cubists never did—nor, for that matter, did any major artist until the second half of the twentieth century. Their work may even directly affect the work of scientists. To quote Peter Weibel's words yet again, "Today art is an offspring of science and technology."

So what is the meaning of this new avant-garde? What are the implications? Are the artists I've described just a disparate group who happen to work with scientists, or are they more? Are they a symptom of something larger? Are they the tip of the iceberg?

There are many ways we can look at the artists I write about here. We can divide them into those who collaborate with scientists and those who are both artist and scientist in their own right. Or we can divide them into those whose work is widely known, such as Antony Gormley, who has a sculpture at CERN, and Aaron Koblin, creative director of the data arts team at Google, who transforms data into witty visual displays, and on the other hand those whose work is still marginal, cult figures maybe known only within this specific world.

Are they the new avant-garde? Though these days there are many conferences and festivals devoted to artsci and many artists turning out stunning artworks, they have to be so described because they are still outside the mainstream. Certainly it is indisputable that what we are looking at here is a brand-new art movement.

More than fifty years have passed since C. P. Snow gave his epoch-defining Rede Lecture in 1959, bewailing the split in intellectual life between what he called the two cultures, with the result that people who considered themselves highly educated were not ashamed that they had no understanding of terms like "mass" or "acceleration" and no knowledge of the second law of thermodynamics. Since then the idea of the two cultures has become part of everyday use. People still think of art and science as being at opposite ends of the spectrum, of scientists as white-coated figures in labs and artists as Salvador Dalí–like characters wielding paintbrushes. They still assume that art depicts nature while science analyzes it, that art brings things together while science pulls them apart.

In fact, both artists and scientists have always been engaged in trying to fathom the reality beyond appearances, the world invisible to our eyes. Leonardo was both artist and scientist, because in his day there was no distinction. It was only with Newton and the first developments of modern physics that a division began to be made between the two. In the centuries that followed, science led to technological development while art took on the role of representing people and landscape, of storytelling and decoration, the creation of beauty. It was not until the first decade of the twentieth century that the two began to converge again.

Now it seems we are finally seeing the coming back together of art and science. Perhaps more than fifty years down the line it is time to discard the two categories that Snow identified, to move beyond his two cultures. Are we witnessing the birth not just of a new art movement but of a whole new culture—a third culture, in which art, science, and technology will fuse? That implies individuals whose understanding of the world includes a merging of art, science, and technology, a blurring of boundaries on the largest of scales, in which these three disciplines no longer function separately.

We might ask whether these artists feel themselves to be part of such a culture, what that means, and whether it is a good thing. Some of the artists and scientists I interviewed argue that, far from art and science coming together, as science grows more advanced we need more intense specialization to pursue research into increasingly complicated and difficult fields. Conversely, many artists told me that for them the distinction between artist and scientist is already meaningless. They regard themselves not as artists or scientists, but as researchers. Ken Perlin, who won an Oscar for the magical animation technology that made films like those produced by Pixar possible, says, "What's the big deal? It's here already." He sees young people coming into media art today who have a mastery of computation, the confidence to make computers do whatever they want. Joichi Ito, director of the MIT Media Lab, claims that "it started here"—in the Media Lab. It's also not news to people in the movie industry. As Rick Sayre of Pixar points out, "motion pictures are by definition art and technology fused together." He continues, "There are artists right now working in digital media, predominantly algorithmic—you might say they are procedural artists, developing algorithms and equations. They tinker with the equations until they like the way the picture looks. It's artwork with a piece of code behind it. That's been going on for more than twenty years." As far as he's concerned, the coming together of art and science is always going to be outside the mainstream. "It is never going to happen in the broad cultural context. But," he adds, "it's already happened in the media world, which is the world of Ars Electronica and the world of visual effects."

Many artists also feel that, as Jonas Loh, creator of the intriguingly sculptural *"Gestalt" of Digital Identity*, puts it, "the border between science and art will melt away, enabling research to be dealt with in a more artistic way. Blurring the line will lead to interesting new developments and to interesting new times." Benedikt Gross, another designer and RCA stalwart who works in data visualization, suggests that programming is going the route of photography. In the 1950s photography was expensive and largely restricted to experts au fait with f-stops and focusing. Now everyone takes photographs with a camera or, more and more often, a phone; photography has become just a tool. Similarly, programming is becoming more and more accessible, done with apps rather than code. There is "no longer a programmer and an artist. They are one and the same."

As far as Michael John Gorman, director of the Science Gallery in Dublin, is concerned, the third culture is already part of our world, though "word is not completely out yet." What is needed, he thinks, is flexible-minded people who can inhabit the worlds of both scientists and artists, who dive across disciplines and feel comfortable swimming in many different environments, "like amphibians."

Not all the artists and scientists I met agree, though. To quote Rolf-Dieter Heuer, the experimental physicist who is director general of CERN, "the questions are so deep today that they demand specialization." He points out the difficulty of working as an individual. "All is teamwork, to function as a creative group." There is so much to learn in any one field that no individual could ever master it all, in which case it is better if we all stick to our own specializations. "You only have one lifetime," as the flamboyant performance artist Stelarc puts it. Nevertheless, this does not preclude collaborations.

Oron Catts, responsible for *Pig Wings* among other biological confections, takes the extreme position that artists need only use scientists and their work. Science should be just a tool, "like paint in a can."

The Russian artists Evelina Domnitch and Dmitry Gelfand, creators of sumptuous environments combining light, physics, chemistry, mysticism, and art, assert that while we may be in at the beginning of a third culture, it is still marginal. The notion of a "culture" goes well beyond that of a "movement," they say, and at the moment this still does not exist.

Ken Arnold, the dynamic head of public programs at the Wellcome Trust, would prefer it if artsci could be eternally emerging and never established. "There's something exciting about it being discovered. As soon as it becomes settled it becomes an orthodoxy and then, frankly, someone needs to give it a good kicking or else it becomes a bit bland." A good point: little by little interdisciplinarity will become just another discipline, the avant-garde will become mainstream, and eventually the process will have to start all over again. Surely this is how knowledge is generated, in an upward spiral, always seeking equilibrium but never attaining it, as new puzzles and new frontiers appear.

Arnold makes another interesting point. "At the micro level

things seem to be blurred"—there seems to be a fusion of art and science—but "on the macro level there is still a mixture, not a compound." In other words, although to practitioners of art and science there seems to be a fusion of the two, the public is not aware of any such fusion as yet. "In the intellectual sphere many people don't define themselves as artist or scientist; but the public sphere is more conservative and shies away from" the notion of a third culture. He adds that "the disagreements between these levels can be quite exciting."

WHAT FORM WILL the third culture take in the long run? What will it mean to our society if these artists and scientists working together really are the vanguard of a brand-new world? It doesn't take a great stretch of the imagination to envisage people in the future working with computers made of not-yet-invented materials, designed according to not-yet-dreamed-of architectures, producing theories that generate images that can be manipulated like equations and which are aesthetic according to some new definition of the term.

Many doubt that this will ever happen. It's madly ambitious to imagine that we can educate students fully in art, science, and technology as these disciplines stand at present, let alone keep up with the ever-increasing amount of new knowledge. But—in answer to such doubts—who could have predicted, even thirty years ago, what the state of art, science, and technology would be today? These fields are already moving closer together. Perhaps in the future there will be such a degree of fusion that a big picture will emerge in which the quantity of knowledge will be manageable, in which many facts and already existing fields will have been absorbed or become irrelevant.

We may question whether we can ever become Renaissance men and women. It's important to remember that during the Renaissance—roughly the fourteenth to the late seventeenth centuries, when the Enlightenment, the Age of Rationalism, began— there might have been less to know, but what was known was often vague. Scholars struggled to unearth knowledge in their efforts to understand their world better. Thinkers like Giordano Bruno and Galileo questioned the truth of given canons of thought and were punished for it.

Today the amount of knowledge is comparatively huge, bogged down with highly specialized, complex subjects. In order to take the first steps toward a third culture, we need to look into our education system, to see how subjects are taught and can be brought together. Our present education system barely acknowledges interdisciplinarity. Schools and universities maintain the traditional departments: physics, chemistry, biology, art. Departments that span disciplines are rare. There are exceptions: media labs such as those at MIT and NYU, art schools such as the School of Visual Arts in New York and the Slade School and Central Saint Martins in London, which recently initiated the first master of arts program in art and science. A number of summer schools in artsci are emerging. The Spanish artist Margarita Cimadevila runs a highly successful summer school in scenic Santiago de Compostela, which she calls "Science & Art: So Similar, So Different." The attendees are mostly high school teachers from Central Europe who return and integrate artsci into their curricula.

Disciplines are also merging. The age of technology has been overtaken by the age of information. Nowadays science and technology can be explored in terms of information content, from the information in a cell to astronomical data transmitted in pixels and the vast amounts of information stored in the cloud. This has also created some of the most advanced forms of artsci—media art, sound art, and data visualization. It is here that interdisciplinarity between the arts and the sciences is beginning to take root.

The most focused trend toward interdisciplinarity is still within the sciences, where departments with hybrid titles like "engineering sciences" are becoming more common. We have biotechnology (biology and technology), nanoelectronics (physics and engineering dealing with the nano level), computational physics (akin to data visualization), and health physics. Eventually specialist areas will disappear and be absorbed into a larger framework with over-arching principles. As a result there will simply be less to learn.

As one would expect, it takes some searching to track down the artworks of the new avant-garde. London's Tate Modern has one major piece, Aaron Koblin's *This Exquisite Forest,* while New York's Museum of Modern Art has had several exhibitions of media art in the past few years thanks to the indefatigable Paola Antonelli, senior curator in the department of art and design. In 2013 Barbara London, associate curator in the department of media and performance at MoMA, put together the first group exhibition to single out sound as a form of artistic expression. One of the artists involved was Tristan Perich, the composer of *1-Bit Symphony,* who contributed a grid of 1,500 speakers, each playing a different tone.

In an upbeat review, Blake Gopnik of the *New York Times* wrote, "Sound art has been on the rise for a decade or two, but it may at last have hit the mainstream." He pointed out that whereas the auction houses Sotheby's and Christie's had not yet sold any sonic art, "they've begun to get good prices for major videos. Sound art's vaguely unworldly air may actually increase its appeal."

Predictably, the establishment art world has given artsci the cold shoulder, just as it did Renoir, Manet, and the Cubists. Even though science, technology, information technology, and software are bedrocks of our twenty-first-century culture, with a few honorable exceptions curators and gallery owners tend to treat artworks based on them with suspicion. Most curators are schooled in traditional art history and are uneasy in the presence of science and technology which they don't fully understand.

Moreover, artsci violates the rules of the art world, in which artworks are commodities. Prime among these is that a work must be unique. Most works of artsci are reproducible. They may need to be repaired or improved as technology advances and end up radically different from the original in look and function. Curators are particularly uncomfortable with computer art generated by algorithms, where the image constantly changes and—worse, from the curator's point of view—the human hand behind it is invisible. To their eyes, such works are not art at all but gimmicks.

There are some notable exceptions. Josiah McElheny and David Weinberg's dazzling glass and aluminum extravaganza *Island Universe* was exhibited in major galleries across Europe and the US, and Keith Tyson's complex and thought-inspiring works in different media are also widely exhibited and sold through galleries.

Just as Cézanne, Manet, and Renoir bypassed the establishment salons and exhibited at the Salon des Refusés and the Salon des Indépendants, and just as the Cubists made use of the services of Daniel-Henry Kahnweiler, so the new avant-garde has established an art world of its own with large venues equipped with all the necessary technology to house the new art: Ars Electronica, the Wellcome Collection, Le Laboratoire, Science Gallery, Documenta, and ZKM. Here they conduct competitions, meet buyers, and gather together to celebrate their work.

If, as seems likely, the new avant-garde is at the forefront of a coming third culture, the establishment art galleries that focus on selling "flat art" that can be hung on walls will find themselves increasingly marginalized. But this too may change as artists come on the scene who grew up surrounded by technology and naturally include it in their art. These future artists may produce sumptuous and decorative science-inspired art. Indeed, the very notion of what art is will change, along with the concept of aesthetics.

In time artsci will be incorporated into art history and art appreciation courses until it too is fully absorbed into the mainstream, in the same way that the works of Debussy, Duchamp, Einstein, Frank Lloyd Wright, James Joyce, Picasso, and Stravinsky, once rejected as scandalous and shocking, have now become part of our heritage.

Thus art, science, and technology as we know them today will disappear, fused into a third culture—leaving the door open for the next, as yet unimaginable, avant-garde.

# NOTES

## Epigraph

xvii  "Art does not reproduce": Klee (1968), 182.

## Preface

xix  "It's just great": Glueck (1966).

xix  "You probably don't remember me": O'Conner (1966).

xx  "A bomb dropped here": Glueck (1966).

## Chapter 1

5  "a good looking bootblack": Stein (1933), 46.

6  "We had no other preoccupation": Parmelin (1969), 106.

6  "M. Maurice Princet preoccupies himself": Salmon (1910), 163. See Miller (2001)

6  "conceived of mathematics like an artist": Salmon (1919), 485.

8  in the face of one of the women: Miller (2001), 113.

8  "terrible picture [that looms] through the chaos": Burgess (1910), 408.

8  "a nightmare"; "overwhelming": Cousins and Seckel (1994), 165–68.

9  "was set free to produce": Seelig (1954), 68.

9  "spellbound": Solovine (1956), x.

9  "This group had a considerable": Einstein to his friend from his patent office days Michele Besso, March 6, 1952, in Speziali (1972), 464–65.

9  "unbearable": From an unpublished 1919 essay by Einstein. See Miller (1998), 137. For an up-to-date Einstein biography see Isaacson (2007).

9  "led to asymmetries that do not": Einstein (1905b), 370.

10  "A consequence of the work on electrodynamics": Einstein, letter written somewhere between June 30 and September 22, 1905, to his friend from his patent office days Conrad Habicht, Einstein (1993), 20–21.

10  Most physicists were convinced: Miller (1998), 325–26.

10 "theorizing out of their depth": Einstein (1923), 484.

11 "overworked" as God: Richardson (1991), 475.

11 "The work of art is born": Kandinsky (1977), 53.

12 "That it was a haystack": Lindsay and Vergo (1982), 363.

12 "spiritual existence": Kandinsky (1977), 52.

12 "The harmony of the new art": Ibid.

13 "with elements borrowed not from": Apollinaire (1913), 69.

13 "We will glorify war": Lista (1973), 98.

14 "Cathedrals made of shit": Richardson (2007), 290.

16 "profound formal distinction": Einstein (1905a), 132. By 1905 Einstein had discovered that light was a wave and particle at the same time—the so-called wave/particle duality. See Miller (1998), pp. 126–28.

18 "Henceforth space by itself": Minkowski (1908), 75.

18 From 1921 onward, themes related to Minkowski's: Hatch (2010).

19 "There are 'made' laws": Mondrian (1937), 353.

20 It did affect the final design: Hatch (2010).

20 "It is necessary to point out": Mondrian (1937), 351.

21 "We do not want to clap the atom": Pauli to Niels Bohr, December 12, 1924, in Hermann, von Meyenn, and Weisskopf (1979), 189.

21 "a simple farm boy with short": Born (1975), 212.

21 "I had the feeling that": Heisenberg (1971), 61.

22 "which appeared very difficult to me": Schrödinger (1926), 735.

22 "The more I reflect on the physical portion": Heisenberg to Pauli, June 8, 1926, in Hermann, von Meyenn, and Weisskopf (1979), 328.

23 "moving around an object": Gleizes and Metzinger (1980), 68.

23 "form to thoughts to an audience": Anderson (1967), 322.

24 "One event can be the cause of another": Parkinson (2008), 55.

24 "Something more real than reality": Richardson (2007), 349.

25 "No, it's sperm": Ibid., 330.

25 "Reality is more than the thing": Ibid., 350.

25 "Human personalities are not measurable": Eddington (1968), 33.

25 "The new geometry of poetic thought": Parkinson (2008), 180.

26 "this harrowing and colossal question": Ibid., 192.

26 "I think it is probable that negative": Dirac (1933), 324–25.

26 "I knew that the theory is correct": From a conversation Einstein had with a student shortly after receiving news of Eddington's result, quoted in Holton (1973), 236–37.

27 "he delights in it because it is beautiful": Poincaré (1958), 8.

27 is eye-catching due to Picasso's violation: Miller (2000), 298–99.

28 "confront us with suggestive dilemmas": Parkinson (2008), 61.

28 "psychologically the modern physicist": Ibid., 62.

28 "modern scientific and artistic thought": Ibid., 61.

28 "today, reason goes so far as to propose": Ibid., 66.

29  "Matta wanted to show the Surrealists up": Ibid., 153.

29  "Matta's first and most important contribution": Ibid., 154.

29  "it tries to poeticize science": Ibid., 160.

30  "In the Surrealist period I wanted to create": Ibid., 216.

31  why the nucleus was stable: Miller (1985) and Miller (2000), 239–43.

31  derived from our everyday world: Miller (2000), 247–49.

31  "A half-assedly thought out semi-visual thing": Gleick (1992), 244.

## Chapter 2

33  "the first art to appear here": Rosenberg (1959), 137.

36  a 90 percent failure rate: Hillman (2007), 68.

37  "never went anywhere": Julie Martin, author interview, October 31, 2011.

37  "one glorious act of mechanical suicide": Klüver (1966), 32.

38  "I can still smell it": Klüver (1998), 5.

38  "During its short life": Lindgren (1969a), 59.

39  Expect the unexpected, was the byword: Perhaps what happened to Duchamp's *The Bride Stripped Bare by Her Bachelors* came to Tinguely's mind. Made up of two panes of glass, it was broken en route from its first public exhibition. Duchamp repaired it but left the cracks intact as an indicator of unpredictability. Some say he thought the cracked version to be an improvement.

39  "Why wasn't I invited?": Klüver (1998), 8.

39  beyond his immediate neighborhood: As told to me by Larry Wright, an assistant and good friend of Rauschenberg.

39  "a ghost bouquet of promises": Rose (1987), 67.

40  "collage of ever-changing bits of music": Klüver and Martin (1991), 85.

40  "is the least individually beguiling": Pippard (1965), 57.

41  "Rauschenberg arrived and we worked frenetically": Klüver and Martin (1991), 86.

41  "as much a part of the work of art": Klüver (1966), 37.

41  "by the artist is not only unavoidable": Ibid.

41  "felt it was a positive use of time": Julie Martin, author interview, October 31, 2011.

42  "new and maybe inhuman objective": Klüver (1966), 37.

42  "Technology is the extension of our nervous system": Ibid., 38.

42  "I'm not so much interested in helping artists": Lingren (1969a), 68.

43  "The artist's work is like that of a scientist": Klüver (1966), 38.

43  "my studio is a sort of laboratory": Laporte (1975), 38.

43  "Many people wanted E.A.T. to be about": Hertz (1995).

43  "What do an artist and a scientist": Julie Martin, author interview.

43  "When I heard last May": Leonard J. Robinson, in transcript of E.A.T.'s first open meeting at the Central Plaza Hotel, November 30, 1966, quoted in Loewen (1975), 54.

43 "I watched Billy having fun": Herb Schneider, interview with Charisse Bardio and Catherine Morris. Unpublished. I thank Jasia Reichardt for access.

44 swinging Greenwich Village art scene: At this time Klüver and Rauschenberg were also involved in a proposed Festival of Art and Technology in Stockholm. They suggested bringing over artists and engineers from Bell Labs because they had already taken part in preliminary meetings. The Swedish organizers insisted on using only their own engineers. Further friction was avoided when the Swedish program folded in summer 1966.

44 well outside company time: Cynthia Pannucci, author interview, January 11, 2012.

44 "At first artists were in control": Herb Schneider, interview with Charisse Bardio and Catherine Morris.

45 "knocked the critics off their rockers": Hillman (2007), 180.

45 "their unhappy wives": Ibid., 68.

45 "just said what he wanted": Julie Martin, author interview.

46 "Good luck": Written September 16, 1966. Hillman (2007), 69–70.

46 "engineers express doubt in front of the artists": Robinson (1967), 16.

48 "[9 Evenings] received, on the whole": O'Doherty (2006), 75.

49 "a new interface between these two areas": Klüver (1966), 34.

49 "Sixty people had questions right away": Julie Martin, author interview.

50 "collaborative relationship between artists and engineers": E.A.T. News 1, no. 2 (June 1, 1967).

50 "politics, and the technical community": E.A.T. News 1, no. 3 (November 1, 1967).

50 "should not be without art": The great German philosopher Immanuel Kant wrote in the Critique of Pure Reason: "Thoughts without content are empty, intuitions without concepts are blind."

51 "via a sophomoric prank": Knowlton (2005), 3.

51 "of, guess what, a nude!": Ibid.

51 "was delighted but worried": Ibid.

52 "the name of Bell Labs with it": Ibid.

52 "Interesting but no big deal": Ken Knowlton, author interview, November 4, 2011.

52 "it was art with a capital A"; "be sure that people know": Ibid.

52 "instead of porn": Ken Knowlton, email to the author, July 17, 2012.

52 "Billy was forever retelling the story": Ibid.

52 "is the topic of the day": Document in Jasia Reichardt's archive.

53 trying to include too many fields: Julie Martin, author interview.

54 "enthusiasm of one man": Reichardt (2008), 76.

54 "edges of things, of how art": Jasia Reichardt, author interview, August 10, 2011.

54 "art's outer periphery": Reichardt (2008), 72.

55 "replace man with a work of abstract art": Quoted at http://olats.org/schoffer /cyspe.htm.

57 "wanted to be artists too"; "meetings were enthusiastic": Jasia Reichardt, author interview.

57 "Look into computers": Reichardt (2008), 77.

57 "to the theories of chance": Ibid.

58 "involvement with the arts": Reichardt (1968), 5.

58 "engineer, mathematician, or architect": Ibid.

58 "scientists masquerading as artists": Willats, interview with Catherine Mason, June 6, 2004, quoted in Mason (2008), 104–5.

59 "a wider consciousness": Ibid., 103.

60 "could encourage exciting innovation": Edmonds (2008), 348.

60 "existential romance, glamour and drama": Ihnatowicz (2008), 111.

61 "science museum for the year 2000": MacGregor (2008), 83.

61 "one day as a landmark": Gosling (1968).

61 "lift a finger to entertain them?": "Fun by Computer," *Evening Standard*, August 2, 1968.

61 "exhibition in the world at the moment": Shepard (1968).

62 "power of new technologies": Mason (2008), 109.

62 "it is prematurely optimistic": Reichardt (1968), 5.

62 found it difficult to obtain funding: See Mason (2008), 106. For a historical revisionist account, see Uselmann (2003).

62 "Thus people who would never have put pencil": Reichardt (1968), 5.

62 "what art is for or about": Shepard (1968).

62 "it will not be worth doing": Melville (1968).

62 "For computer scientists whose work": Mason (2008), 100.

63 "This exhibition is not"; "Many artists today": Hultén (1968a).

63 "in planning for such a world": Hultén (1968b).

64 "The idea becomes a machine that makes the art": Lewitt (1967).

64 "forms of experimental art": Shanken (2002b), 433.

64 "intensive artistic experimentation": Ibid.

## Chapter 3

67 "I can still remember him": Noll (2011), 4.

67 "computer art": Ibid.

68 "only the art world": Ibid.

68 "Béla always felt that his patterns": Ibid.

69 Noll exhibited variations: Ibid., 7.

69 "very first major public showing": Ibid.

69 "the artist will simply 'create'": Preston (1965).

70 "Some guests were nervous": Candy and Edmonds (2002), 6.

71 "artificial art": Klütsch (2007), 421.

71 "The drawings were not very exciting": Bowlin (2007), 4.

71 "With hindsight, the digital computer": A. Michael Noll, email to the author, January 30, 2012.

72 "The generation of random numbers": Coveyou (1998), 178.

72 soothing, and abstract: Noll (1966), 8–9.

72 Both pictures were actually conceived by humans: Ibid., 9.

72 "artistic merit is not": Ibid.

73 It was the brainchild: Mason (2008), 10.

73 "The Group's avant-garde artistic": Ibid., 16.

73 "influenced by Charles J. Biederman's": Mason (2008), 24.

74 "The world is as it is": Hamilton (2003), 60–62.

74 "There is something to gain": Whyte (1968), 1.

75 "Popular (designed for a mass audience)": Hamilton, as quoted in his obituary in the *Daily Telegraph*, September 13, 2011.

75 "that it caused an uproar": Victor Pasmore to Richard Yeomans, October 10, 1983, in Yeomans (1987).

76 Jokingly he recalls having been: Roy Ascott, author interview, December 6, 2012.

76 "It can be argued that": Ascott (2008), 9.

76 "kinetic art king": Roy Ascott, author interview.

76 "Cybernetics was everywhere": Eléonore Schöffer, author interview, May 21, 2012.

77 "Eureka experience": Shanken (2002a).

77 "is a proper study for the artist": Ascott (1964), 100–101.

77 Then they were released into: Pethick (2006), 1.

78 Yet there was a positive aspect too: Gustave Metzger, Third Manifesto, July 3, 1961. This was meant to advertise his famous performance on the South Bank where he appeared in military attire and attacked an array of nylon sheets with a spray gun filled with acid, destroying them.

78 greatly inspired by Metzger: Gere (2002), 100.

78 "cultures being leveled": Brian Eno, speaking at Art and the Mind festival, Winchester, UK, March 7, 2004. Quoted in Mason (2008), 68.

79 "Murder of the Art Schools": Patrick Heron, "The Murder of Art Schools," *Guardian*, June 22, 1971.

79 "creativity is not totally in the hands of the artist": Ernest Edmonds, author interview, June 21, 2011.

80 "clear emphasis on composition": Ibid.

80 "non-constrained than it appeared to be": Ibid.

80 With the rather basic computers: They used a Honeywell 200 which was comparable to the IBM 1401, at least one model behind the IBM 7090, the powerhouse mainframe of the day.

80 "it may no longer be necessary": Ibid.

80 "concept of art can survive artists as we know them": Ibid., 12.

80  "We live in an industrialized": Ihnatowicz (1985).

81  To back up this approach: Brown (2003), 2.

81  "Our aim is to get people (the press)": Gardner (1969).

81  *Senster* went way beyond: Philips engineers had also assisted Schöffer in some of his works, providing sophisticated electronic circuitry.

81  "In the quiet of the early morning": Zivanovic (2008), 103.

82  "For me, Ihnatowicz and Cohen": Brown (2008), 277.

82  "I met my first computer"; "I was simply grabbed": Harold Cohen, email to the author, June 6, 2011.

84  He studied at Manchester College of Art: Paul Brown, author interview, September 20, 2011.

85  "The audience was enraptured": Ibid.

85  "amplification of minute turbulent events": Ibid.

85  "into the computational domain": Ibid.

85  "Now I could take enormous risks": Susan Collins, author interview, October 10, 2012.

85  "computer facility for fine art": Ibid.

85  "I had such a clear vision": Ibid.

86  "art and technology": Ibid.

87  "computer as a control system": Ernest Edmonds, author interview, June 21, 2011.

87  "control issues as conceptual": Ibid.

87  "a poem is something more"; "essence of software": Ibid.

87  "Calder walked into Mondrian's studio": Ibid.

87  To him, the work he created: Ibid.

87  "If you want to paint": Ibid.

88  "Man, you're a real artist": Roy Ascott, author interview.

88  "technology of thought transfer": Ibid.

88  "telematic embrace": Ascott (1990).

88  "real power of the computer": Gerfried Stocker, author interview, September 3, 2012.

89  "consciousness is not generated": Roy Ascott, author interview, June 12, 2012.

## Chapter 4

91  "They have computer tools already": Bruce Wands, author interview, November 3, 2011.

92  "I saw the future": Ibid.

92  One of his works: Wands (2006), 21.

92  "preconceived notions as science being logical": Bruce Wands, author interview, November 3, 2011.

93  "To me improvisation": Bruce Wands, email to the author, February 13, 2013.

93  "the artist has control over a wide range": Ibid.

94  "where high resolution live action": Ibid.

94  "primarily intuitive": Ibid.

95  "Take Jane Austen": Ibid.

95  "Needless to say": Ken Perlin, author interview, November 1, 2011.

95  describes him as a genius: Bruce Wands, author interview.

95  "opted for mathematics because": Ken Perlin, author interview, November 1, 2011.

95  "where it's at": Ibid.

96  "Einstein was an artist": Ibid.

96  "There is a willed ignorance"; "Great discoveries": Ibid.

96  "I'm a researcher": Ibid.

96  "For me, science and technology": Rick Sayre, author interview, January 7, 2013.

97  "doing shoestring visual effects": Ibid.

97  "hive mind": Ibid.

97  "rattle off things as aesthetic touch points": Ibid.

97  "Pixar can try something": Ibid.

97  "I don't have a precise definition of aesthetics": Ibid.

98  "It was the most complex of any skin": Ibid.

98  It was the robot: See the Ars Electronica archive site: http://90.146.8.18/en/archives/prix_archive/prix_projekt.asp?iProjectID=2474.

98  "peripheral areas, particularly underground things": Rick Sayre, author interview, January 7, 2013.

98  "wretched hive of scum and villainy": Ibid.

99  "we met Rick Sayre": Gargaj (2007).

99  "a key part in how greenery"; "Fresh blood is a great thing": Rick Sayre, author interview.

99  the program in Media Arts: http://web.mit.edu/catalog/degre.archi.media.html.

99  "place where the future is lived": http://web.mit.edu/files/overview.pdf.

100  "I always retained the notion of neoteny": Joichi Ito, author interview, September 3, 2011.

100  "because I felt I could": Ibid.

100  "Ito, 44, is recognized as one": http://web.Mit.edu/newsoffice/2011/ito-media-lab-director.html?tmpl+component&print=1.

101  "He can position the lab at the edge": Markoff (2011).

101  "not very collaborative"; "Process, not end result"; "is not a tech shop for artists": Ibid.

101  "People spend very little time sitting"; "Shut up and build it": Ibid.

102  "problem-seeking, as opposed to": Neri Oxman, author interview, October 18, 2011.

102  "aesthetic fascination with"; Ibid.

102  "era of glamour geeks": http://www.future-ish.com/2009/05/design-idol-sceleb-neri-oxman.html.

102  "produces unbelievable patterns"; "tries to spec out algorithms": Neri Oxman, author interview.

102  "now, with the Internet"; "I miss the old world": Ibid.

103  "*Beast* lets go of boundaries": Ibid.

103  "A da Vinci stage has to be reached": Ibid.

104  "iPad fever. You build"; "doesn't believe that that is": Ibid.

104  "translation between disciplines": Ibid.

104  "The Media Lab is an idea factory": Michael Bove, author interview, October 20, 2011.

104  "What if": Ibid.

104  "Look, I don't know if this is art": Ibid.

105  "totally radical way of making large scale": Paula Dawson, email to the author, September 10, 2013.

106  "Apart from the fact that Paula is creative": Michael Bove, email to the author, August 28, 2012.

106  "it's all about creativity here": Joe Paradiso, author interview, October 20, 2011.

107  "Today art is an offspring of science and technology": Peter Weibel, author interview, January 25, 2013.

109  Science was too well established: Ibid.

109  "ideas about space, time, and relativity": Ibid.

109  "in dialogue with scientists": Ibid.

109  "spooky actions-at-a-distance": Einstein to Max Born, March 3, 1947, in Born (1971), 158.

109  "could be wonderful media art": Peter Weibel, author interview.

110  "It exists only thanks to the media": Peter Weibel, author interview, February 15, 2013.

110  "Aesthetics is a general medium"; "For me aesthetics has nothing to do with beauty"; "Many of my works are not directly": Peter Weibel, email to the author, January 29, 2013.

111  "It is ridiculous that art historians": Peter Weibel, author interview.

111  "how photography deeply influenced painting": Ibid.

111  "I give art a strong theoretical, political, and social urgency"; "theory-dependent because it wants": Cook (2000).

111  "just about anything goes"; "We are interested in using design": Gaver (2011).

111  "highly committed to playing": Quoted in the RCA's catalogue, available at: http://www.rca.ac.uk/Default.aspx?ContentID=159827&GroupID=161712&CategoryID=36692&Contentwithinthissection&More=1.

111  "have a high tech savvy"; "They can be pure artists": Fiona Raby, author interview, November 18, 2011.

112 "technological dreams"; "mess up all those very nicely": Ibid.

112 "I don't think we ever recovered!": Fiona Raby, email to the author, February 6, 2013.

112 "no playfulness in England": Fiona Raby, author interview.

112 "Can designers be involved in biology?": Ibid.

113 "Beautiful flowers, mind-altering weeds"; "a controlled ecosystem of entertainment": Benqué (2010).

113 "They all sit at the same table": "While the designers may know": Fiona Raby, author interview.

114 "that of a provocateur"; "tampers with nature"; "seductiveness of technology": Alexandra Daisy Ginsberg, author interview, May 15, 2012.

115 "Private industry will finance"; "art will become independent"; "will be an empire of data": Peter Weibel, author interview.

115 "Today art is an offspring of science and technology": Peter Weibel, author interview.

## Chapter 5

116 "Art does not reproduce": Klee (1968).

116 "In a time when so many artists": Parsons (1998). "Jim Miller" is a pseudonym for Bill James and Lisa Miller.

117 "a child of the space age": Paul Friedlander, author interview, March 30, 2012.

117 "felt squeamish about the label": Ibid.

119 "I feel that I have been true to Barbour's ideas": Friedlander (2006).

119 "Only as an artist am I able": Julian Voss-Andreae, email author interview, August 26, 2011.

120 "It was then that I realized that I needed": Ibid.

120 "probably all intellectually conjured, brain-born art": Ibid.

121 "drew without the intellect interfering": Ibid.

121 "intrigued by the time when relativity and quantum physics": Ibid.

121 "honest and strong dynamic"; "quantum dance": Julian Voss-Andreae, email author interview, August 26, 2011.

121 "To me, form and function are always a unit": Ibid.

121 "I feel a strong excitement about work": Ibid.

122 "Is it possible to create a sonic rainbow?": Domnitch and Gelfand (2004), 391.

122 "The *Camera Lucida* project began": Ibid., 392.

123 "came out of a series of very long": Evelina Domnitch and Dmitry Gelfand, author interview, September 9, 2012.

123 "they offered no buzz"; "What is light?": Ibid.

124 "recorded music or recorded art"; "the narrative is too restrictive": "live cinema": Ibid.

124 "The image was animated"; "art became a means": Ibid.

124  "iridescent depth"; Domnitch and Gelfand (2011).

125  "there was no theoretical model": Evelina Domnitch, email to the author, September 25, 2012.

125  went on to propose one: Yasui et al. (2008).

125  "Raoul claimed that his encounter": Evelina Domnitch and Dmitry Gelfand, author interview.

125  "Aesthetics is to be emotionally moved": Ibid.

125  "We always hope the unexpected will appear": Ibid.

125  "I'm an optimist"; I want to see just over the horizon": Nathan Cohen, author interview, June 6, 2011.

126  "I almost became a geologist": Ibid.

126  "look for ways to share": Ibid.

126  "need to be sensitive"; "ongoing and organic": Ibid.

126  "that space which exists between two and three dimensions": "Where does the picture end": Ibid.

126  "part of the real world": Ibid.

128  "finding patterns in what we see": Nathan Cohen, email to the author, March 16, 2013.

128  "has made possible new ways": Cohen (2013).

128  "technology and fine art": Quoted ibid.

128  "make a journey toward understanding": Nathan Cohen, author interview.

128  "The history of Renaissance art was primed": Ibid.

128  "heartbeat of the primordial universe": Robert Fosbury, voice recording for the author, March 17, 2013.

129  "Symbolic meaning—it was the": Ibid.

130  "observing the night sky": Antonella Nota, email to the author, March 15, 2013.

130  allowed her "to reestablish": Ibid.

131  "interpret science with design": Vanessa Harden, author interview, August 3, 2011.

131  "distant cosmological phenomena": http://www.vanessaharden.com/Urban-Sputnik.

131  "was crucial throughout the project": Andrew Jaffe, email to the author, September 6, 2011.

132  "enabled/forced me to think"; "I think the effect has been minimal": Ibid.

133  "The Entire Universe on a Dimmer Switch": Spears (2006).

134  "the act of looking at a reflective object": http://whitecube.com/artists/josiah_mcelheny/.

134  "Dark Matter rap": For the complete lyrics see http://www.astronomy.ohio-state.edu/~dhw/.

134  "David's view of cosmology"; "two-way dialogue in which Weinberg": Josiah McElheny, author interview, November 3, 2011.

135  "More people saw *Island Universe*": http://arxiv.org/pdf/1006.1013v1.pdf.

135 "big gap in technical level": David Weinberg, email to the author, September 5, 2011.

135 helping him to see: Ibid.

135 Weinberg knew from the start; "This was the deal"; "my colleagues are sometimes miffed": Ibid.

137 We called the work *Anschaulichkeit*: On the Kantian notion of *Anschaulichkeit* and quantum physics, see Miller (2000).

140 CERN "always welcomed artists": Renilde Vanden Broeck, author interview, October 5, 2012.

141 "not merely art as illustration": Ken McMullen, author interview, December 7, 2011.

141 The London Institute suggested: Ibid.

141 wrote in his diary: Benson (2001a), 12.

141 "Artists today are beginning to realize": Benson (2001b).

141 "The laws of physics are not going to go away": CERN press office, available at: press.web.cern.ch/press-releases/2000/06/signatures-invisible.

142 "Signatures of the Invisible is a groundbreaking initiative": Maiani (2001), 3.

142 "There is fusion": Grandjean (2000).

142 "Not a straightforward illustrative": McMullen (2001), 4.

142 "a kind of archetype": Deacon (2001), 22.

143 "The mathematical, the simple, and the complex": Sand (2001), 32.

144 "There can never be": March-Russell (2001), 6.

144 "crumple theory": Sexton (2001), 11.

146 "what we see is somebody": Ken McMullen, author interview.

146 "so that anyone can get the drift": Ibid.

146 "What would C. P. Snow make of all this?": Pile (2001).

147 "A conversation with a theoretical physicist": Campbell-Johnston (2001), 21.

147 "For me it was a no-brainer:" See http://cdsweb.cern.ch/record/1204806.

148 "great arts for great science" Koeli (2012), available at: arts.web.cern.ch.

149 "inspiration partner"; "speed dating": Ariane Koek, author interview, October 4, 2011.

149 "don't get too close": Ibid.

149 "in collaboration with the local police": Julius von Bismarck, author interview, September 1, 2012.

149 "inspiration partner"; "I was more interested"; "international galleries are interested already": Ibid.

149 Then they all had to critique: See cdsweb.cern.ch/record/1481328.

151 "What do I want to do?"; "von Bismarck brings humor": Ibid.

151 A number of physicists: See, too, Butterworth (2012).

151 "After three months of brain work": Ibid.

152 "It's all about credentials"; "Get a great artist"; "Use the data"; "That'll be astounding": Joe Paradiso, author interview.

152 "plenty of space for installations": Luis Álvarez-Gaumé, author interview, October 7, 2011.

152  "hopefully they will leave something behind"; "artists with abstract ideas": Ibid.

152  "the fusion of mathematics and reality"; "a comatose god running its own universe": Keith Tyson, quoted at: http://onestoparts.com/review-keith-tyson-panta-rhei-pace-london.

153  "world as information": Keith Tyson, author interview, May 2, 2012.

153  "Stuff happens at the edge of a cloud"; "dichotomy between science and art"; exclusive categories: Ibid.

153  "like good artists, good scientists": Ibid.

154  "It, like, knocked me out"; "artistic encouragement was in the genes": Steve Miller, author interview, November 20, 2012.

154  "The habitual gestures": Quoted in Heiferman (2007), 54.

155  "totally enchanted with that epoch": Steve Miller, author interview.

155  "how it happened in Cubism"; "changes in technology"; "Man, I wanted"; "reinvent new genres": Ibid.

155  "use science to look at pathology": Ibid.

155  "somebody else's piece of paint": Heiferman (2007), 54.

155  "a blob of paint": Steve Miller, author interview.

155  "Technology allows you to penetrate": Ibid.

156  "beauty of these images is that they are":"Hiroshima or bloody fetuses": Heiferman (2007), 55.

156  "It was a new way to take on": Katz (1999).

156  "new technologies to reinvent": Steve Miller, author interview.

156  "All of a sudden"; "We really have windows": Katz (1999).

157  whereas today all artists": Steve Miller, author interview.

158  "brought into radiological labs all sorts of things": Ibid.

159  "They invited me to do whatever": Ibid.

159  "Steve showed up at the site": Steve Adler, email to the author, December 9, 2012.

159  "investigation of matter"; "how to tell the story": Steve Miller, author interview.

159  "time line of human development, from mud pies": Ibid.

161  "Silk-screened on canvas, Dr. MacKinnon's": Hargrave (2007).

161  "If you read this review through the lens": Steve Miller, email to the author, December 14, 2012.

161  "an affirmation of life": McQuaid (2007).

161  "beauty is, for many, a forbidden word"; "it's the result of a process": Heiferman (2007), 58.

161  "scientists cooperate with me"; "cross-pollination that can loosen everybody up": Steve Miller, author interview.

161  scientists "appreciate art but": Ibid.

162  "to present complicated information to an audience": Ibid.

162  "Steve's influence on my work at that time": Steve Adler, email to the author, December 9, 2012.

162  "been aware of the creative intellect": "that both art and science play equal roles": Ibid.

162  with the Industrial Revolution; "I see my motivation": Ibid.

162  "Expertise demands discipline"; "There is so much to learn": Steve Miller, author interview.

163  "own laboratory in the garden": Antony Gormley, email author interview, April 12, 2012.

164  "three-dimensional drawing in space": Quoted at http://www.antonygorm ley.com/sculpture/item-view/id/252.

164  "little conference": Antony Gormley, email author interview.

164  "I would be very honored to be able to donate": Ibid.

165  "an attempt to materialize the place at the other side": http://arts.web.cern .ch/works/feeling-material-xxxiv.

165  "led to a reassessment of the standard view": Antony Gormley, email author interview.

165  the highest density cloud: Antony Gormley, email to the author, August 21, 2013.

165  Gormley's objects relate to topology: http://www.antonygormley.com/ resources/essay-item/id/122.

165  "Collaboration is alive and well"; Antony Gormley, email author interview.

165  "I don't think the two cultures"; Ibid.

166  "in particular Roger Penrose"; "Art does a similar thing": Ibid.

166  "Art does not reproduce": Klee (1968) 182.

166  "I hope that scientists": Rolf-Dieter Heuer, author interview, October 5, 2011.

166  "questions are so deep"; "difficulty to act as an individual": Ibid.

167  "The solitary worker": Ibid.

167  "functionality rather than beauty"; "but this goes along with": Ibid.

167  "If it functions well": Ibid.

## Chapter 6

168  "world's greatest art scandal": Taylor (2010).

168  preeminent Pollock scholar: O'Connor (1967).

170  "similar to Pollock's work": Richard Taylor, author interview, September 13, 2011.

170  "Suddenly the secrets of Jackson Pollock": Taylor (2002), 118.

171  "keep the viewer alert"; "I think that one": Taylor et al. (2011), 11.

172  "There was a reviewer": Rouché (1950).

172  "A striking visual similarity": Taylor (2005), 2.

173  Researchers have found that : See Taylor et al. (2011).

173  "To fake a Pollock": Richard Taylor, author interview.

173  "Perhaps it may even": Taylor (2002), 121.

173  "gift + purchase": Abbott (2006), 648.

174 "I was completely blown away"; "there are too many things": Kennedy (2005).

174 "restraint of trade": Peers (2005).

175 "rethink its involvement": Kennedy (2005).

175 "If Ellen Landau's opinion": Ibid.

175 "Unlike the authors of the Pollock": Ibid.

175 Three had appeared on Alex Matter's: Kennedy (2006a).

176 "made me sign": Richard Taylor, email to the author, September 19, 2012.

176 Anticipating accusations of bribery: Kennedy (2006a).

176 "the unprecented nature": Taylor (2010), 2.

176 In fact, by commissioning him: Richard Taylor, email to the author, September 20, 2012.

176 "Thus the PKF": Richard Taylor, email to the author, September 19, 2012.

176 which published them: Kennedy (2006a). A more detailed report appeared in Abbott (2006).

176 "impeded scholarly debate": Kennedy (2006a).

177 "very new and contested"; "a full-scale catalogue": Ibid.

177 "reinforced his initial doubts": Ibid.

177 One Pollock scholar claimed: Kennedy (2006a).

177 "But this was not an ideal world": Richard Taylor, email to the author, September 19, 2012.

177 "you've just lost them": Taylor (2010), 3.

177 paintings were not fractal: Kennedy (2006b).

177 "Oh my God"; "mathematics had an incredible": Richard Taylor, author interview.

177 "Jones-Smith, along with": Jones-Smith and Mathur (2006).

177 "I am pleased they have": Griffith (2006).

178 If Jones-Smith and Mathur's: see Taylor, Micolich, and Jonas (2006).

178 "have done is a simple trick": Rehmeyer (2007), 124.

178 "This is like taking": Richard Taylor, email to the author, September 27, 2012.

178 "Nature liked controversy": Richard Taylor, email to the author, September 28, 2012.

178 Aware of the controversy: Peers (2005).

179 "began aggressively pursuing": Steven Litt, email to the author, September 26, 2012.

179 "Orion uses microscopy": http://www.orionanalytical.com/.

180 Matter admitted that the paintings: Litt (2007a).

180 "Robi paints": See Cernuschi and Landau (2007), 3, for a photograph of the wrapping.

180 There were a lot of: Landau (2007), 27, 84.

180 "Hang in there": Litt (2007b).

180 "these are Pollocks": Ellen Landau, email to Albert Albano, February 22, 2006, in Litt (2007b).

180 "going to lead to the fact that the works": Mark Borghi, email to Albert Albano, February 22, 2006, in Litt (2007b).

180 "a strategy for a graceful": Ibid.

180 In an email to Landau: Litt (2007b).

180 "Despite the paint analysis": Ibid.

181 "troubled by both Martin's"; "learned that fractal analysis is invalid"; "personal and artistic": Ibid.

181 "dubious and unproven as a way": Jeremy Epstein, in Peers (2005).

181 lawsuit was "very negotiable": Litt (2008).

181 He told the *Cleveland Plain Dealer* that he would release: Litt (2007a).

182 The disputed paintings were exhibited: http://www.bc.edu/bc_org/avp/cas/artmuseum/exhibitions/archive/pollock-matters/index.html.

182 "still other avenues of exploration"; "Robi paints": Litt (2007a).

182 "neither she nor her sister recalled paints": Harvard University Art Museums (2007), 9.

182 "is not as hard and fast": Cernuschi and Landau (2007), 6.

182 Landau also mentioned a fingerprint": Ibid.

182 This had been mentioned in an article: Biro (2007), 157.

183 "As a scientist and a scholar": Litt (2008).

183 "which has not been published": Newman (2008), 129.

183 "demonstrably flawed": Landau (2007), 5.

183 "I think they took a fairly simplistic": Minkel (2007).

184 Taylor explains that their color separation: Richard Taylor, email to the author, September 27, 2012.

184 "rightly embarrassed": Ibid.

184 "no reliable way to distinguish": Minkel (2007).

184 "The entire art world was present": F. V. O'Connor, email to the author, September 27, 2012.

184 "graciously declined": Flescher (2008), 8.

185 "These pictures are not a new bunch": Karmel (2008), 17.

185 "does not mean that Pollock created the painting": Newman (2008), 23.

185 boards were manufactured in the 1970s: Martin (2008), 32.

185 "are patently *inconsistent* with the claimed": Ibid., 35.

185 "red herring": "Q & A," 37.

186 "fractal analysis technique": Richard Taylor, email to the author, September 27, 2012.

186 "can still see the expression on the face": F. V. O'Connor, email to the author, September 27, 2012.

186 "patently wrong"; "fine art and other cultural property": Litt (2007b).

186 results were "irrefutable"; "utmost seriousness": Ibid.

186 "ranges from unlikely": Ibid.

187 "underscore the central importance"; "extremely improbable": Ibid.

187 Landau removed the paragraph: Litt (2010).

187 Perhaps Pollock stared at them: Taylor et al. (2011), 2.

188 "identifying the artist's 'hand'": Richard Taylor, email to the author, September 24, 2006.

188 "throw a narrow beam of light": Taylor (2002), 121.

## Chapter 7

189 "We are witnessing the birth": De Menezes (2007), 215.

191 "art is too important to be left": Carnie (2002).

191 "Each work is approached differently": Marron (2011).

192 "My artistic production gets inspiration": Angheleddu (2011).

192 "a major component in scientific discovery": Dowson (2011).

193 "Our work tries to retrace lines": Yonetani and Yonetani (2011).

193 "how to define one's personality": Aldworth (2010).

193 "scan might look like it could show": Aldworth (2011).

193 "Where am I"; "You will look into my brain": Ibid.

195 it "invites analogies": Kemp (2010).

195 "ephemeral, oscillating brain activity": Morton Kringelbach, email to the author, August 28, 2011.

198 "avoids either the sensationalism of gore": Miller (2011).

198 "I'm fascinated by the mystery of our status as biological beings": Ibid.

198 "semi-living sculptures": Ibid., 233.

199 "Wet biology art practices are engaged": Catts and Zurr (2007), 232.

199 "what it is to be human, a quest in which": Oron Catts, author interview, June 30, 2011.

199 "The function of art"; "areas of incompatibility": Ibid.

199 "what connects artists and scientists": Stelarc, author interview, June 22, 2011.

200 "In the 21st century we have become captivated" Sellars (2011).

200 "what it means to be a body": Stelarc, author interview.

201 "be no more biological death": Ibid.

202 "This exhibit is an adventure": Carter (2011)

202 "Wherever science is leading us": Reisz (2011).

202 "intriguing, because at first glance": Sowels (2011).

202 "Scientists are leading the way": Lewis (2011).

202 I chaired three debates: See http://www.artandscience.org.uk/debates/.

202 "Miller spoke about an emerging third avenue of art"; "Perhaps as scientists increasingly collaborate": O'Callaghan (2011).

205 nom de guerre: ORLAN (2004), 9.

205 "Very boring": ORLAN, author interview, February 21, 2012.

205 "I am both man and woman": ORLAN (2004), 9.

205 "I must have control over": ORLAN, author interview.

206 "beef implants," bones from a cow: Ibid.

206   "Cells expand in time and space": Ibid.

207   "This particular work is caught": ORLAN (2008), 89.

208   "biotechnology is taken out of the laboratory": Ibid.

208   "transgenic art": Kac (2007), 163.

208   "Kac's point of view was that he saw": Roy Ascott, author interview, June 12, 2012.

208   "to *enact* critical views": Kac (2007), 164.

210   "My methodologies in creating": "With them I learned": Marta de Menezes, email author interview, August 10, 2012.

210   "I always refer to my work": Ibid.

211   "Art uses every other discipline": Suzanne Anker, author interview, November 2, 2011.

211   "Chromosomes can't tell me who I am": Ibid.

211   "how parts fit together": Ibid.

211   "It was the first time I saw the philosophical": Ibid.

212   "too hard to be an artist in New York City": Ibid.

212   "think about manipulating nature": Ibid.

213   "anything to do with the natural world": "bioart or science is an umbrella term"; "That's because artists believe": Ibid.

213   "My interest is not in being a rookie scientist": Ibid.

213   "Things that are going on in labs": Martineau (2012).

214   "present the technology with problems it can't do": Suzanne Anker, author interview.

215   "can bring other things to"; "can come into a lab": Ibid.

215   "artist might think up"; "the scientists changed their tune": Ibid.

215   "Going back into one's ghetto": Ibid.

216   "Much of my early work": Grande (2007).

216   "While Haeckel drew": Brandon Ballengée, author interview, November 7, 2011.

216   "are a 'sentinel' species": Grande (2007).

216   "Beuysian idea": Brandon Ballengée, author interview.

217   "relates to ecological problems not necessarily": Ibid.

217   "eco-artists [can] transform the society": Grande (2007).

217   don't consider his work genuine art; "Frogs sell better in Europe": Brandon Ballengée, author interview.

217   "'eco-artists' whose art may be unrecognizable as art": Triscott and Pope (2010), 7.

217   "If only the government": Sato and Kazan (2008).

218   "They called the cops": Joe Davis, author interview, September 1, 2012.

218   "DNA can encode any information": Sato and Kazan (2008).

219   The sounds were transmitted for several minutes: Ibid.

220   metallizing non-recombinant bacteria: Joe Davis, email to the author, August 5, 2013.

221 "often I speak a language that nobody"; "Artists should create things": Joe
Davis, author interview.

221 "The concept of a garden": Jun Takita, author interview, August 18, 2011.

221 "Most of all, the little fireflies": Takita (2008), 141.

221 "because without science there is no way": Jun Takita, author interview.

221 "artistic action" is necessary: Takita (2008), 141.

222 "a direct extension of the body": Ibid.

223 "against the conventions"; "in which light creates a strange relief": Ibid.

223 "this is the expression of man's": Quoted at http://juntakita-artworks
.blogspot.co.uk/.

223 "beauty in proportion"; "how to do the work"; "Process rather than result":
Jun Takita, author interview.

224 "We are interested in blurring the boundaries": Marion Laval-Jeantet and
Benôit Mangin, author interview, September 3, 2011.

224 "I had the feeling of being": Quoted at http://www.designboom.com/weblog/
cat/10/view/16123/art-oriente-objet-may-the-horse-live-in-me.html.

224 experiment cannot happen twice; "explores the new hybrid organisms":
Jens Hauser, email to the author, November 3, 2012.

226 "no scientists work alone": Marion Laval-Jeantet and Benôit Mangin, author
interview.

226 "Why restrict one's mind?": Ibid.

226 "We know that's bullshit": Ibid.

226 "Space for freedom is shrinking"; "like the idea that it's never ending": Ibid.


## Chapter 8

229 "What does it mean if you really": Bernhard Leitner, author interview, June
23, 2012.

230 "reigning aesthetic of pulverization": Ross (2009), 429.

230 "re-envisioned sound itself": Blume (2008), 15.

230 "This subconscious detachment"; "a spark-like confluence of thought":
Bernhard Leitner, author interview.

231 "I could not have developed my work"; "I needed a city that is really pulsating"; "where most things happened": Ibid.

231 "theoretical investigations"; "sound as architectural material"; "no one
talked about sound": Ibid.

231 "empirical studies": Blume (2008), 15.

231 "corporeality of listening": Bernhard Leitner, author interview.

233 "I hear with my knee": Ibid.

233 "In the sound chair": Ibid.

233 "a kind of holistic thinking"; "It was a medical research project": Blume
(2008), 19.

234 "medical research following an empirical": Bernhard Leitner, author interview.

234 "The art market"; "White Cube and sound art just don't work": Ibid.

235 "We, O+A, listen to everything"; "hallelujah" moment: O+A (2009), 63.

235 "jaw-droppingly profound": Ibid, 64.

236 was completely transformed: Ibid.

236 "We were a big hit": Sam Auinger, author interview, September 4, 2011.

236 "how music changes have a lot": Ibid.

236 "besides the most famous Mr. Bach": Ibid.

236 "Morton Feldman"; "could organize time in a way": Ibid.

237 "everything perfect, not one note too much": Ibid.

237 the "church becomes a sound box": Ibid.

237 "amazing"; "unbelievably distorted barrage"; "brain and the precondition": Ibid.

237 "to go deeper into sound": Ibid.

238 sonic commons: O+A (2009), 67.

238 "unite people to listen": Sam Auinger, author interview.

238 "Is it possible to hear Bonn's history": Ibid.

238 "Spaces are reanimated through the energy of sound": Ibid.

239 "personal perceptions and struggle with form"; "time and quality of research": Ibid.

239 "knows where to begin": Ibid.

239 "Completely uninspiring"; they should be able to arrange the room: Ibid.

239 "It's a question of aesthetics": Ibid.

239 "We can close our eyes": David Toop, author interview, July 13, 2012.

239 "three most influential British": Lamb (2012).

240 "catastrophic"; "desire to cross boundaries"; Music turned out to be "the right": David Toop, author interview.

240 "But actually it was the flute": Ibid.

241 "any device can makes sound": Ibid.

241 "In 1971, there was no such thing": Ibid.

241 "no coherent language for texture": Ibid.

241 "based on straight physics": Ibid.

242 "to work as simply as possible"; "works with the computer and bamboo"; "a polarity of means": Ibid.

242 "Working with a computer, I can create"; Ibid.

242 "Technology is extremely important"; "There are many wonderful examples"; "sound making": Ibid.

243 "goes back to the sixties, to Stockhausen"; "in live performances"; Ibid.

244 "putting sounds under a microscope": Toop (2011).

244 "We are all David Toop now": Reynolds (2012).

244 "Toop's writing always meant for me": Cascella (2012).

245 "you'd be most likely to find on Bjork's": Lamb (2012).

245 "relentless cycle of compromises": Gross (1997).

245 "Intuitional"; "Intuition is when you're not thinking"; "I am always thinking about aesthetics": David Toop, author interview.

245 "I have a split personality": Ibid.

245 "The analytical faculty": Ibid.

245 "methods of working have changed": Ibid.

246 "I don't use any notation in the conventional sense": David Toop, email to the author, December 23, 2012.

246 "directed improvisation": David Toop, author interview.

246 "today demands proof"; "what a human being can produce spontaneously"; "a tendency to apply logic": Ibid.

247 Nature's own music: Jo Thomas, author interview, September 2, 2012.

247 "ambient sounds such as the synchrotron machinery": Ibid.

248 "I was obsessed with the ways"; "not really interested in playing tunes"; "incredible sound studios": Ibid.

248 "how sound moved": Ibid.

248 "any music by anyone": Ibid.

248 "sees sound"; "paint and draw music": Ibid.

249 malfunctions in the machinery: Ibid.

249 "We always thought incredibly big"; "This is a universe and this is": Ibid.

249 "nice union of ideas"; "very creative when they worked with me"; "interested in an exchange"; "A work has a belief in itself"; "transfer of knowledge": Ibid.

250 "Always a bit of a surprise": Paul Prudence, author interview, September 11, 2011.

251 "I learned animation packages"; "tension between how much randomness": Ibid.

251 "This creates an element of tension"; "I can control the amount": Ibid.

251 "old conventional ideas of story": Ibid.

251 "Quite belittling": Ibid.

251 "a bit of a conundrum"; "Was it important to know": Ibid.

252 "We are all David Toop": Ibid.

252 "I don't interact with scientists": Ibid.

252 He has even lectured: Prudence (2012).

252 "Media art, that's the catchall term"; "I'm not a category person": Paul Prudence, author interview.

252 "Even the art/sci thing confuses me": Ibid.

252 "Aesthetics is complex"; "It's not just about new media": Ibid.

253 "aesthetic sensibility"; "It's some sort of formal way of putting together"; "It doesn't come out of nothing": Ibid.

253 "If I had all day every day": Tod Machover, author interview, October 20, 2011.

253 "It ideally balanced complexity": Machover (2012), 400.

254 "a virtuoso at computer manipulation of sound"; "of extroverted eclecticism": Rothstein (1991).

254 "morph the cello sound": Troxell (2011), 7–8.

255 "a repertoire of colors": Machover (2005), 7.

255 "If you're immune to raw awesomeness": Troxell (2011), 3.

255 "Visual art is easier to grasp": Tod Machover, author interview.

255  "Tod Machover is one of the strangest composers alive today": Troxell (2011), 2–3.

256  "music and medicine as a sort of prescription"; "Music can tell you something": Tod Machover, author interview.

257  "I love this, I love that"; "There are so many ideas"; "A new language is needed": Ibid.

257  "Nothing interested me more than music": Robert Rowe, author interview, November 30, 2012.

257  "an excellent way to learn": Ibid.

257  "computer didn't know what it was doing": Ibid.

258  "in a very demanding environment"; "make the computer musically aware": Ibid.

258  "my long-time interest": Ibid.

258  "Indeed it does"; "It's hard-core computer science": Ibid.

259  "The computer is (at least in my music)": Robert Rowe, email to the author, December 27, 2012.

259  "Are there fundamental rhythmic": Robert Rowe, author interview.

260  The neuroscientist's input includes: Ibid.

260  "But you have to start somewhere": Ibid.

260  "It was a valiant attempt"; "is to reduce the distance": Ibid. See also Wilson (2010).

261  "more and more people are thinking computationally"; "truly creative thinking"; Robert Rowe, author interview.

261  "Music is and will always remain": Robert Rowe, NYU faculty profiles, available at: http://steinhardt.nyu.edu/profiles/faculty/robert_rowe.

261  "wanted to see what they are": Tristan Perich, author interview, November 30, 2012.

262  "It is possible to create": Ibid.

263  "With electronic sound": Ibid.

263  "Just two notes oscillating"; "phasing process. Things go in and out": Ibid.

263  "You can see the aesthetic": Ibid.

263  "Simple stuff is the most pristinely": Ibid.

263  "Electronic music seemed": Ibid.

264  "There's a difference between process": Hallett (2009).

264  "with music in mind": Tristan Perich, author interview.

## Chapter 9

266  "It may be the best": Tufte (1983), 40.

267  "landscape of data": Mike Phillips, author interview, May 14, 2012.

267  *A Mote it is*: *Hamlet*, Shakespeare, Act 1, Scene 1, line 112.

269  "makes it a powerful": Ibid.

269  "Then the computer came along": Mike Phillips, author interview.

269   "heroes of mine"; "digital art today makes the mistake": Ibid.

269   "I sometimes dreamed in BASIC": Ibid.

270   a title he had to negotiate for fiercely: Ibid.

270   "in the computing department I talked art": Ibid.

271   "Not conclusive": Ibid.

271   "There are not enough collaborations"; "a lot of collaborations have not": "Speed dating"; "doesn't root properly": Ibid.

272   always tried to steer clear: Ibid.

272   "But at a certain age"; "They engaged with it": Ibid.

272   "Whenever I fly over data fields": Ibid.

272   "When scientists have data"; "But not artists": William Latham, author interview, August 7, 2012.

272   Books on computing codes: Todd and Latham (1992).

273   an extraordinarily ambitious brief: http://www.amazon.co.uk/Evolutionary-Art-Computers-Stephen-Todd/dp/012437185X/ref=sr_1_fkmr0_1?s=books&ie=UTF8&qid=1345968753&sr=1-1-fkmr0.

273   "mixture of art and science": William Latham, author interview.

273   "apparent mismatch between": Ibid.

273   "With just a simple set": Ibid.

274   "brainstorming, plagiarizing the natural world": Ibid.

274   "The artist became like a gardener": Ibid.

274   "feel that [Mutator] provides": Todd and Latham (1992), 105.

275   "the art world was never": William Latham, author interview.

275   "drug-fueled raves": Ibid.

276   "symmetry, elegance, and balance"; "tech-inspired art"; "They looked organic": Ibid.

276   "peer disdain"; "Scientists don't want to be seen": Ibid.

276   "What I'm trying to do"; "Art is running out of track"; "then art becomes a research project": Ibid.

277   "The most stimulating debates": Ibid.

277   "75 percent artist, 25 percent scientist": Ibid.

277   "It took three years": Erik Guzman, author interview, December 10, 2012.

277   "over the hump"; "Got the whole thing done": Ibid.

278   "Very few artists understand": Ibid.

278   "humans reacting to spaces"; "Anyone who sets"; "he really understood weight": Ibid.

278   "digital as material": Ibid.

278   "What are you doing with my": Ibid.

278   "big enough that people can see it"; "Beyond apps, computers, iPads": Ibid.

279   "It's like the space race"; "I wish I was sharp enough": Ibid.

279   "advantage of a culture": Ibid.

279   "Chelsea for signals": Ibid.

279   "Most galleries go for flat paintings"; "Yeah, make this really cool thing"; Ibid.

279 "new aesthetic forms": Jonas Loh, author interview, December 19, 2012.

280 "made programming very easy": Ibid.

280 an "interesting data set that communicates"; Ibid.

281 "What is produced is art": Ibid.

281 "new sort of angle": Ibid.

282 "Absolutely!" says Loh: Ibid.

282 "Communicating data in an interesting": Ibid.

282 "artist can only aid scientists"; "scientists find it hard": "Artists use technologies": Ibid.

282 "The border between science and art"; "Blurring the line will lead": Ibid.

282 "The amount of data around these days": Benedikt Gross, author interview, October 10, 2012.

283 "there were no books on computer design": Ibid.

283 he and three colleagues: Gross et al. (2009).

283 "it would be cool to program"; "A logical next step"; "Programming to create a visual output": Benedikt Gross, author interview.

284 "Visualization of data always interested me": Ibid.

284 "Visual design started out producing logos": Ibid.

285 "As a designer I don't care": Ibid.

285 "Programming is going to be"; "we are in a time like photography"; "no longer a programmer and an artist"; "coding style": Ibid.

285 "Art raises the question"; "I think I'm somewhere in between": Ibid.

285 "The diversity is huge"; "We play around with technology": Ibid.

286 "I believe that computation can reproduce": Perry (2011).

286 "A graphics package is like a language": Perry (2011). The Kandinsky quote is from Kandinsky (1977), 27.

286 "dove right in"; "There was no before": Scott Draves, author interview, December 14, 2012.

286 "at home and alone": Ibid.

287 cool but "pretty exotic"; "still didn't think of himself as an artist": Ibid.

287 "creating art of out math": Perry (2011).

287 "to reveal the beauty contained in them": Draves and Draves (2010).

287 "Holy shit, this is art"; "This was really what I wanted to do": Scott Draves, author interview.

288 "much more complicated": Ibid.

289 "What is life in the abstract sense?": Ibid.

289 "struggling with the definition of information"; "they kept trying to quantify": Scott Draves, email to the author, December 14, 2012.

290 "a difference which makes a difference": Bateson (1972), 459.

290 "It was a sign to me": Scott Draves, email to the author, December 14, 2012.

290 "I want to be surprised by the computer": Scott Draves, author interview.

290 "You're right"; "pick out the best ones": Ibid.

290 "I have a love relationship with technology"; "although collaborative": Ibid.

290 "there's something going on": Ibid.

291 "in principle computers can think"; "the arc of history": Ibid.

291 "This is really getting spooky": Ibid.

291 Draves "hates science fiction in which": Ibid.

291 he replies, "Both"; "Everybody wants to be cool": Ibid.

291 He also recently produced a work he calls *243*: See http://scottdraves.com/about.html.

292 "The art world doesn't care about technology": Scott Draves, author interview.

292 "don't understand the material and can be fooled": Scott Draves, author interview.

292 "Anything that's digital is copyable"; "Barriers to acceptance are still": Ibid.

292 "The electronic world has created its own"; "It's over, we've won": Scott Draves, email to the author, December 26, 2012.

292 "we'll have to wait another twenty years": Ibid.

292 "We have to get past irony": Scott Draves, author interview.

292 "I create art with data": Aaron Koblin, author interview, December 19, 2012.

293 "visually exciting while also related": Ibid.

293 "I'm 50 percent artist, 50 percent nerd": Koblin (2011).

294 "those systems and [wondered] how": Aaron Koblin, author interview.

294 "doing artwork about live data": Ibid.

294 "I was excited about visuals": Ibid.

295 "didn't meet sheeplike criteria": Ibid.

295 "I try to reembody data into a context": Ibid.

295 "If I find something interesting": Ibid.

296 "time-based and so opens up"; "allow interfaces enabling you"; "pull out a pattern like the Russell thing": Ibid.

296 "cleanly designed"; "telling a story in its form": Ibid.

296 "the emotional test": Ibid.

296 "is tangible [while] what scientists": Ibid.

296 to "think in terms of"; "in the worlds of science and engineering": Ibid.

296 "there is the communication problem": Ibid.

297 "pushing technology to the limits": Ibid.

297 "help esoteric scientific pursuits"; "cutting checks for those working": Ibid.

297 "I have been in and out of favor"; "currently galleries have a love of relics"; "It's more for collectors": Ibid.

297 "I've been studying art for about": Ibid.

297 "A lot of people in the 'maker culture'"; "Any distinction between art and science": Ibid.

297 "New terms continuously arise": Ibid.

297 "I do see art as experimentation": Aaron Koblin, email to the author, January 4, 2013.

298 "At Google, I think we have a healthy mix": Ibid.

299 "users create short animations": See http://www.exquisiteforest.com/concept. The image first appeared in *Nature*: Boyack et al. (2006).

299 "The image was constructed": Paley (2010).

300  "Today such a representation"; My driving goal"; "how "seductive they are"; "playful meaning": W. Bradford Paley, author interview, December 12, 2012.

300  "meaning or sense of the information": Ibid.

301  "Photography was very formative": Ibid.

301  "visually engaging": Ibid.

301  "Studying the Cubists led me to wonder": Ibid.

301  "Dad wouldn't pay for an art history": Ibid.

302  "Personally and emotionally, my driving goal"; "select it and draw it out of the world": Ibid.

302  "the meaning from information process": Ibid.

302  "The layperson could take this": Ibid.

302  "I am a designer": Ibid.

302  were still too many overlaps: Ibid.

303  It is "not meant to pull": Ibid.

303  "How do you make a mark": Ibid.

303  "As a curator": Ibid.

303  "If I create something and someone": Ibid.

303  curated by Christiane Paul at the Whitney Museum: See http://artport.whitney .org/commissions/codedoc/paley.shtml, and Mirapaul (2002).

304  a form of creative writing: Mirapaul (2002).

304  "artifacts out there in the world": Bradford Paley, author interview.

304  "Neither. I'm a toolmaker": Ibid.


## Chapter 10

305  Artsci is "not only sexy"; "Artists are like scavengers": Gerfried Stocker, author interview. September 3, 2012.

305  "Today art is an offspring": Peter Weibel, author interview.

305  "The exhibition's overarching theme": Ouroussoff (2008).

306  "Artists could not afford not to be": Ken Arnold, author interview, December 2, 2011.

307  "physical, visual, material aspects": Ibid.

308  "one of the most exciting": Attenborough (1994).

308  "Creativity, that crucial element": Laurence Smaje, author interview, August 9, 2012.

308  "science was right": Smaje (1998), 46.

308  "Another way was pictorial art": Laurence Smaje, author interview.

308  "Why not have an exhibition?": Ibid.

309  "The sensory organ of the inner"; "Because it works": Matthew Holley, email to the author, August 28, 2012.

310  "a genuinely innovative exhibition": *Artists and Illustrators*, September 1995.

310  "an extremely intelligent": Pertusini (1995), 33.

310  "scientists can play an important": Going Public (1996), 13.

310  "We began to wonder": Smaje (1998), 47.

310  They realized that there had to be: The artists were Heather Ackroyd and Dan Harvey and the scientists were Howard Thomas and Helen Ougham.

311  "We wouldn't be doing": http://www.viewingspace.com/genetics_culture/pages_genetics_culture/gc_w02/gc_w02_ackroyd_harvey.htm.

311  "[Sci-Art] was an idea": Terry Trickett, email to the author, January 21, 2012.

311  "Will they come?": Clare Matterson, email to the author, November 7, 2012.

312  "artists grappling with big ideas": Ken Arnold, author interview.

312  "to look at things"; "a mutually shared": Ibid.

312  "types of artists"; "scientists interested in working": Ibid.

312  "The process of science, owing to": Ibid.

312  "ruined [their] career": Ibid.

313  "unorthodox in career structure": Ibid.

313  "Collaboration is bloody difficult"; "Never, ever, ever": Ibid.

313  Arnold mentions the "myth": Ibid.

313  "scientific ideas are everyone's"; "Artistic ideas are not": Ibid.

314  "halfway through"; "gift of knowledge": Ibid.

314  "if both parties are clear"; "power of collaboration"; "people who serially do this": Ibid.

314  "The Wellcome sits"; "In the permanent site": Ibid.

314  "In two clicks we can go": Ibid.

315  "the art of my generation": Gerfried Stocker, author interview.

315  "a catchall for"; "encompasses everything"; "blurring distinctions": Ibid.

315  "role of art": Ibid.

315  "Artists can make a living": Ibid.

315  because it's so new: Ibid.

316  "in the nature of science–media": Ibid.

316  This is just surface: Ibid.

316  "We put our hands behind"; at the surface; "how one should think": Ibid.

316  "no one from art history"; "in between [and are] eager": Ibid.

316  "new aesthetics of the future": Ibid.

316  "has improved drastically": Ibid.

316  They are "hybrids": Ibid.

317  "a role model for art"; "is a performative museum": Weibel and Reidel (2010), 6–7.

317  "Machinery, Materials, and Men": Peter Weibel, author interview.

318  "Since the beginning of time": David Edwards, author interview, August 19, 2011.

318  "Le Laboratoire is based": Ibid.

318  "wonderful naivete"; "all is possible"; "no distinction between art": Ibid.

318  "a kind of experimentation": Ibid.

319  "As time has gone on": Ibid.

319  "too much institutionalization": Ibid.

319  "One of the challenges"; "We shouldn't be too careful": Ibid.

319  "process of artscience needs to be better": Ibid.

320  "Instantly geeks up": http://www.wired.com/reviews/2012/12/le-whaf/.

320  "like little kids"; "artists sometimes ask": David Edwards, author interview.

320  "Science Gallery is a platform": Michael John Gorman, author interview, December 11, 2012.

320  "The exhibitions are not really": Ibid.

321  "got the art/science bug": Ibid.

321  "For me the exciting"; "infanticize science": Ibid.

321  "bringing [both artists and scientists] out"; "public can come": Ibid.

321  "Wellcome likes a bit of competition": Ibid.

321  "huge amount of bad stuff": Ibid.

321  "Wouldn't it be interesting": Ibid.

322  acting as a "platform": Ibid.

322  "We don't sell work": Ibid.

322  "In a weird"; "It's popular, it's sexy": Ibid.

322  it's "already happened": Ibid.

323  "like amphibians": Ibid.

323  "The distance between idea": Dianne Harris, email to the author, December 20, 2011.

323  "to tap unseen forces": Dianne Harris, author interview, December 19, 2011.

324  "tap into the unseen": Ibid.

324  "But some artists just want": Ibid.

325  "in a more metaphysical way": Ibid.

325  "a quest for knowledge": Tony Langford, author interview, December 19, 2011.

325  "less metaphysical": Dianne Harris, author interview.

325  "doesn't necessarily belong": Ibid.

325  "work is experiential and performative": Watson (2010).

326  "It was not just": Robert Devcic, author interview, April 3, 2012.

326  "likes to be involved": work could "generate discussion": Ibid.

326  "people want to know more": Ibid.

326  "public feel out of"; "People gravitate to them": Ibid.

327  "to understand and assimilate": Nicola Triscott, author interview, March 14, 2012.

327  "a way could be found to commission": Ibid.

327  "When I began talking": Ibid.

327  "lots of goodwill": Ibid.

327  "Narrow?"; "Just about everything": Ibid.

328  "to open people's eyes": Ibid.

328  "But I realized": Ibid.

328  "finding artists by in-depth": Ibid.

328  "credit where credit is due": Ibid.

328  "extraordinary breakthroughs"; "really brilliant scientists": Ibid.

328  "Everywhere is a hotbed": Ibid.

## Chapter 11

329  "throw some narrow beam": Taylor (2002), 121.

329  "in a sense": Zeki (2009).

330  "The scientist does not study": Poincaré (1958), 8.

330  "asymmetries that are not inherent": Einstein (1905b).

331  "Look into computers": Reichardt (2008), 77.

331  "aesthetic fascination with forms": Neri Oxman, author interview.

331  "tries to spec out": Ibid.

331  "minimalism in my means": Ernest Edmonds, author interview.

331  "Writing computer programs": Ibid.

332  physicist and magician of light: Paul Friedlander, email to the author, September 10, 2011.

332  "In a time when so many": Quoted in Parsons (1998).

332  "simpler is better": Ken Perlin, author interview.

332  "goes along with beauty": Rolf-Dieter Heuer, author interview.

332  "If it functions well": Ibid.

332  "To me, form and function"; "I cannot separate": Julian Voss-Andreae, email author interview, August 26, 2011.

332  "I don't have a precise": Rick Sayre, author interview.

333  "new aesthetic forms": Jonas Loh, author interview.

333  "I want to give up control": Scott Draves, author interview.

334  "symmetry, elegance, and balance": William Latham, author interview.

334  "Is it exciting? Does it resonate?": Aaron Koblin author interview.

334  "Personally and emotionally": Bradford Paley, author interview.

334  "For me aesthetics starts with"; "aesthetics is a general medium": Peter Weibel, email to the author, January 29, 2013.

335  "Aesthetics is to be": Evelina Domnitch and Dmitry Gelfand, author interview.

335  "aesthetic emerges"; "an aesthetics of the ephemeral": Ibid.

335  "Aesthetics is intuition": David Toop, author interview.

335  "A work has a belief"; "transfer of knowledge": Jo Thomas, author interview.

335  "Aesthetics is complex"; "It's some sort of formal": Paul Prudence, author interview.

335  "Aesthetics feeds into": Robert Rowe, author interview.

336  "What else do we need to know": Ibid.

336  "I let the music play me": Bruce Wands, author interview, November 3, 2011.

336  "algorithms to produce": Ibid.

336  "You can see their aesthetic": Tristan Perich, author interview.

337 "Simple stuff is the most"; Ibid.

337 "strategy of seduction": Oron Catts, author interview.

337 "how to do the work": Jun Takita, author interview.

338 "art of painting new structures": Apollinaire (1913), 68. See also Miller (2001), 165.

339 "not to capture appearances": Ibid.

339 "new aesthetics of the future": Gerfried Stocker, author interview.


## Chapter 12

340 "Everything is possible": Salmon (1922), 45.

341 "Today art is an offspring": Peter Weibel, author interview.

343 "What's the big deal?": Ken Perlin, author interview.

343 "it started here": Joichi Ito, author interview.

343 "motion pictures are by definition"; "There are artists right now"; "it's already happened": Rick Sayre, author interview.

343 "the border between science": Jonas Loh, author interview.

343 "no longer a programmer": Benedikt Gross, author interview.

344 "word is not completely out yet"; "like amphibians": Michael John Gorman, author interview.

344 "the questions are so deep"; "All is teamwork": Rolf-Dieter Heuer, author interview.

344 "You only have one lifetime": Stelarc, author interview.

344 "like paint in a can": Panel discussion at Art & Science: Merging Art & Science to Make a Revolutionary New Art Movement, July 8, 2011, GV Art, London.

344 "culture" goes well beyond: Evelina Domnitch and Dmitry Gelfand, author interview.

344 "There's something exciting about": Ken Arnold, author interview.

344 "At the micro level"; "In the intellectual sphere"; "disagreements can be": Ibid.

347 "Sound art has been"; "they've begun to get": Gopnik (2013).

# BIBLIOGRAPHY

Abbott, Alison. 2006. "The Hands of a Master." *Nature* 439 (February 9): 648–50.

Aldworth, Susan. 2010. "Fine Art Visiting Speakers Programme." http://research.ncl.ac.uk/fineartvisitingspeakers/programme.html.

———. 2011. In *Art & Science: Merging Art & Science to Make a Revolutionary New Art Movement.* http://www.artandscience.org.uk/wp-content/uploads/2011/07/GV-Art-Art-Science-e-catalogue.pdf.

Anderson, Mogens. 1967. "An Impression." In S. Rozental, *Niels Bohr: His Life and Work as Seen by His Friends and Colleagues*, 321–24. New York: Wiley.

Angheleddu, Davide. 2011. In *Art & Science: Merging Art & Science to Make a Revolutionary New Art Movement.* http://www.artandscience.org.uk/wp-content/uploads/2011/07/GV-Art-Art-Science-e-catalogue.pdf.

Antonelli, Paola, ed. 2011. *Talk to Me: Design and Communication between People and Objects.* New York: Museum of Modern Art.

Apollinaire, Guilluame. 1913. *Les Peintres cubists: Mediations esthétiques.* Paris: Figuière. Reprinted with an introduction and annotation by L. C. Breunig and J.-C. Chevalier. Paris: Hermann, 1980. The page reference is to the 1980 edition.

Ascott, Roy. 1964. "The Construction of Change." In Edward A. Shanken, ed., *Cybernetics to Telematics: The Art, Pedagogy and Theory of Roy Ascott.* Berkeley: University of California Press, 2007.

———. 1990. ""Is There Love in the Telematic Embrace?" *Art Journal* 49, no. 3 (Autumn): 241–47.

———. 2008. "Creative Cybernetics: The Emergence of an Art Based on Interaction, Process, and System." In Paul Brown et al., *White Heat Cold Logic.*

Attenborough, David. 1994. http://www.metstudio.com/exhibition_designers/portfolio_science4live.html.

Ballengée, Brandon. 2010. *Malamp: The Occurrence of Deformities in Amphibians.* London: The Arts Catalyst.

Bateson, Gregory. 1972. *Steps to an Ecology of Mind.* Chicago: University of Chicago Press.

Benqué, David. 2010. "Acoustic Botany." http://www.davidbenque.com/projects/
    acoustic-botany.

Benson, Michael. 2001a. "How It Went." *Signatures of the Invisible: Documents 1.1.*
    Exhibition catalogue. London: Atlantis Gallery.

——. 2001b. "CERN Project Brings Science and Art Together." *CERN Courier*, April
    30.

Biro, Paul. 2007. "Fingerprinting Jackson Pollock?" In Ellen G. Landau and Claude
    Cernuschi, *Pollock Matters*, 155–58. Chicago: University of Chicago Press.

Blume, Eugen. 2008. "Eugen Blume in Conversation with Bernhard Leitner." In
    *Museum for Contemporary Art: Bernhard Leitner*, 5–22. Berlin: National
    Museums of Prussian Cultural Heritage.

Born, Max. 1971. Translated by Irene Born, *The Born–Einstein Letters: Correspon-
    dence between Albert Einstein and Max and Hedwig Born from 1916 to 1955.*
    New York: Walker and Company.

——. 1975. *My Life: Recollections of a Nobel Laureate.* New York: Scribners.

Bowlin, Rhonda. 2007. "The Zuse Z64 Graphomat as Utilized by Artist Frieder
    Nake." In ITGM–705 Interactive Design and Media Applications, Unit 2—
    Art Review 1: Early Digital Artifact. Available at: http://issuu.com/rhonda
    bowlin/docs/rbowlin_artreview01.

Boyack, Kevin, Dick Flavans, and W. Bradford Paley. 2006. "A Map of Science,"
    *Nature* 444, 21/28 December, 985.

Brown Paul. 2003. "The Idea Becomes a Machine: AI and Alife in Early Brit-
    ish Computer Arts." Available at: http://www.paul-brown.com/WORDS/
    CR2003.pdf.

——. 2008. "From Systems Art to Artificial Life." In Paul Brown et al., *White Heat
    Cold Logic*, 275–79.

——, and Phil Husbands. 2010. "Not Intelligent by Design." In Hazel Gardiner and
    Charlie Gere, eds., *Art Practice and Digital Design*, 61–92. Farnham, UK: Ash-
    gate. Although Brown and Husbands cowrote this chapter, they delineated
    who wrote what.

Brown, Paul, Charlie Gere, Nicholas Lambert, and Catherine Mason, eds. 2009.
    *White Heat Cold Logic : British Computer Art 1960–1980.* Cambridge, MA:
    MIT Press.

Burgess, Gelett. 1910. "The Wild Men of Paris." *Architectural Record* 27, no. 5 (May):
    401–14.

Burnham, Jack. 1971. "Notes on Art and Information." In Judith Benjamin Burn-
    ham, ed., *Software, Information Technology: Its New Meaning for Art*, 10–14.
    New York: Jewish Museum

Butterworth, Jon. 2012. "Does art-from-science really add anything?" http://www
    .theguardian.com/science/life-and-physics/2012/jun/18/art-from-science-
    copenhagen.

Campbell-Johnston, Rachel. 2001. "They Blinded Me with Science." *The Times*,
    March 21.

Candy, Linda, and Ernest Edmonds, eds. 2002. *Explorations in Art and Technology.* London: Springer-Verlag.

Carnie, Andrew. 2002. "Scientists and Artists Must Rub Shoulders,"*Guardian*, March 17. Available at: http://www.guardian.co.uk/education/2002/mar/17/arts.highereducation.

Carter, Sally. 2011. "The Emergence of Art-Science." *British Medical Journal*, August 13. Available at: http://www.bmj.com/content/343/bmj.d5133.

Cascella, Daniela. 2012. "This Sonorous Writing of Nowhere. (Or: To Go Back to Words and Sounds. To Forget about Deleuze)." *En abîme.* Available at: https://enabime.wordpress.com/2012/03/08/this-sonorous-writing-of-nowhere-or-to-go-back-to-words-and-sounds-to-forget-about-deleuze/.

Cattrell, Annie. 2011. In *Art & Science: Merging Art & Science to Make a Revolutionary New Art Movement.* Available at: http://www.artandscience.org.uk/wp-content/uploads/2011/07/GV-Art-Art-Science-e-catalogue.pdf.

Catts, Oron, and Ionat Zurr. 2007. "Semi-Living Art." In Eduardo Kac, ed., *Signs of Life: Bio Art and Beyond*, 231–47. Cambridge, MA: MIT Press.

Cernuschi, Claude, and Ellen G. Landau. 2007. "Introduction." In Ellen G. Landau and Claude Cernuschi, *Pollock Matters*, 3–8. Chicago: University of Chicago Press.

Chipp, Herschel B. 1968. *Theories of Modern Art: A Source Book by Artists and Critics.* Berkeley: University of California Press.

Clemens, David. 1968. "Scene." *Daily Mirror*, August 9.

Cohen, Nathan. 2013. "New Perspectives." *ACM Computers in Entertainment.* Available at: http://dl.acm.org/pub.cfm?id=J912.

Cook, Sarah. 2000. "Interview with Peter Weibel, Chairman and CEO of the Zentrum fur Kunst und Medientechnologie in Karlsruhe, Germany." http://www.crumbweb.org/getInterviewDetail.php?id=9&op=3.

Cornock, Stroud, and Ernest Edmonds. 1973. "The Creative Process Where the Artist Is Amplified or Superseded by the Computer." *Leonardo* 6: 11–16.

Cousins, Judith, and Hélène Seckel. 1994. "Chronology of *Les Demoiselles d'Avignon.*" In William Rubin, Judith Cousins, and Hélène Seckel, eds., *Les Demoiselles d'Avignon*, 145–212. New York: Museum of Modern Art.

Coveyou, Robert R. 1998. Quoted in Ivars Peterson, *The Jungles of Randomness: A Mathematical Safari.* New York: Wiley.

David, D. M. 1968. "Art and Technology." *Art in America* 56 (January): 28–47.

Davis, Joe. 2007. "Cases for Genetic Art." In Eduardo Kac, ed., *Signs of Life: Bio Art and Beyond,* 249–66. Cambridge, MA: MIT Press.

Dayal, Geeta. 2010. "Carsten Nicolai." *Frieze* (June 28). Available at: http://www.frieze.com/shows/review/carsten_nicolai/.

Deacon, Richard. 2001. In *Signatures of the Invisible: Works, 2.1.* Exhibition catalogue. London: Atlantis Gallery.

De Menezes, Marta. 2007. "Art: In Vivo and in Vitro." In Eduardo Kac, ed., *Signs of Life: Bio Art and Beyond*, 215–29. Cambridge, MA: MIT Press.

Dirac, P. A. M. 1933. "Theory of Electrons and Positrons." Nobel lecture presented December 12, 1933. Available at: http://www.nobelprize.org/nobel_prizes/ physics/laureates/1933/dirac-lecture.pdf.

Domnitch, Evelina, and Dmitry Gelfand 2004. "*Camera Lucida*: A Three-Dimensional Sonochemical Observatory." *Leonardo* 37: 391–96.

———. 2011. "Memory Vapor." http://www.portablepalace.com/memoryvapor.html.

Dowson, Katherine, 2011. In *Art & Science: Merging Art & Science to Make a Revolutionary New Art Movement*. Available at: http://www.artandscience.org.uk/ wp-content/uploads/2011/07/GV-Art-Art-Science-e-catalogue.pdf.

Draves, Scott, and Isabel Walcott Draves. 2010. "The Flame Algorithm and Its Open Source Culture." *Computer Graphics* 44 (August 3). Available at: http://www .siggraph.org/publications/newsletter/volume-44-number-3/the-flame-algorithm-and-its-open-source-culture.

Eddington, Arthur Stanley. 1968. *The Nature of the Physical World*. Ann Arbor: University of Michigan Press.

Edmonds, Ernest. 2008. "Constructive Computation." In Paul Brown et al., *White Heat Cold Logic*, 345–59.

Edwards, David. 2008. *Artscience: Creativity in the Post-Google Generation*. Cambridge, MA: Harvard University Press.

Einstein, Albert. 1905a. "Über einen die Erzeugung und Verwandlung des Lichtes betreffenden heuristischen Gesichtspunkt [On a Heuristic Viewpoint on the Production and Transformation of Light]." *Annalen der Physik* 17: 132–48.

———. 1905b. "Zur Elektrodynamik bewegter Körper [On the Electrodynamics of Moving Bodies]." *Annalen der Physik* 17: 891–921; quotation is from the English translation in Arthur I. Miller, *Albert Einstein's Special Theory of Relativity: Emergence (1905) and Early Interpretation (1905–1911)* (New York: Springer-Verlag, 1998).

———. 1923. "Fundamental Ideas and Problems of the Theory of Relativity." In *Nobel Lectures: Physics, 1901–1921*, 482–90. Amsterdam: Elsevier.

———. 1993. *Collected Papers of Albert Einstein*, vol. 5. Princeton: Princeton University Press. Page reference is to the English translation.

Fahlström, Öyvind. 1966. "Armory." Getty Research Institute, Department of Special Collections, archive 94003.1.42:20–21 [Quoted in Hillman (2007), 64].

Flescher, Sharon. 2008. "Introduction." *IFAR Journal: International Foundation for Art Research* 10, no. 1: 8–9.

Friedlander, Paul. 2006. "Timeless Universe: 2006." Exhibition catalogue, Sala Parpalló. Available at: http://www.paulfriedlander.com/text/salal2006.htm.

———. 2011. "Interview: Once a Physicist." *Physics World*. Available at: http://www .iop.org/careers/working-life/profiles/page_57653.html.

Gardner, James. 1969. James Gardner to Robin Clark, November 18, commissioning Clark to "write the story" of *Senster*. Quoted in Mason (2008), 256, note 27.

Gargaj. 2007. "Demoscene Outreach Tour 2007." http://hugi.scene.org/online/hugi34/hugi%2034%20-%20demoscene%20reports%20gargaj%20demo-scene%20outreach%20tour%202007.htm.

Gaver, Bill. 2011. "Designing Interactions: Interview with Fiona Raby and Tony Dunne." http://www.designinginteractions.com/interviews/DunneandRaby.

Gere, Charlie. 2002. *Digital Culture*. London: Reaktion.

Gertner, John. 2012. *The Idea Factory: Bell Labs and the Great Age of American Innovation*. New York: Penguin.

Gleick, James. 1992. *Genius: The Life and Science of Richard Feynman*. New York: Pantheon.

Gleizes, Albert, and Jean Metzinger. 1980. *Du "Cubisme."* Paris: Edition Figuière: 1912; Saint-Vincent-sur-Jabron: Éditions PRÉSENCE. Page reference is to the 1980 reprint.

Glueck, Grace. 1966. "Arts and Engineering Are Mixing It Up at the Armory." *New York Times*, October 14: 87.

———. 1989. "Howard Wise, 86, Dealer Who Helped Technology in Art." *New York Times*, obituary, September 8. Available at: http://www.nytimes.com/1989/09/08/obituaries/howard-wise-86-dealer-who-helped-technology-in-art.html.

Going Public. 1996. "Finding the Art in Science." In *Going Public: An Introduction to Communicating Science, Engineering and Technology*. Available at: http://www.berr.gov.uk/files/file14581.pdf.

Gopnik, Blake. 2013. "Did You Hear That? It Was Art." *New York Times*, August 5. Available at: http://www.nytimes.com/2013/08/04/arts/design/museums-embrace-works-made-of-sound.html?ref=design.

Gosling, Nigel. 1968. "Man in an Automated Wonderland." *Observer*, August 4.

Grande, John K. 2007. "Bio-Art with Brandon Ballengée." http://ecologicalart.org/branbal.html.

Grandjean, Emmanuel. 2000. "Au CERN, douze artistes vont titiller les particules [At CERN, a dozen artists are going to titillate the particles]." *Tribune de Genéve*, July 7.

Griffith, Susan. 2006. "Case Western Reserve University Physicists Refute Analysis of Jackson Pollock's Paintings." News Center, Case Western Reserve, November 30. Available at: http://blog.case.edu/case-news/2006/11/30/pollock.

Gross, Benedikt, with Hartmut Bohnacker, Julia Laub, and Claudius Lazzeroni. 2009. *Generative Gestaltung: Entwerfen. Programmieren. Visualisieren [Generative Design: Visualize, Program, and Create with Processing]*. Mainz: Schmidt Hermann Verlag. English translation, Princeton: Princeton Architectural Press, 2012.

Gross, Jason. 1997. "Interview with David Toop." *Perfect Sound Forever*. Available at: http://www.furious.com/perfect/toop.html.

Hadley, Katherine. 1968. "Serendipity with Cybernetics." *Hampstead and Highgate News*, August 9.

Hallett, Nick. 2009. "Tristan Perich." *BOMBSITE*. Available at: http://bombsite .com/issues/999/articles/3361.

Hamilton, Richard. 2003. Interview with Hans-Ulrich Obrist. "Pop Daddy." *Tate*, no. 4, April 1. 60–62. Available at: http://www.tate.org.uk/context-comment/ articles/pop-daddy-richard-hamilton-early-exhibition.

Hargrave, Katie. 2007. "Steve Miller@Rose Art Museum." Available at: http:// www.bigredandshiny.com/cgi-bin/BRS.cgi?section=review&issue=72&arti cle=ROSE_ART_MUSEUM_816954.

Harvard University Art Museums. 2007. "Technical Analysis of Three Paintings Attributed to Jackson Pollock." Available at: www.harvardartmuseums.org/ sites/harvardartmuseums.org/files/straus_pub_pollock.pdf.

Hatch, John G. 2010. "Some Adaptations of Relativity in the 1920s and the Birth of Abstract Architecture." *Nexus Network Journal* 12, no. 1: 131–47.

Heiferman, Marvin. 2007. "Every Body a Spectacle: An Interview with Steve Miller." In *Steve Miller: Spiraling Inward*. 53–61. Brandeis, MA: Rose Art Museum.

Heisenberg, Werner. 1971. *Physics and Beyond: Encounters and Conversations*. Translated by Arnold J. Pomerans. New York: Harper & Row.

Hermann, Arno, Karl von Meyenn, and Victor F. Weisskopf. 1979. *Wolfgang Pauli: Scientific Correspondence with Bohr, Einstein, Heisenberg, a.o., Volume I: 1919–1929*. New York: Springer-Verlag.

Hertz, Garnet. 1995. "The Godfather of Technology and Art: An Interview with Bill Klüver." April 19. www.conceptlab.com/interviews/kluver.html.

Hillman, Susanne Naima. 2007. "Robert Rauschenberg, Robert Whitman and Billy Klüver: From '9 Evenings' to 'Experiments in Art and Technology.'" Unpublished PhD thesis, Rutgers, the State University of New Jersey.

Hochfield, Sylvia. 2008. "The Blue Print." *ARTnews*, June 1. Available at: www.art news.com/2008/06/01/the-blue-print/.

Holton, Gerald. 1973. *Thematic Origins of Scientific Thought: Kepler to Einstein*. Cambridge, MA: Harvard University Press.

Hultén, Pontus. 1968a. "Introductory Panel—Wall Label," for *The Machine: As Seen at the End of the Mechanical Age* at Museum of Modern Art, New York, November 27—February 9, 1969. Available at: http://www.moma.org/ docs/press_archives/4149/releases/MOMA_1968_July-December_0081 .pdf?2010.

———. 1968b. Museum of Modern Art, New York, press release, November 27. Available at: http://www.moma.org/docs/press_archives/4149/releases/ MOMA_1968_July-December_0081.pdf?2010.

Ihnatowicz, Edward. 1985. "Cybernetic Sculpture." Unpublished, undated (c. 1985). Quoted in Mason (2008), 79.

Ihnatowicz, Richard. 2008. "Forty Is a Dangerous Age: A Memoir of Edward Ihnatowicz." In Paul Brown et al., *White Heat Cold Logic*, 111–17.

Isaacson, Walter. 2007. *Einstein, His Life and His Universe*. New York: Simon & Schuster.

Jones-Smith, Katherine, and Harsh Mathur. 2006. "Revisiting Pollock's Drip Paintings." *Nature* 444, E9–E10 (November 30).

Kac, Eduardo. 2007. "Life Transformation—Art Mutation." In Eduardo Kac, ed., *Signs of Life: Bio Art and Beyond*, 163–84. Cambridge, MA: MIT Press.

Kandinsky, Wassily. 1977. *Concerning the Spiritual in Art*. Translated by M. T. H. Sadler. New York: Dover Publications. First published in 1914. Translated from W. Kandinsky, *Über das Gestige in der Künst*. Munich: R. Piper & Co., 1912.

Karmel, Pepe. 2008. "What's the Matter with Pollock: Issues of Connoisseurship and Abstraction." *IFAR Journal: International Foundation for Art Research* 10, no. 1: 10–17.

Katz, Carissa. 1999. "Steve Miller: Seeing Through His Subjects." *East Hampton Star*, September 23. Available at: http://stevemiller.com/pdf/EastHampton-99.pdf.

Kemp, Martin. 2001. "Suggestive Signatures." *Nature* 410 (March 22): 414.

———. 2010. "A Flowering of Pleasure and Pain." *Nature* 465 (May 20): 265. Available at: http://www.kringelbach.dk/papers/Nature_Kemp2010.pdf.

Kennedy, Randy. 2005. "Is This a Real Jackson Pollock?" *New York Times*, May 29. Available at: http://www.nytimes.com/2005/05/29/arts/design/29kenn.html?pagewanted=print.

———. 2006a. "Computer Analysis Suggests Paintings Are Not Pollocks." *New York Times*, February 9. Available at: http://www.nytimes.com/2006/02/09/arts/design/09poll.html.

———. 2006b. "The Case of Pollock's Fractals Focuses on Physics." *New York Times*, December 2. Available at: http://www.nytimes.com/2006/12/02/books/02frac.html.

Klee, Paul. 1968. "Creative Credo." In Herschel B. Chipp, *Theories of Modern Art: A Source Book by Artists and Critics* (Berkeley: University of California Press, 1968), 182–86. Originally published in Paul Klee, *Schöpferische Konfession*, ed. Kasimir Edschmid. Berlin: Erich Reiss, 1920 (*Tribune der Kunst und Zeit*, no. 13). Translation by Norbert Guterman from *The Inward Vision: Watercolors, Drawings and Writings by Paul Klee* (New York: Abrams, 1959), 5–10.

Klütsch, Christoph. 2007. "Computer Graphic—Aesthetic Experiments between Two Cultures." *Leonardo* 40, no. 5: 421–25.

Klüver, Billy. 1966. "'The Great Northeastern Power Failure' [1966]." In Randall Packer and Ken Jordan, eds., *The Great Northeastern Power Failure*, 33–38. New York: Norton. Reprint edition, 2001.

———. 1998. "The Engineer as Catalyst: Billy Klüver on Working with Artists." Interview with Paul Miller. *IEEE Spectrum* 35, no. 7 (July): 20–29. Available at: http://paulspen.com/the-engineer-as-catalyst-billy-kluver-on-working-with-artists. All references are to the online source.

———, and Julie Martin. 1991. "Four Difficult Pieces." *Art in America* 79: 81–138.

Knowlton, Ken. 2005. www.kenknowlton.com/pages/04portrait.htm.

Koblin, Aaron. 2011. http://www.youtube.com/watch?v=cAsEeY-QE1U.

Lamb, Maggoty. 2012. "Maggoty Lamb Interviews Visionary Music Critic David Toop." *Guardian*, March 23. Available at: http://www.guardian.co.uk/music/2012/mar/23/maggoty-lamb-interviews-david-toop.

Landau, Ellen G. 2007. "Action/Reaction: Friendship of Herbert Matter and Jackson Pollock," In Ellen G. Landau and Claude Cernuschi, *Pollock Matters*, 9–57. Chicago: University of Chicago Press.

Laporte, Geneviève. 1975. *Sunshine at Midnight*. Translated by Douglas Cooper. Worthing, UK: Littlehampton Book Services, Ltd.

Latham, William. Early 1990s. *Secret Passions*. http://www.youtube.com/watch?v=AN6ngsckRZs.

Leahy, James. 2003. "Science and Its Signs: An Interview with Ken McMullen." http://sensesofcinema.com/2003/28/ken_mcmullen/.

Lewis, Helen. 2011. "Written in the Body." http://www.newstatesman.com/blogs/helen-lewis-hasteley/2011/08/art-body-tissue-pieces.

Lewitt, Sol. 1967. "Paragraphs on Conceptual Art." *Artforum* (June). Available at: http://www.tufts.edu/programs/mma/fah188/sol_lewitt/paragraphs%20on%20conceptual%20art.htm.

Lindgren, Nilo. 1969a. "Art and Technology: I. Steps toward a New Synergism." *IEEE Spectrum*, April: 59–68.

——. 1969b. "Art and Technology: II. A Call for Collaboration." *IEEE Spectrum*, May: 46–56.

Lindsay, Kenneth, and Peter Vergo. 1982. *Kandinsky: Complete Writings on Art*. London: G. K. Hall.

Lista, Giovanni. 1973. *Futurisme, Manifestes, proclamations, documents*. Lausanne: L'Âge d'homme.

Litt, Steven. 2007a. "Shadows of Doubt Loom over Pollock-Style Works." *Cleveland Plain Dealer*, October 23. Available at: http://blog.cleveland.com/pdextra/2007/10/shadows_of_doubt_loom_over_pol.html.

——. 2007b. "E-mails Show Adviser to Owner of Alleged Pollocks Suspected They Weren't Genuine." *Cleveland Plain Dealer*, December 21. Available at: http://blog.cleveland.com/ent_impact_arts/print.html?entry=/2007/12/pollock.html.

——. 2008. "Even as Exhibit Opens, Controversy Simmers." *Cleveland Plain Dealer*, February 4. Available at: http://lynndunham.blogspot.co.uk/2008/02/art-even-as-exhibit-opens-pollock.html.

——. 2010. "Collector Eugene Thaw Speaks on the Pursuit of Quality." *Cleveland Plain Dealer*, March 6. Available at: http://www.cleveland.com/arts/index.ssf/2010/03/collector_eugene_victor_thaw_s.html.

Loewen, Norma. 1975. *Experiments in Art and Technology: A Descriptive History of the Organization*. Unpublished PhD thesis, New York University.

MacGregor, Brent. 2008. "Cybernetic Serendipity Revisited." In Paul Brown et al., *White Heat Cold Logic*, 83–93.

Machover, Tod. 2005. "Tech Night to Premier Machover Work." Interview in *MIT Tech Talk*, June 1, 7.

———. 2012. "Future Opera for Robots and People Too." In *Innovation Perspectives for the 21st Century*, 399–408. Madrid: Tf. Editores.

Maiani, Luciano. 2001. "Introduction." In *Signatures of the Invisible: Documents 1.1.* Exhibition catalogue. London: Atlantis Gallery.

March-Russell, John. 2001. "Extracts from a Conversation in the CERN Canteen on 3 March 2000 between John March Russell and Ken McMullen." In *Signatures of the Invisible: Works 2.1.* Exhibition catalogue. London: Atlantis Gallery.

Markoff, John. 2011. "M.I.T. Media Lab Names a New Director." *New York Times*, April 25.

Marron, David. 2011. *Art & Science: Merging Art & Science to Make a Revolutionary New Art Movement*. Available at: http://www.artandscience.org.uk/wp-content/uploads/2011/07/GV-Art-Art-Science-e-catalogue.pdf.

Martin, James. 2008. "What Materials Tell Us about the Age and Attribution of the Matter Paintings." *IFAR Journal* (International Foundation for Art Research) 10, no. 1: 25–35.

Martineau, Lisa. 2012. "Suzanne Anker: Doyenne of Bio Art." *NYCity Woman*, May 24. http://www.nycitywoman.com/features/suzanne-anker-doyenne-bio-art.

Mason, Catherine. 2008. *A Computer in the Art Room: The Origins of British Computer Arts 1950–1980*. Norfolk, UK: J. J. G. Hindrigham.

McMullen, Ken. 2001. "History." In *Signatures of the Invisible: Documents 1.1.* Exhibition catalogue. London: Atlantis Gallery.

McQuaid, Cate. 2007. "Building Blocks." *Boston Globe*, October 14. Available at: http://stevemiller.com/press-articles/building-blocks/.

Melville, Robert. 1968. "Signalling the End." *New Statesman*, August 9.

Miller, Arthur I. 1985. "Werner Heisenberg and the Beginning of Nuclear Physics." *Physics Today* 38 (November): 60–68.

———. 1998. *Albert Einstein's Special Theory of Relativity: Emergence (1905) and Early Interpretation (1905–1911)*. New York: Springer-Verlag; Reading, MA: Addison-Wesley, 1981.

———. 2000. *Insights of Genius: Imagery and Creativity in Science and Art*. Cambridge, MA: MIT Press.

———. 2001. *Einstein, Picasso: Space, Time, and the Beauty That Causes Havoc*. New York: Basic Books.

———. 2010. *137: Jung Pauli and Pursuit of a Scientific Obsession*. New York: Norton.

———. 2011. *Art & Science: Merging Art & Science to Make a Revolutionary New Art Movement*. Available at: http://www.gvart.co.uk/press/GV%20Art%20-%20Art%20&%20Science%20-%20e-catalogue.pdf.

Minkel, J. R. 2007. "Pollock or Not? Can Fractals Spot a Fake Masterpiece?" *Scientific American*, October 31. Available at: http://www.scientificamerican.com/article.cfm?id=can-fractals-spot-genuine.

Minkowski, Hermann. 1908. "Raum und Zeit [Space and Time]," translated by W. Perrett and G. B. Humphrey, *The Principle of Relativity*, 75–91. New York: Dover Publications, 1923.

Mirapaul, Matthew. 2002. "Arts Online: Secrets of Digital Creativity Revealed in Miniatures." *New York Times*, September 16. Available at: http://www.nytimes.com/2002/09/16/arts/arts-online-secrets-of-digital-creativity-revealed-in-miniatures.html.

Mondrian, Piet. 1937. "Figurative Art and Nonfigurative Art." In Herschel B. Chipp, *Theories of Modern Art: A Source Book by Artists and Critics*, 349–62. Berkeley: University of California Press, 1968. The essay was originally published in 1937 in John Leslie Martin, Naum Gabo, and Ben Nicholson, eds., *Circle*, 41–56. London: Faber & Faber. Page references are to Chipp, 1968.

Nake, Frieder. 2002. "Personal Recollections of a Distant Beginning." In Linda Candy and Ernest Edmonds, eds., *Explorations in Art and Technology*. London: Springer, 6–7.

Netzer, Nancy. 2007. "Director's Preface." In Ellen G. Landau and Claude Cernuschi, *Pollock Matters*, 1–2. Chicago: University of Chicago Press.

——, and Michele Derrick. 2007. "Scientific Examination of the Paint on Nine Matter Paintings." In Ellen G. Landau and Claude Cernuschi, *Pollock Matters*, 105–29. Chicago: University of Chicago Press.

Newman, Richard. 2008. "An Overview of the Scientific Analysis of the 'Matter Paintings.'" *IFAR Journal* 10, no. 1: 24.

Noll, A. Michael. 1966. "Human or Machine: A Subjective Comparison of Piet Mondrian's 'Composition with Lines' (1917) and a Computer-Generated Picture." *Psychological Record* 16: 1–10.

——. 2011. "Early Years at Bell Labs (1961–1971)." Unpublished. Courtesy of A. Michael Noll.

O+A [Bruce Odland and Sam Auinger]. 2009. "Reflections on the Sonic Commons." *Leonardo Music Journal* 19: 63–68.

O' Callaghan, Tiffany. 2011. "Escape Artists: Breaking Out of the Lab." http://www.newscientist.com/blogs/culturelab/2011/07/escape-artists-breaking-out-of-the-lab.html.

O'Conner, Patrick. 1966. "500 People on Stage, 'Theatre and Engineering' Less Than Pleasing." *Jersey Journal*, October 17: 11.

O'Connor, Francis. 1967. *Jackson Pollock*. New York: Museum of Modern Art.

O'Doherty, Brian. 1966. "New York: 9 Armored Nights." In Catherine Morris, ed., *9 Evenings Reconsidered: Art, Theater, and Engineering*. Cambridge: MIT List Visual Arts Center, 2006.

ORLAN. 2004. *ORLAN*. Paris: Flammarion.

———. 2008. "Harlequin Coat." In Jens Hauser, ed., *Sk-interfaces: Exploring Borders—Creating Membranes in Art Technology and Society,* 83–89. Liverpool: Liverpool University Press.

Ouroussoff, Nicolai. 2008. "The Soul in the New Machines." Review of "Design and the Elastic Mind" at the Museum of Modern Art. *New York Times,* February 22. Available at: http://www.nytimes.com/2008/02/22/arts/design/22elas.html?pagewanted=all.

Paley, W. Bradford. 2010. http://wbpaley.com/brad/mapOfScience/index.html.

Parkinson, Gavin. 2008. *Surrealism, Art and Modern Science.* New Haven: Yale University Press.

Parmelin, Hélène. 1969. *Picasso Says.* Translated by Christine Trollope. London: Allen and Unwin.

Parsons, Paul. 1998. Quoted in "How to Make LightWork of the Invisible: A British Aartist Is Shedding Light on the Dark Side of the Universe," *Guardian,* April 9. Available at: http://www.paulfriedlander.com/text/guardian.html.

Peers, Alexandra. 2005. "Summer Mystery: Who Painted Those Pollocks." *Wall Street Journal,* August 9. Available at: http://online.wsi.com/article/SB115507231837630311.html.

Perry, Phoenix. 2011. "Scott Draves." *TRIANGULATION.* http://www.triangulationblog.com/2011/01/scott-draves.html.

Pertusini, Angela. 1995. "An Ear for Art." *The Big Issue,* April. Reprinted in *Look Hear,* exhibition catalogue, 1995, 10. This document is an exhibition assessment by the Wellcome Trust.

Pethick, Emily. 2006. "Degree Zero." *Frieze,* no. 101 (September). Available at: www.frieze.com/issue/article/degree_zero/.

Phillips, Mike. 2012. *Spectre.* http://digitalekunst.ac.at/index.php?id=266.

Picabia, Francis. (undated). *La Pomme des Pins.* Paris: Saint-Raphaël.

Pierce, John R. 1967. "Remarks by John R. Pierce." In Julie Martin, ed., *E.A.T. News* 1 (November 1).

Pile, Stephen. 2001. "Can Art Explain Science?" *Daily Telegraph,* March 10.

Pippard, Lucy. 1965. "New York Letter." *Art International* 9: 57.

Poincaré, Henri. 1958. *The Value of Science.* Translated by G. B. Halsted. New York: Dover.

Preston, Stuart. 1965. "Art ex Machina." *New York Times,* April 18, 23.

Prudence, Paul. 2012. Lecture and workshop at the Emergent Digital Practices Department, University of Denver, May. Available at: http://www.du.edu/ahss/edp/news.html#prudence.

"Q & A: Are They Pollocks: What Science Tells Us about the Matter Paintings." *IFAR Journal* 10, no. 1 (2008): 36–43.

Reas, Casey. 2009. "Aaron Koblin Interview." http://wiki.processing.org/w/Aaron_Koblin_Interview.

Rees, Jane. 2008. "Cultures in the Capital." *Nature* 451 (February 21): 891.

Rehmeyer, Julie J. 2007. "Fractal or Fake: Novel Art-Authentication Method Is Challenged." *Science News* 171 (February 24): 122–24.

Reichardt, Jasia, ed. 1968. *Cybernetic Serendipity; The Computer and the Arts*. New York: Studio International.

———. 1971. *Cybernetics, Art, and Ideas*. London: Studio Vista and Greenwich.

———. 2008. "In the Beginning. . . ." In Paul Brown et al., *White Heat Cold Logic*, 71–81.

Reisz, Matthew. 2011. "The Pick—Art & Science: Merging Art & Science to Make a Revolutionary New Art Movement." *Times Higher Education*, July 7. Available at: http://www.timeshighereducation.co.uk/story.asp?sectioncode=26&storycode=416758.

Reynolds, Simon. 2012. "We Are All David Toop Now." *The Wire* 338 (April).

Richardson, John, with the collaboration of Marilyn McCully. 1991. *A Life of Picasso, Volume I: 1881–1906*. New York: Random House.

———. 1996. *A Life of Picasso: Volume II, 1907–1917. The Painter of Modern Life*. New York: Random House.

———. 2007. *A Life of Picasso. The Triumphant Years 1917–1922*. New York: Random House.

Robinson, Leonard J. 1967. "What Really Happened at the Armory." Unpublished paper, March 16. Quoted in Loewen (1975), 68.

Rose, Barbara. 1987. *Rauschenberg*. New York: Vintage.

Rosenberg, Harold. 1959. "Tenth Street: A Geography of Modern Art." *Arts News Journal* 28: 120–43.

Ross, Alex. 2009. *The Rest Is Noise: Listening to the Twentieth Century*. New York: Harper Perennial.

Rothstein, Edward. 1991. "Yo-Yo Ma and His New 'Hyper' Cello." *New York Times*, August 17. Available at: http://www.nytimes.com/1991/08/17/arts/review-music-yo-yo-ma-and-his-new-hyper-cello.html.

Rouché, Burton. 1950. "Unframed Space." *New Yorker*, August 5, 16.

Salmon, André. 1910. "Courier de ateliers." *Paris Journal 4* (10 May).

———. 1919. "Les Origines et Intentions du Cubisme." *Demain* #68 (May 10), (April 26), 485–89.

———. 1922. *Propros d'Atelier*. Paris: Cré.

Sand, Monica. 2001. In *Signatures of the Invisible: Works, 2.1*. Exhibition catalogue. London: Atlantis Gallery.

Sato, Rebecca, and Casey Kazan. 2008. "The Brilliantly Weird World of MIT's 'Mad' Scientist." *Daily Galaxy*. http://www.dailygalaxy.com/my_weblog/2008/03/the-brilliantly.html.

Schrödinger, Erwin. 1926. "Uber das Verhältnis Heisenberg-Born-Jordanschen Quantenmechanik zu der meinen [On the Relation between Heisenberg-Born-Jordan's Quantum Mechanics and Mine]." *Annalen der Physik* 70: 734–56.

Seelig, Carl. 1954. *Albert Einstein und die Schweiz*. Zürich: Europa Verlag.

Sellars, Nina. 2011. In *Art & Science: Merging Art & Science to Make a Revolutionary New Art Movement*. Available at: http://www.artandscience.org.uk/wp-content/uploads/2011/07/GV-Art-Art-Science-e-catalogue.pdf.

Sexton, Ian. 2001. "The Magic of Light." In *Signatures of the Invisible: Works, 2.1.* Exhibition catalogue. London: Atlantis Gallery.

Shanken, Edward A. 1997. "Telematic Embrace: A Love Story? Roy Ascott's Theories of Telematic Art." http://telematic.walkerart.org/timeline/timeline_shanken.html.

———. 2002a. "Cybernetics and Art: Cultural Convergence in the 1960s." In Bruce Clarke and Linda Dalrymple Henderson, eds., *From Energy to Information.* Palo Alto: Stanford University Press.

———. 2002b. "Art in the Information Age: Technology and Conceptual Art." *Leonardo* 35, no. 4: 433–38.

Shepard, Michael. 1968. "Machine and Mind." *Sunday Telegraph*, August 11.

Smaje, Laurence. 1998. "Funding Partnerships: Nightmare or Panacea?" *Science and Public Affairs*, Winter: 45–49.

Snow, C. P. 1969. *The Two Cultures and a Second Look.* Cambridge, UK: Cambridge University Press. All quotations are from his 1959 Rede Lecture.

Solovine, Maurice. 1956. *Albert Einstein: Lettres à Maurice Solovine.* Paris: Gauthier-Villars.

Sowels, Katia. 2011. "The Merging of Art and Science." *Independent*, September 16. http://blogs.independent.co.uk/2011/09/16/the-merging-of-art-science-2/.

Spears, Dorothy. 2006. "The Entire Universe on a Dimmer Switch." *New York Times*, May 7. Available at: http://www.nytimes.com/2006/05/07/arts/design/07spea.html?pagewanted=all.

Speziali, Pierre. 1972. *Albert Einstein, Michele Besso: Correspondance, 1903–1955.* Paris: Hermann.

Stein, Gertrude. 1933. *The Autobiography of Alice B. Toklas.* New York: Random House.

Stelarc. 2010. "Excess and Indifference: Alternate Body Architectures." In Hazel Gardiner and Charlie Gere, eds., *Art Practice in a Digital Culture*, 93–116. Burlington, VT: Ashgate.

———. 2011. In *Art & Science: Merging Art & Science to Make a Revolutionary New Art Movement.* http://www.artandscience.org.uk/wp-content/uploads/2011/07/GV-Art-Art-Science-e-catalogue.pdf.

Takita, Jun. 2008. "Light only light." In Jens Hauser, ed., *Sk-interfaces: Exploring Borders—Creating Membranes in Art Technology and Society*, 141–43. Liverpool: Liverpool University Press.

Taylor, Richard. 2002. "Order in Pollock's Chaos." *Scientific American*, December: 116–21.

———. 2005. "Fractal Expressionism: Can Science Be Used to Further Understanding of Art?" http://materialscience.uoregon.edu/taylor/art/fractal_taylor.html.

———. 2010. "The Curse of Jackson Pollock: The Truth behind the World's Greatest Art Scandal." *Oregon Quarterly* 89, no. 4 (Summer). Available at: http://www.oregonquarterly.com/summer2010/10-pollock.php.

———, A. P. Micolich, and D. Jonas. 2006. "Taylor et al. Reply." *Nature* 144, E10–E11 (November 30).

——, B. Spehar, P. Van Donkelaar, and C. M. Hagerhall. 2011. "Perceptual and Physiological Responses to Jackson Pollock's Fractals." *Frontiers in Human Neuroscience* 5 (June): 1–13.

Todd, Stephan, and William Latham. 1992. *Evolutionary Art and Computers*. New York: Academic Press.

Toop, David. 2011. "Flat Time/Sounding—An interview with David Toop at Whitechapel Gallery" (March 30). http://www.youtube.com/watch?v=rSNcH59UdR8.

Triscott, Nicola, and Miranda Pope. 2010. "An Itinerant, a Messenger and an Explorer: The Work of Brandon Ballengée." In Ballengée (2010), 5–7.

Troxell, Cooper. 2011. "A Glimpse at Music from Beyond the Singularity." Interview with Tod Machover. http://www.kickstarter.com/blog/a-glimpse-at-music-from-beyond-the-singularity.

Tufte, Edward R. 1983. *The Visual Display of Quantitative Information*. Cheshire, CT: Graphics Press.

Usselmann, Rainer. 2003. "The Dilemma of Modern Art: Cybernetic Serendipity at the ICA London." http://www.rainerusselmann.net/2008/12/dilemma-of-media-art-cybernetic.html.

Wands, Bruce. 2006. *Art of the Digital Age*. London: Thames & Hudson.

Watson, Daniella. 2010. Interview with Dianne Harris. I thank Dianne Harris for a transcript.

Weibel, Peter, and Christiane Reidel, eds. 2010. *ZKM: Center for Art and Media*. Karlsruhe: ZKM.

Whyte, Lancelot Law, ed. 1968. *Aspects of Form: A Symposium on Form in Nature and Art*. 2nd ed. London: Lund Humphries.

Wilson, Chris. 2010. "I'll Be Bach." *Slate*, May 19. http://www.slate.com/articles/arts/music_box/2010/05/ill_be_bach.html.

Wilson, Stephen. 2010. *Art + Science Now: How Scientific Research and Technological Innovation Are Becoming Key to 21st-Century Aesthetics*. London: Thames & Hudson.

Yasui, Kyuichi, et. al. 2008. "The Range of Ambient Radius for an Active Bubble in Sonoluminescence and Sonochemical Reactions." *Journal of Chemical Physics* 128: 184705–12. Available at: http://dx.doi.org/10.1063/1.2919119.

Yeomans, Richard. 1987. *The Foundation Course of Victor Pasmore and Richard Hamilton 1954–1966*. PhD thesis. Institute of Education, University of London.

Yonetani, Ken, and Julia Yonetani. 2011. *In Art & Science: Merging Art & Science to Make a Revolutionary New Art Movement*. http://www.artandscience.org.uk/wp-content/uploads/2011/07/GV-Art-Art-Science-e-catalogue.pdf.

Zeki, Semir. 2009. "Statement on Neuroesthetics." Neuroesthetics Web.24 (November). http://www.neuroesthetics.org/statement-on-neuroesthetics.php.

Zivanovic, Aleksandar. 2008. "The Technologies of Edward Ihnatowicz." In Paul Brown et al., *White Heat Cold Logic*, 95–110.

# ILLUSTRATION CREDITS

## Chapter 2

2.1: *Variations V* (1965). John Cage, David Tudor, Gordon Mumma (foreground); Carolyn Brown, Merce Cunningham, Barbara Dilley (background). Photographer: Hérve Gloaguen. Courtesy of John Cage Trust.

2.2: *Studies in Perception I,* 5' x 12', 1966. Courtesy of Leon Harmon and Ken Knowlton.

2.3: Courtesy of ADAGP.

2.4: © Cybernetic Serendipity.

2.5: © Cybernetic Serendipity.

## Chapter 3

3.1: De Agostini/Getty Images.

3.2: Copyright © 1965 A. Michael Noll. Created in 1964.

3.3: Cellulose on glass, wood. Collection of the artist. Courtesy of the artist.

3.4: Courtesy of the artist.

3.5: Paul Brown, *Swimming Pool,* 1997. Giclée print, 75 x 50 cm. Courtesy of the artist.

3.6: *Seascape, Folkestone, 25th October 2008 at 11:41 am.* © Susan Collins.

## Chapter 4

4.1: Plotter drawing on paper, 30.5 x 30.5 cm. Courtesy of the artist.

4.2: In collaboration with Prof. W. Craig Carter (MIT) and Stratasys. Prototype for a Chaise Lounge 2008–10, acrylic composites. Museum of Science, Boston. Neri Oxman, architect and designer.

4.3: Courtesy of Joseph Paradiso.

4.4: Photograph © Peter Weibel.

4.5: Courtesy of Tobie Kerridge, Nikki Stott, and Ian Thompson.

## Chapter 5

5.1: Courtesy of the artist. Steel, 100" x 44" x 20" (2.50 x 1.10 x 0.50 m). Location: City of Moses, WA.

5.2: Photograph © Dmitry Gelfand. Acoustically induced light emissions in xenon-saturated sulphuric acid.

5.3 (a): Nathan Cohen.

5.3 (b): Nathan Cohen.

5.4: © Tim Roth.

5.5: Courtesy of the artist. A project by Vanessa Harden and Dr. Dominic South-gate in collaboration with Dr. Roberto Trotta, Dr. David Clements, and Dr. Andrew Jaffe from the Imperial College Astrophysics Group. Photo Credit—Kevin Hill.

5.6: © Josiah McElheny. Courtesy Andrea Rosen Gallery, New York. Photo by Tom Powel. Chrome-plated aluminum, electric lighting, hand blown glass, steel cable and rigging 16 feet diameter.

5.7: Courtesy of the artist. Installation. Material: Stainless steel wire.

5.8: Courtesy of the artist.

5.9: Courtesy of the artist.

5.10: Courtesy of the artist and Alexander Levy, Berlin.

5.11: Collection Paul Brennan, New York.

5.12: Antony Gormley. *Feeling Material XXXIV*, 2008. 5mm square section mild steel bar, 155 x 244 x 153 cm. Collection of CERN, Geneva, Switzerland. Photograph by Stephen White, London. © the artist.

5.13: Courtesy of the artist.

## Chapter 7

7.1: Courtesy of the artist and GV Art gallery, London.

7.2: Courtesy of the artist and GV Art gallery, London.

7.3: Courtesy of the artist and GV Art gallery, London.

7.4: Courtesy of the artist and GV Art gallery, London. 20 giclée prints, 175 x 140 cm.

7.5: Courtesy of the Artist and GV Art gallery, London. In collaboration with Professor Morton L. Kringelbach.

7.6: Courtesy of the Artist and GV Art gallery, London. In collaboration with Dr. Matthew Sellars.

7.7: Courtesy of the Artist, Dominik Mersch Gallery, Sydney, and GV Art gallery, London.

7.8: Courtesy of Scott Livesy Galleries, Melbourne.

7.9: Courtesy of the artist.

7.10: Live Butterflies with modified wing patterns. Courtesy Marta de Menezes.

7.11: Courtesy of the artist.

7.12: Courtesy of the artist.

7.13: Courtesy of the artist.

7.14: Courtesy of the artist. Performance by *Art Orienté* (Laval-Jeantet and Mangin), Kapelica Gallery, Ljubljana, February 22, 2011. Image by Miha Fras.

## Chapter 8

8.1: © Atelier Leitner.

8.2: Copyrights: Andreas Langen and bonn hoeren, 2012.

8.3: Courtesy of the artist. Photograph by Jana Chiellino.

8.4: Courtesy of the artist. Photograph by Stephen Kill FBIPP BSc, Chief Photographer, STFC Rutherford Appleton Laboratory, Harwell Science and Innovation Campus.

8.5: Courtesy of the artist. *Paul Prudence performing Cyclotone in 2013.*

8.6: Courtesy MIT Media Lab.

8.7: Courtesy Jonathan Williams.

8.8: Courtesy D. Yee.

## Chapter 9

9.1: *A Mote*, 2009, microscope digital image, image courtesy of the artist.

9.2: *AFM Mote*, 2009, AFM digital image, image courtesy of the artist.

9.3: Software: Mutator, FormGrow, and Winsom. Produced at the IBM UK Scientific Centre. The image is copyright William Latham.

9.4: Credit line image: Studio NAND 2012. Credit line project: Moritz Stefaner, Drew Hemment, Studio NAND 2012. Courtesy of the artist.

9.5: Courtesy of the artist. *Speculative Sea Level Explorer* by Benedikt Gross. Data: "SRTM30 Plus" by *Scripps Institution of Oceanography*, San Diego. http://benedikt-gross.de/log/2013/04/speculative-sea-level-explorer/.

9.6: Courtesy of the artist.

9.7: Courtesy of Aaron Koblin and Chris Milk.

## Insert

Insert 1: Courtesy of the artist and Site Gallery, Sheffield. Installation, Site Gallery, Sheffield: two computers, data projectors, Perspex screens, and cameras.

Insert 2: Photograph by Oliver Strewe.

Insert 3: Courtesy of the artist. *String Theory II* is on permanent display at the Rasmuson Centre, Anchorage, Alaska. 6 meters tall, base 1.75 meters in diameter. Illumination high power video projector with artist's computer-generated kinetic images.

Insert 4: © Keith Tyson. Mixed Media on Watercolor Paper. 157 cm x 126 cm.

Insert 5: Courtesy of Galerie Albert Benamou, Paris.

Insert 6: Courtesy of the artist and GV Art gallery, London. *Pig Wings*, 2000–2002 by the Tissue Culture & Art Project (Oron Catts, Ionat Zurr in collaboration with Guy Ben-Ary). Medium: Gold-coated pig bone marrow stem cells, differentiated into bone tissue, and degradable polymer scaffold. Size: approx 4 x 1 x 0.4 in cm, each set.

Insert 7: Courtesy of the artist and GV Art gallery. This project has been assisted by the Australian government through the Australia Council, its arts funding and advisory body.

**Australian Government**

Australia Council for the Arts

Insert 8: Eduardo Kac, *GFP Bunny*, 2000, transgenic artwork. Alba, the fluorescent rabbit. Courtesy Black Box Gallery, Copenhagen.

Insert 9: Courtesy of the artist and Ronald Feldman Fine Arts, New York, NY. Photograph of cleared and stained multi-limbed Pacific tree frog from Aptos, California, in scientific collaboration with Stanley K. Sessions. Unique digital C-print on watercolor paper. 46 x 34 inches.

Insert 10: Courtesy of Joonjae Moon.

Insert 11: By Scott Draves and the Electric Sheep.

Insert 12: Courtesy of the artist.

Insert 13: Courtesy of W. Bradford Paley. Research and node layout by Kevin Boyack and Dick Klavans; data from Thompson ISI; information layering, graphics, and typography by W. Bradford Paley. Copyright © 2006–2008 W. Bradford Paley, all rights reserved.

Insert 14: Courtesy of W. Bradford Paley. Research and node layout by Kevin Boyack and Dick Klavans; data from Thompson ISI; information layering, graphics, and typography by W. Bradford Paley. Copyright © 2006–2008 W. Bradford Paley, all rights reserved.

# INDEX

Page numbers in *italics* refer to illustrations.